In Vitro Toxicology

Second Edition

In Vitro Toxicology

Second Edition

Edited by
Shayne Cox Gad

CRC Press
Taylor & Francis Group
Boca Raton London New York

CRC Press is an imprint of the
Taylor & Francis Group, an **informa** business

CRC Press
Taylor & Francis Group
6000 Broken Sound Parkway NW, Suite 300
Boca Raton, FL 33487-2742

Visit the Taylor & Francis Web site at
http://www.taylorandfrancis.com

and the CRC Press Web site at
http://www.crcpress.com

To my mother, Norma Jean Cox Gad,
who taught me the value in, and
to care for, all life.

Contributors

Daniel Acosta: College of Pharmacy, University of Cincinnati Medical Center, Cincinnati, Ohio

Katherine L. Allen: IDEC Pharmaceuticals, La Jolla, California

Florence G. Burleson: BRT-Burleson Research Technologies, Raleigh, North Carolina

Gary R. Burleson: BRT-Burleson Research Technologies, Raleigh, North Carolina

Ai-Lean Chew: Department of Dermatology, University of California, San Francisco, California

Elaine M. Faustman: Department of Environmental Health, University of Washington, Seattle, Washington

Shayne Cox Gad: Gad Consulting Services, Raleigh, North Carolina

P. D. Gautheron: Laboratoires Merck Sharp and Dohme-Chibret, Riom, France

Carol E. Green: SRI International, Menlo Park, California

Saadia Kerdine: Immunotoxicology Group, INSERM U461, Faculté de Pharmacie Paris XI, Chatenay-Malabry, France

Herve Lebrec: Immunotoxicology Group, INSERM U461, Faculté de Pharmacie Paris XI, Chatenay-Malabry, France

Thomas A. Lewandowski: Department of Environmental Health, University of Washington, Seattle, Washington

Howard I. Maibach: Department of Dermatology, University of California, San Francisco, California

Ann D. Mitchell: Genesys Research, Incorporated, Research Triangle Park, North Carolina

Marc Pallardy: Immunotoxicology Group, INSERM U461, Faculté de Pharmacie Paris XI, Chatenay-Malabry, France

Rafael A. Ponce: Department of Environmental Health, University of Washington, Seattle, Washington

Kenneth Ramos: Department of Physiology and Pharmacology, College of Veterinary Medicine, Texas A&M University, College Station, Texas

J. F. Sina: Merck Research Laboratories, West Point, Pennsylvania

Joan B. Tarloff: University of the Sciences in Philadelphia, Pennsylvania

Janis Teichman: SRI International, Menlo Park, California

Stephen G. Whittaker: Department of Environmental Health, University of Washington, Seattle, Washington

Patricia D. Williams: SRA Life Sciences, Falls Church, Virginia

Contents

Preface

Toxicology has made tremendous strides in the sophistication of the models used to identify and understand the mechanisms of agents that can harm or kill humans and other higher organisms. Initially, other people were used as surrogates for monarchs or others. Other animals then came to be used, and until recently, this use, while becoming increasingly refined, also came to serve as the "gold standard" against which truth (at least in regulatory, legal, and economic senses) was judged.

Nonanimals or *in vitro* models timely started to gain significant use in the 1960s. Because of concern about animal welfare, economics, and the need for greater sensitivity and understanding of mechanisms, interest in *in vitro* models has increased. As our technology has advanced, such interest has deepened.

As the contents of this volume demonstrate, an extensive body of *in vitro* models now exists for use in either identifying or understanding most forms of toxicity. The availability of *in vitro* models spans both the full range of endpoints (irritation, sensitization, lethality, mutagenicity, and developmental toxicity) and the full spectrum of target organ systems (skin, eye, heart, liver, kidney, nervous system, etc.). This volume devotes chapters to each of these specialty areas from a perspective of presenting the principal models and their uses and limitations. All of these chapters have been extensively revised and updated since the first edition of this volume appeared.

Chapters that overview the principles involved in the general selection and use of models, and that address the issues of safety concerns and regulatory acceptance of these methods, are also included.

By the time this book sees print, as in any such volume, portions will again be dated but not obsolete. The authors and I hope this will provide a sound basis for broad understanding and utilization of these models.

Shayne Cox Gad

In Vitro Toxicology

Second Edition

Introduction

Shayne Cox Gad

GAD Consulting Services, Raleigh, North Carolina

Toxicology, in the sense used in this volume, is the science concerned with identifying and understanding the mechanisms of agents adversely affecting the health of humans, other animals, and living portions of the environment. Most of it, however, is concerned with those man-made chemical agents adversely affecting the health of humans.

The current test methods designed and used to evaluate the potential of man-made materials to cause harm to the people making, transporting, using, or otherwise coming into contact with them continue to hold a unique and ambivalent place in our society. On the one hand, our society is not only critically dependent on technologic advances to improve or maintain standards of living, but it is also intolerant of risks, real or potential, to life and health that are seemingly avoidable. On the other hand, the traditional tests (with both their misuse and misunderstanding of their use) have served as the rallying point for those individuals concerned about the humane, ethical, and proper use of animals. This concern has caused all testing using animals to come under question on both ethical and scientific grounds, and it has provided a continuous stimulus for the development of alternatives and innovations.

Since 1980, tremendous progress has been made in our understanding of biology down to the molecular level. This progress has translated into many modifications and improvements in *in vivo* testing procedures that now give us tests that (1) are more reliable, reproducible, and predictive of potential hazards in humans, (2) use fewer animals, and (3) are considerably more humane than are earlier test forms. At the same time, a multitude of *in vitro* test systems have been proposed, developed, and "validated" to at least some extent. Yet the perception persists that little has changed in how toxicology testing is performed.

It is hoped that this volume will make more people aware of both the current range of techniques available and the means and extent of their application. More importantly, it is hoped that the whole process involved in testing will continually be modified, so that only what needs to be done will be and that those tests that are done will answer the desired questions in a manner maximizing efficiency, effectiveness, scientific quality, and dependability while limiting any discomfort or suffering in animals.

The entire product safety assessment process, in the broadest sense, is a multistage process in which none of the individual steps is overwhelmingly complex, but the integration of the whole process involves fitting together a large complex pattern of pieces. This volume as a whole seeks to address the questions of the integration of current state-of-the-art *in vitro* methodologies into the product safety assessment process [4].

1.1 DEFINITIONS

Various terms are used to describe the different kinds of testing and research performed by the model systems used. By and large, *in vivo* (though technically implying the use of living organisms) is used to denote the use of intact higher organisms (vertebrates).

In vitro, meanwhile, is used to describe those tests using other than intact vertebrates as model systems. These tests include everything from lower organisms (planaria and bacteria) to cultured cells and computer models. The next section looks in more detail at the different "levels" of *in vivo* models and at their advantages and disadvantages.

In between clearly *in vivo* and *in vitro* models (and overlapping both of them) are the "alternatives." This term has a different meaning to different people. In its broadest sense, it incorporates everything that reduces higher animal usage and suffering in the existing traditional test designs. This definition includes use of the following range of situations:

1. A reduced volume of test material in a rabbit eye irritation test
2. An "up-and-down" method or a limited test design to characterize lethality in the rat
3. Earthworms instead of rats or mice for lethality testing
4. Fish instead of rats or mice for carcinogenicity bioassays
5. Computerized structure activity models for predicting toxicity
6. True *in vitro* models

This volume, however, will concentrate on *in vitro* models.

1.2 LEVELS

As the definitions above illustrate, many approaches to having a predictive model for use in toxicology are available. One way to classify these approaches is presented in Table 1.1, which looks at the different levels of models by their complexity. Each level of approach has advantages and disadvantages, some of which are specific to the concerns and viewpoints of the user. Each of these levels represents a different approach to a problem set.

TABLE 1.1. Levels of models for toxicity testing and research

Level/model	Advantages	Disadvantages
In Vivo (intact higher organism)	Full range of organismic responses	Costs Ethical/animal welfare concerns Species-to-species variability
Lower organisms (earthworms, fish)	Range of integrated organismic responses	Frequently lack responses of higher organism Animal welfare concerns
Isolated organs	Intact, yet isolated tissue and vascular system Controlled environmental and exposure conditions	Donor organisms still required Time consuming and expensive No intact organismic responses Limited length of viability
Cultured cells	No intact animals directly involved Ability to carefully manipulate system Low costs Wide range of variables can be studied	Instability of system Limited enzymatic capabilities and viability of system No or limited integrated multicell or organismic responses
Chemical/biochemical systems	No donor organism problems Low cost Long-term stability of preparation Wide range of variables can be studied Specificity of response	No *de facto* correlation to *in vivo* system Limited to investigation of single defined mechanism
Computer simulations	No animal welfare concerns Speed and low per-evaluation cost	Problematic predictive value beyond narrow range of structures Expensive to establish

Several approaches to *in vitro* toxicity or target organ models are available. The first and oldest approach is that of the isolated organ preparation. Perfused and superfused tissues and organs have been used in physiology and pharmacology since the late 19th century. A vast range of these approaches are available, and a number of them have been widely used in toxicology (ref. 10 presents an excellent overview). Almost any endpoint can be evaluated in most target organs (the central nervous system being a notable exception), and these are closest to the *in vivo* situation and, therefore, generally the easiest to extrapolate or conceptualize from.

Those things that can be measured or evaluated in the intact organism can also largely be evaluated in an isolated tissue or organ preparation. The drawbacks or limitations of this approach are also compelling, however.

An intact animal generally produces one tissue preparation. Such a preparation is viable generally for a day or less before it degrades to the point of losing utility. As a result, such preparations are useful as screens only for agents having rapidly reversible (generally pharmacologic or biochemical) or acute mechanisms of action. They are superb for evaluating mechanisms of action at the organ level for agents acting rapidly, but they are generally not useful for evaluating cellular effects or for evaluating agents acting over a course of more than a day.

The second approach is to use tissue or organ culture. Such cultures are attractive because of maintaining the ability for multiple cell types to interact in at least a near-physiological manner. They are generally not as complex as the perfused organs, but they are stable and useful over a longer period of time, somewhat increasing their utility as screens. They are truly a middle ground between the perfused organs and the cultured cells. Good models performing in a manner representative of the *in vitro* organ are available only for relatively simple organs (such as the skin and bone marrow).

The third and most common approach is that of cultured cell models. These models can be either primary or transformed (immortalized) cells, but the former have significant advantages in use as predictive target organ models. Such cell culture systems can be used to identify and evaluate interactions at the cellular, subcellular, and molecular level on an organ- and species-specific basis [1]. The advantages of cell culture are as follows: (1) Single organisms can generate multiple cultures for use, (2) these cultures are stable and useful for protracted periods of time, and (3) effects can be studied precisely at the cellular and molecular level. The disadvantages are that isolated cells cannot mimic the interactive architecture of the intact organ, and they will respond over time in a manner that becomes decreasingly representative of what happens *in vivo*. An additional concern is that, with the exception of hepatocyte cultures, the influence of systemic metabolism is not factored in unless extra steps are taken. Stammati et al. [15] and Tyson and Stacy [16] present some excellent reviews of the use of cell culture in toxicology. Any such cellular systems would be more likely to be accurate and sensitive predictors of adverse effects if their function and integrity were evaluated while they were operational. For example, cultured nerve cells should be excited while being exposed and evaluated.

1.3 HISTORY

The key assumptions underlying modern toxicology are as follows: (1) Other organisms can serve as accurate predictive models of toxicity in humans, (2) selection of an appropriate model to use is key to accurate prediction in humans, and

(3) understanding the strengths and weaknesses of any particular model is essential to understanding the relevance of specific findings to humans. The nature of models and their selection in toxicologic research and testing have only recently become the subject of critical scientific review. Usually in toxicology when we refer to "models" we really are referring to test organisms, though, in fact, the ways in which parameters are measured (and which parameters are measured to characterize an endpoint of interest) are also critical parts of the model (or, indeed, may actually constitute the "model").

Though principles for test organism selection have been accepted, these have not generally been the final basis for such selection. It is a fundamental hypothesis of both historical and modern toxicology that adverse effects caused by chemical entities in higher animals are generally the same as those induced by those entities in humans. Many critics point to individual exceptions to this and conclude that the general principle is false. Yet, as our understanding of molecular biology advances and we learn more about the similarities of structure and function of higher organisms at the molecular level, the more it becomes clear that the mechanisms of chemical toxicity are largely identical in humans and animals. This increased understanding has caused some of the same people questioning the general principle of predictive value to in turn suggest that our state of knowledge is such that mathematical models or simple cell culture systems could be used just as well as intact animals to predict toxicities in humans. This last suggestion also misses the point that the final expressions of toxicity in humans or animals are frequently the summation of extensive and complex interactions at cellular and biochemical levels. Zbinden [18] published extensively in this area, including an advanced defense of the value of animal models. Lijinsky [9] reviewed the specific issues about the predictive value and importance of animals in carcinogenicity testing and research. Though it was once widely believed (and still is believed by many animal rights activists) that *in vitro* mutagenicity tests would entirely replace animal bioassays for carcinogenicity, this is clearly not the case on either scientific or regulatory grounds. Though differences in the responses of various species (including humans) to carcinogens exist, the overall predictive value of such results (when tempered by judgment) is clear. At the same time, well-reasoned use of *in vitro* or other alternative test model systems is essential to the development of a product safety assessment program that is both effective and efficient [5].

The subject of intact animal models (and of their proper selection and use) has been addressed elsewhere by this author [6] and will not be further addressed here. However, alternative models using other than intact higher organisms are seeing increasing use in toxicology for a number of reasons.

1.3.1 The Four R's

The first and most significant factors behind the interest in so-called *in vitro* systems have clearly been political—an unremitting campaign by a wide spectrum of individuals concerned with the welfare and humane treatment of laboratory animals [14]. The historical beginnings of this campaign were in 1959, when Russell and Burch [13] first proposed what have come to be called the three R's of humane animal use in research: replacement, reduction, and refinement. These R's have served as the conceptual basis for reconsideration of animal use in research.

Replacement means using methods that do not use intact animals in place of those that do. For example, veterinary students may use a canine cardiopulmonary-resuscitation simulator, Resusci-Dog, instead of living dogs. Cell cultures may replace mice and rats that are fed new products to discover substances poisonous to humans. In addition, using the preceding definition of animal, an invertebrate (e.g., a horseshoe crab) could replace a vertebrate (e.g., a rabbit) in a testing protocol.

Reduction refers to the use of fewer animals. For instance, changing practices allow toxicologists to estimate the lethal dose of a chemical with as few as one-tenth the number of animals used in traditional tests. In biomedical research, long-lived animals, such as primates, may be used in multiple sequential protocols, assuming they are not deemed inhumane or scientifically conflicting. Designing experimental protocols with appropriate attention to statistical inference can lead to either a decrease or an increase in the number of animals used. Through coordination of efforts among investigators, several tissues may be simultaneously taken from a single animal. Reduction can also refer to the minimization of any unintentionally duplicative experiments, perhaps through improvements in information resources.

Refinement entails the modification of existing procedures so that animals are subjected to less pain and distress. Refinements may include administration of anesthetics to animals undergoing otherwise painful procedures, administration of tranquilizers for distress, humane destruction before recovery from surgical anesthesia, and careful scrutiny of behavioral indices of pain or distress, followed by cessation of the procedure or the use of appropriate analgesics. Refinements also include the enhanced use of noninvasive imaging technologies that allow earlier detection of tumors, organ deterioration, or metabolic changes and the subsequent early euthanasia of test animals.

Progress toward these first three R's has been previously reviewed [5]. However, a fourth R, responsibility, has been introduced that was not in Russell and Burch's initial proposal. To toxicologists, this is the cardinal R. They may be personally committed to minimizing animal use and suffering and to doing the best possible science of which they are capable, but at the end of it all, toxicologists must stand by their responsibility to be conservative in ensuring the safety of the people using or exposed to the drugs and chemicals produced and used in our society.

During the past decade, issues of animal use and care in toxicologic research and testing have become one of the fundamental concerns of both science and the public. Are our results predictive of what may or may not be seen in humans? Are we using too many animals, and are we using them in a manner that gets the answer we need with as little discomfort on the part of the animal as possible? How do we balance the needs of humans against the welfare of animals?

In 1984, the Society of Toxicology (SOT) held its first symposium and addressed scientific approaches to these issues. The last such symposium for SOT was in 1988. Each year has brought new regulations, attempts at federal and state legislation, and demonstrations directly affecting the practice of toxicology. Increasing amounts of both money and scientific talent have been dedicated to progress in this area. At the same time, the public clearly supports animal use in research when they see a need and benefit, which is shown in Table 1.2.

TABLE 1.2. Public opinion on animal use in research

A 1989 survey conducted for the American Medical Association [2] sampled almost 1,500 households and found that:

- 64% opposed organizations attempting to stop the use of animals in research testing.

- 77% thought animal research was necessary to progress in medicine.

- Other polls have given the same results regarding medical research or general issues of animal research and testing. A majority has been found, however, to oppose animal testing of cosmetics, regarding it as unwarranted [3].

During the same time frame, interest and progress in the development if *in vitro* test systems for toxicity evaluations have also progressed. Early reviews by Hooisma [8], Neubert [11], and Williams et al. [17] record the proceedings of conferences on the subject, but Rofe's 1971 review [12] was the first found by this author. Though it is hoped that in the long term some of these (or other) *in vitro* methods will serve as definitive tests in place of those using intact animals, at present, it appears more likely that their use in most cases will be as screens. Goldberg and Frazier [7] give an overview of the general concepts and status of *in vitro* alternatives. The first edition of this work captured the practical status of the field in the early 1990s.

1.4 DRIVING FORCES

A number of reasons drive toxicology toward a broader use of *in vitro* test systems. These reasons can generally be summarized as political, financial, and technological.

The political reasons are the need to deal with the pressures of the animal welfare movement and its influence on the public and regulators. The economic reasons are based on the rapidly increasing costs of laboratory animals and their upkeep, which translates into spiraling costs for traditional *in vivo* models. The technologic reasons encompass all requirements for having better (i.e., in this case, more predictive of effects in humans) and faster answers.

With increasing scientific need for alternatives to animal experimentation and increasing perception of the potential scientific, ethical, and commercial value of *in vitro* techniques in toxicology, scientific effort in this area increased dramatically over the past few years. It is essential, however, that the potential for the reduction on (or avoidance of) whole-animal experiments should not force irrational acceptance of invalidated tests. It is also important that the value of some *in vitro* approaches should be recognized as complementary to whole-animal experiments at the current state of our knowledge.

The most important advantage of *in vitro* tests, with a potential not yet realized, is that they allow comparisons of the effects of cellular and organ exposure to drugs and chemicals to be extrapolated across the species to include humans through the use of human cell cultures from necropsy or biopsy material. In other words, such techniques have the potential to allow the toxicologic evaluation of compounds in animals and humans on an equal basis, which cannot be achieved in the classic *in vivo* toxicologic testing. The difficulties experienced by many laboratories in obtaining human tissue cannot be ignored, however, and are a serious impedance to progress in this area. The comparison across species may be extended further to the establishment of cultured cell lines that can allow scientists in different laboratories to compare results and permit the necessary standardization fundamental to good scientific practice.

It remains the opinion of the author that it is essential to establish standards of methodology that will allow parallel assessment by independent laboratories of *in vitro* parameters of closest relevance to the *in vivo* situation. Certainly much is at stake, concerning both safety and the expensive commercial risk, in interpretations of *in vitro* data that may "kill" a perfectly valid development compound or allow an unacceptably toxic compound to proceed, with ultimate adverse effects in humans.

This issue leads inevitably to questions on the predictive value of positive or negative results *in vitro*, but the weighing given to such tests can only be established with the experience of time and hard data, by relating *in vitro* observations to proven *in vivo* effects with well-studied compounds. It is clearly important that the aim of all laboratories should be to establish *in vitro* endpoints that will bear the closest possible relationship to responses obtained *in vivo*. If this result is not achieved, *in vitro* parameters will not gain the required scientific and regulatory acceptance in relation to their relevance to ultimate safety *in vivo* in humans.

Although a predominance of work in the field of genotoxicity testing has occurred, *in vitro* approaches are moving into the field of immunotoxicology and will inevitably expand to the study of all potential target organs. However, the continuing debate concerning the validity of many genotoxicity tests, particularly regarding their ability to predict potential carcinogens, emphasizes the rigorous evaluation that must be applied to *in vitro* approaches if they are to gain acceptance by the scientific community.

Following the thalidomide tragedy and the establishment of regulatory authorities to consider new drugs and other chemicals, the pharmaceutical and other industries set up more formal toxicologic evaluation procedures. "In-house" toxicology departments were soon supplemented by contract research organizations, and a series of standard "regulatory" tests became an international requirement for all new compounds that might be taken by humans (and other animals) either deliberately or accidentally. The objective of these stereotyped studies was to identify the nature of the toxicity of a compound and to assess the potential risks by extrapolation from the toxic responses at various dose levels to the therapeutic dose in humans, with the highest "no-effect" dose in animals being used as the basis of the so-called "therapeutic ratio."

Although the therapeutic ratio is a valid concept, one of the fundamental guiding principles of toxicologists is that they are not trying to demonstrate that a potential drug or other chemical is nontoxic. The fact that all chemicals are toxic has been well recognized for centuries.

If we accept that all chemicals have some potential hazard, it follows that toxicologists are not looking at a compound to see whether it is toxic, but to find out the degree of toxicity and the nature of the toxicity. What is important in the development of drugs is the ration between the therapeutic and the toxic doses or blood levels and between the desired and the unwanted effects. The "regulatory" tests achieve this balance with a degree of certainty.

Another important guiding principle for the toxicologist is that toxicology is essentially a predictive science. We study the nature of the toxic effect to assess the risk to humans. Unfortunately, sufficient instances of toxic effects have occurred only after wide exposure of humans to compounds, which have previously fully satisfied international regulatory requirements with regard to animal testing. This has raised questions in the minds of many toxicologists as to the predictability of "standard" tests for many substances.

The toxicologist is going to administer increasingly higher doses of a compound to experimental animals to identify target organs or other limiting toxicity. After doing this procedure in collaboration with colleagues in other disciplines, the toxicologist has to make risk assessments and contribute to the development and selection of other candidate compounds. Once target organ or limiting toxicity is identified, mechanistic studies are required, and it is here that *in vitro* techniques are now becoming widely and increasingly used.

Subject to the validation discussed above, these tests must surely take their place alongside, for example, biochemical or pharmacologic tests *in vitro* at the subcellular, cellular, or organ level, which are currently used together with *in vivo* tests in forming an overall, and more complete, scientific picture of a new test compound.

Often, whole-animal studies may not be appropriate for mechanistic studies because in many cases the adverse effect becomes apparent only after long periods of chronic dosing.

It has often been possible to demonstrate that risk to humans is not likely once the mechanism of action of the adverse effect in the experimental animals has been understood. Sometimes, the species specificity of toxic effects has been confirmed by using cell cultures from several species, including humans. When the mechanism of an adverse effect seen in whole-animal studies has been studied and is considered to be perhaps predictive of risk in humans, the compound may need evaluation in short-term models designed to detect potential adverse effects when only small amounts of compound are available. *In vitro* techniques are frequently the most appropriate means of doing this. In both economic terms and use of animals, such early comparative tests with different compounds in *in vitro* systems must be attractive. The extension of this technique to multicompound screening is a matter of individual research and development strategy. Such short-term models may also be used in drug design, because when playing "molecular roulette," potency data in biologic or pharmacologic assays provide only part of the information required by the medical chemist. A modern, cost-effective approach to drug design must take into account toxicologic potential as well as inherent biologic activity.

It is anticipated that the progress of the science of toxicology in the pharmaceutical, agrochemical, and other similar industries will lead increasingly to mechanistic approaches to toxicology and increasing use of *in vitro* techniques and models.

Contributors to this book have covered almost every area of toxicology and have used a full range of *in vitro* techniques ranging from mammalian cell lines (including human) at one end of the spectrum to abattoir material (eyes) and classic pharmacologic-isolated organ techniques (hearts) at the other. Whether chicken eggs used for studies on chorioallantoic membranes or whole-rat embryos are *in vitro* or *in vivo* are moot points, but they certainly represent humane alternatives to the use of whole animals and provide elegant investigational tools and models for toxicologic study.

It is clear that this volume should provide a valuable reference for scientists involved in the toxicologic investigation and evaluation of potential new drugs, agrochemicals, food additives, and so on. It should interest graduate and postgraduate students and research workers in toxicology because this subject becomes an integral part of the training of toxicologists, particularly because the individual chapters not only cover the philosophy and strategy of the use of *in vitro* models, but they also give attention to detailed methodology.

References

1. Acosta D, Sorensen EMB, Anuforo DC, et al. An *in vivo* approach to the study of target organs toxicity of drugs and chemicals. *In Vitro Cell Dev Biol Anim* 1985;21:495–504.

2. American Medical Association. Public support for animals in research. *Ann Med News* 1989;June 9.

3. Cowley G, Hager M, Drew L, et al. The battle over animal rights. *Newsweek* 1988;Dec 26.

4. Gad SC. A tier testing strategy incorporating *in vitro* testing methods for pharmaceutical safety assessment. *Hum Innov Alter Anim Exp* 1989;3:75–79.

5. Gad SC. Recent developments in replacing, reducing and refining animal use in toxicologic research and testing. *Fundam Appl Toxicol* 1990;15(1):8–16.

6. Gad SC, Chengelis CP. *Animal Models in Toxicology*. New York: Marcel Dekker, 1992.

7. Goldberg AM, Frazier JM. Alternatives to animals in toxicity testing. *Sci Am* 1989;261:24–30.

8. Hooisma J. Tissue culture and neurotoxicology. *Neurobehav Toxicol Teratol* 1982;4:617–622.

9. Lijinsky W. Importance of animal experiments in carcinogenesis research. *Environ Mol Mutagen* 1988;11:307–314.

10. Mehendale HM. Application of isolated organ techniques in toxicology. In: Hayes AE, ed. *Principles and Methods of Toxicology*. New York: Raven, 1989;699–740.

11. Neubert D. The use of culture techniques in studies on prenatal toxicity. *Pharmacol Ther* 1982;18:397–434.

12. Rofe PC. Tissue and culture toxicology. *Food Cosmet Toxicol* 1971;9:685–696.

13. Russell WMS, Burch RL. *The Principles of Humane Experimental Technique*. London: Methuen, 1959.

14. Singer P. *Animal Liberation: A New Ethic for Our Treatment of Animals*. New York: Random House, 1975.

15. Stammati AP, Silano V, Zucco F. Toxicology investigations with cell culture systems. *Toxicology* 1981;20:91–153.

16. Tyson CA, Stacy NH. *In vitro* screens from CNS, liver and kidney for systemic toxicity. In: Mehlman M, ed. *Benchmarks: Alternative Methods in Toxicology*. Princeton, NJ: Princeton Scientific, 1989:111–136.

17. Williams GM, Dunkel VC, Ray VA, eds. Cellular systems for toxicity testing. *Ann NY Acad Sci* 1983;407.

18. Zbinden G. *Predictive Value of Animal Studies in Toxicology*. Carshalton, U.K.: Centre for Medicines Research, 1987.

2

General Principles for *In Vitro* Toxicology

Shayne Cox Gad

GAD Consulting Services, Raleigh, North Carolina

As introduced in the last chapter, *in vitro* methods actually encompass a broad range of techniques and models for use in toxicity testing. These techniques have varying degrees of reliability and acceptance. Some techniques may be directly substituted in place of existing *in vivo* models, and other techniques are currently suitable only as screens or adjunct tests [1,9]. The continuing challenge to the practicing toxicologist is the appropriate and timely selection and use of new models and methodologies [11]. The essential starting place for such decisions is a clear statement and understanding of the objective behind any testing program, along with an understanding of the entire safety assessment process.

The entire product safety assessment process, in the broadest sense, is a multistage process in which none of the individual steps is overwhelmingly complex, but the integration of the whole process involves fitting together a large complex pattern of pieces. The most important part of this product safety evaluation program is, in fact, the initial overall process of defining and developing an adequate data package on the potential hazards associated with the product life cycle (the manufacture, sale, use, and disposal of a product and associated process materials). To do this process, one must ask a series of questions in a highly interactive process, with many of the questions designed to identify or modify their successors. First, what is the objective of the testing (i.e., what question is being asked) being conducted?

Required here are (1) an understanding of the way in which a product is to be made and used and (2) an awareness of the potential health and safety risks associated with the exposure of humans associated with these processes or the product's use. Such an understanding and awareness is the basis of a hazard and toxicity profile. Once such a profile is established, the available literature should be searched to determine what is already known.

Taking into consideration this literature information and the previously defined exposure profile, a tier approach (Fig. 2.1) has traditionally been used to generate a list of tests or studies to be performed. What goes into a tier system is determined by (1) regulatory requirements imposed by government agencies, (2) the philosophy of the parent organization, (3) economics, and (4) available technology.

TIER TESTING			
Testing tier	Mammalian Toxicology	Genetic Toxicology	Remarks
0	Literature review	Literature review	Upon initial identification of a problem, database of existing information and particulars of use of materials are established
1	Cytotoxicity screens Dermal sensitization Acute systemic toxicity Lethality screens	Ames test *In vitro* SCE *In vitro* cytogenetics Forward mutation/CHO	R&D material and low volume chemicals with severely limited exposure
2	Subacute studies Metabolism Primary dermal irritation Eye irritation	*In vivo* SCE *In vivo* cytogenetics	Medium volume materials and/or those with a significant chance of human exposure
3	Subchronic studies Reproduction Developmental toxicology Chronic studies Mechanistic studies		Any materials with high volume or a potential for widespread or long-term human exposure or one which gives indications or specific long-term effects

FIG. 2.1. The usual way of characterizing the toxicity of a compound or product is to develop information in a tiered manner. More information is required (i.e., a higher tier level is attained) as the volume of production and potential for exposure increase. A common scheme is shown.

How such tests are actually performed is determined on one of two bases. The first (and most common) basis is the menu approach: selecting a series of standard design tests as "modules" of data, and then modifying the design of each module to meet the specifics of the particular case. The second basis is an interactive/iterative approach, in which strategies are developed and studies are designed based both on needs and on what has been learned to date about the product. This process has been previously examined in some detail [4,7]. Our interest here, however, is the specific portion of the process involved in generating data (namely, the test systems), and we are also interested in how *in vitro* systems may be incorporated.

2.1 TEST SYSTEMS: CHARACTERISTICS, DEVELOPMENT, AND SELECTION

Any useful test system must be sufficiently sensitive that the incidence of false negatives is low. Clearly, a high incidence of false negatives is intolerable. In such a situation, large numbers of dangerous chemical agents would be carried through extensive additional testing only to find that they possess undesirable toxicologic properties after the expenditure of significant time and money. On the other hand, a test system that is overly sensitive will give rise to a high incidence of false positives, which will have the deleterious consequence of rejecting potentially beneficial chemicals. The "ideal" test will fall somewhere between these two extremes and thus provide adequate protection without unnecessarily stifling development.

The "ideal" test should have an endpoint measurement that provides data, such that dose-response relationships can be obtained. Furthermore, any criterion of effect must be sufficiently accurate in the sense that it can be used to reliably resolve the relative toxicity of two test chemicals, which produce distinct (in terms of hazard to humans), yet similar responses. In general, it may not be sufficient to classify test chemicals into generic toxicity categories. For instance, a test chemical falling into an "immediate" toxicity category, yet borderline to the next, more severe toxicity category, should be treated with more concern than is a second test chemical falling at the less toxic extreme of the same category. Therefore, it is essential for any credible test system to be able to both place test chemicals in an established toxicity category and rank materials relative to others in the category.

The endpoint measurement of the "ideal" test system must be objective, which is important so that a given test chemical will give similar results when tested using the standard test protocol in different laboratories. If it is not possible to obtain reproducible results in a given laboratory over time or between various laboratories, the historical database against which new test chemicals are evaluated will be time/laboratory dependent. If this condition is the case, significant limitations on the application of the test system will occur because it could potentially produce conflicting results. From a regulatory point of view, this possibility would be highly undesirable. Along these lines, it is important for the test protocol to incorporate internal standards to serve as quality controls. Thus, test data could be represented using a reference scale based on the test system response to the internal controls. Such normalization, if properly documented, could reduce intertest variability.

From a practical point of view, several additional features of the "ideal" test should be satisfied. *In vitro* alternatives to current *in vivo* test systems basically should be designed to evaluate the observed toxic response in a manner as closely predictive of the outcome of interest in humans as possible. In addition, the test should be fast enough that the turnaround time for a given test chemical is reasonable for the intended purpose (rapid for a screen, timely for a definitive test). Obviously, the speed of the test and the ability to conduct tests on several chemicals simultaneously will determine the overall productivity. The test should be inexpensive, so that it is economically competitive with current testing practices. Finally, the technology should be easily transferred from one laboratory to another without excessive capital investment (relative to the value of the test performed) for test implementation.

It should be kept in mind that though some of these practical considerations may appear to present formidable limitations for any given test system at the present time, the possibility of future developments in testing technology could overcome these obstacles. In the real-world environment, these practical considerations are grounds for consideration of multiple new candidate tests on the basis of competitive performance. The most predictive test system in the universe of possibilities will never gain wide acceptance if it takes years to produce an answer or costs substantially more than are other test systems only marginally less predictive.

The point is that these characteristics of the "ideal" test system provide a general framework for evaluation of alternative test systems in general. No test system is likely to be "ideal." Therefore, it will be necessary to weigh the strengths and weaknesses of each proposed test system to reach a conclusion on how "good" a particular test is.

In both theory and practice, *in vivo* and *in vitro* tests have potential advantages. Tables 2.1 and 2.2 summarize their advantages. How then might the proper tests be selected, especially in the case of the choice of staying with an existing test system or adopting a new one? The next section will present the basis for selection of specific tests.

TABLE 2.1. Rational for using *in vivo* test systems

1. They provide evaluation of actions/ effects on intact animal and organ/tissue interactions.

2. Either neat chemicals or complete formulated products (complex mixtures) can be evaluated.

3. Either concentrated or diluted products can be tested.

4. They yield data on the recovery and healing processes.

5. They are the required statutory tests for agencies under such laws as the Federal Hazardous Substances Act (unless data are already available), Toxic Substances Control act, Federal Insecticides, Fungicides, and Rodenticides (FIFRA), Organization for Economic Cooperation (OECD), and the Food and drug Administration (FDA) laws.

6. Quantitative and qualitative tests with scoring system are generally capable of ranking materials as to relative hazards.

7. They are amenable to modifications to meet the requirements of special situations (such as multiple dosing or exposure schedules).

8. They have an extensive available database and cross-reference capability for evaluation of relevance to human situation.

9. They involve the ease of performance and relative low capital costs in many cases.

10. Tests are generally both conservative and broad in scope, providing for maximum protection by erring on the side of overprediction of hazard to humans.

11. Tests can be either single endpoint (such as lethality, corrosion, etc.) or shot-gun (also called multiple endpoint, and includes such test systems as a 13-week oral toxicity study).

TABLE 2.2. Limitations of *in vivo* testing systems serving
as a basis for seeking *in vitro* alternatives for toxicity tests

1. Complications and potential confounding or masking findings of *in vivo* systems.

2. *In vivo* systems may only assess short-term site of application or immediate structural alterations produced by agents. Specific *in vivo* tests may only be intended to evaluate acute local effects (however, this may be a purposeful test system limitation).

3. Technician training and monitoring are critical (particularly due to the subjective nature of evaluation).

4. *In vivo* test in animals do not perfectly predict results in humans if the objective is to exclude or identify severe-acting agents.

5. Structural and biochemical differences between test animals and humans make extrapolation from one to the other difficult.

6. Lack of standardization of *in vivo* systems.

7. Variable correlation with human results.

8. Large biological variability between experimental units (i.e., individual animals).

9. Large, diverse, and fragmented databases which are not readily comparable.

2.1.1 Considerations in Adopting New Test Systems

Conducting toxicologic investigations in two or more species of laboratory animals is generally accepted as being a prudent and responsible practice in developing a new chemical entity, especially one that is expected to receive widespread use and to have exposure potential over human lifetimes. Adding a second or third species to the testing regimen offers an extra measure of confidence to the toxicologist and the other professionals responsible for evaluating the associated risks, benefits, and exposure limitations or protective measures. Although undoubtedly broadening and deepening a compound's profile of toxicity, the practice of enlarging on the number of test species is an indiscriminate scientific generalization, as has been demonstrated in multiple points in the literature (as reviewed in ref. 8). Moreover, such a tactic is certain to generate the problem of species-specific toxicoses. These toxicoses are defined as toxic responses or inordinately low biologic thresholds for toxicity evident in one species or strain, whereas all other species examined are either unresponsive or strikingly less sensitive. Species-specific toxicoses usually imply that different metabolic pathways for converting or excreting xenobiotics are involved or that anatomic differences are involved. The investigator confronting such findings must be prepared to address the all-important question: Are humans likely to react positively or negatively to the test agent under similar circumstances? Assuming that numeric odds prevail and that humans automatically fit into the predominant category would be scientifically irresponsible, whether on the side of being safe or at risk. Such a confounded situation can be an opportunity to advance more quickly into the heart of the search for predictive information. Species-specific toxicoses can frequently contribute toward a better understanding of the general case if the underlying biologic mechanism either causing or enhancing toxicity is defined, especially if it is discovered to uniquely reside in the sensitive species.

The designs of our current tests appear to serve society reasonably well (i.e., significantly more times than not) in identifying hazards that would be unacceptable. However, the process can just as clearly be improved from the standpoint of both improving our protection of society and doing necessary testing in a manner that uses fewer animals in a more humane manner.

Substantial potential advantages exist in using an *in vitro* system in toxicologic testing; these advantages include (1) isolation of test cells or organ fragments from homeostatic and hormonal control, (2) accurate dosing, and (3) quantitation of results. It should be noted that, in addition to the potential advantages, *in vitro* systems per se also have a number of limitations that can contribute to there being unacceptable models. Findings from an *in vitro* system that either limit their use in predicting *in vivo* events or make them totally unsuitable for the task include wide differences in the doses needed to produce effects or differences in the effects elicited. Some reasons for such findings are detailed in Table 2.3.

Tissue culture has the immediate potential to be used in two very different ways in industry. First, it can be used to examine a particular aspect of the toxicity of a compound in relation to its toxicity *in vivo* (i.e., mechanistic or explanatory studies). Second, it can be used as a form of rapid screening to compare the toxicity of a group of compounds for a particular form of response. Indeed, the pharmaceutical industry has used *in vivo* test systems in these two ways for years in the search for new potential drug entities.

The theory and use of screens in toxicology has previously been reviewed by this author [4–6]. Mechanistic and explanatory studies are generally called for when a traditional test system gives a result that is unclear or for which the relevance to the real-life human exposure is doubted. *In vitro* systems are particularly attractive for such cases because they can focus on well-defined single aspects of a problem or pathogenic response, free of the confounding influence of the multiple responses of an intact higher level organism. Note, however, that first one must know the nature (indeed, the existence) of the questions to be addressed. It is then important to devise a suitable model system that is related to the mode of toxicity of the compound.

Currently, much controversy exists over the use of *in vitro* test systems: Will they find acceptance as "definitive test systems," or only be used as preliminary screens for such final tests? Or in the end, will they not be used at all? Almost certainly, all three of these cases will be true to some extent. Depending on how the data generated are to be used, the division between the first two is ill-defined at best.

Before trying to definitively answer these questions in a global sense, each of the endpoints for which *in vitro* systems are being considered should be overviewed and considered against the factors outlined up to this point.

TABLE 2.3. Possible interpretations when in vitro
data do not predict results of *in vivo* studies

1. Chemical is not absorbed at all, or it is poorly absorbed, in *in vivo* studies.

2. Chemical is well absorbed but subject to "first-pass effect" in the liver.

3. Chemical is distributed so that less (or more) reaches the receptors than would be predicted on the basis of its absorption.

4. Chemical is rapidly metabolized to an active or inactive metabolite that has a different profile of activity or different duration of action than does the parent drug.

5. Chemical is rapidly eliminated (e.g., through secretory mechanism).

6. Species of the two test systems used are different.

7. Experimental conditions of the *in vitro* and *in vivo* experiments differed and may have led to different effects than expected. These conditions include factors such as temperature or age, sex, and strain of animal.

8. Effects elicited *in vitro* and *in vivo* by the particular test substance in question differ in their characteristics.

9. Tests used to measure responses may differ greatly for *in vitro* and *in vivo* studies, and the types of data obtained may not be comparable.

10. The *in vitro* study did not use adequate controls (e.g., pH, vehicle used, volume of test agent given, samples taken from sham-operated animals), resulting in "artifacts" of method rather than results.

11. *In vitro* data cannot predict the volume of distributional central or peripheral compartments.

12. *In vitro* data cannot predict the rate constants for chemical movement between compartments.

13. *In vitro* data cannot predict the rate constants of chemical elimination.

14. *In vitro* data cannot predict whether linear or nonlinear kinetics will occur with specific dose of a chemical *in vivo*.

15. Pharmacokinetic parameters (e.g., bioavailability, peak plasma concentration, half-life) cannot be predicted based solely on *in vitro* studies.

16. *In vivo* effects of chemical are caused by an alteration in the higher order integration of an intact animal system, which cannot be reflected in a less complex system.

2.2 TARGET ORGAN TOXICITY MODELS

This final model review section addresses perhaps the most exciting potential area for the use of *in vitro* models—as specific tools to evaluate and understand discrete target organ toxicities. Here, the presumption is that a reason to believe (or at least suspect) exists that some specific target organ (nervous system, lungs, kidney, liver, heart, etc.) is or may be the most sensitive site of adverse action of a systemically absorbed agent. From this starting point, a system that is representative of the target organ's *in vivo* response would be useful in at least two contests.

First, as with all of the other endpoints addressed in this chapter, a target organ predictive system could serve as a predictive system (in general, a screen) for effects in intact organisms, particularly humans. As such, the ability to identify those agents with a high potential to cause damage in a specific target organ at physiologic concentrations would be extremely valuable.

The second use is largely specific to this set of *in vitro* models, which is to serve as tools to investigate, identify, or verify the mechanisms of action for selective target organ toxicities. Such mechanistic understandings then allow for one to know if such toxicities are relevant to humans (or to conditions of exposure to humans), to develop means to either predict such responses while they are still reversible or to develop the means to intervene in such toxosis (i.e., first aid or therapy), and, finally, to potentially modify molecules of interest to avoid unwanted effects while maintaining desired properties (particularly important in drug design).

In the context of these two uses, the concept of a library of *in vitro* models [5,6] becomes particularly attractive. If one could accumulate a collection of "validated," operative methodologies that could be brought into use as needed (and put away, as it were, while not being used), this would represent an extremely valuable competitive tool. The question becomes one of selecting which systems/tools to put into the library, and how to develop them to the point of common utility.

Additionally, one must consider what forms of markers are to be used to evaluate the effect of interest. Initially, such markers have been exclusively either morphologic (in that a change in microscopic structure occurs), observational (is the cell/preparation dead or alive, or has some gross characteristic changed?), or functional (does the model still operate as it did before?). Recently, it has become clear that more sensitive models do not generate just a single endpoint type of data, but rather a multiple set of measures providing in aggregate a much more powerful set of answers.

A wide range of target–organ-specific models have already been developed and used. Their incorporation into a library-type approach requires that they be evaluated for reproducibility of response, ease of use, and predictive characteristics under the intended conditions of use. These evaluations are probably at least somewhat specific to any individual situation. The remaining chapters in this volume address each of these applications in some detail.

2.3 SENSITIVITY, SPECIFICITY, AND PREDICTIVE VALUE

Two of the key issues that must be confronted when considering any new test system are predictive value and sensitivity. These issues (along with the general scientific requirement of reproducibility) are points that must be evaluated for an *in vitro* system. Both of these characteristics are essential for a test system to be able to identify situations (in statistical terms, "population elements") that are different in some specified manner.

Sensitivity determines how much "power" a test has: How much different does an endpoint need to be before it is identified as different (or an effect is detected)? Predictive value determines how selective a test is in determining that an effect is present. A highly specific test will detect only "real" effect with a high level of confidence. These two characteristics are not independent. That is, changes in one characteristic result in changes in the other, all other factors being held constant. This relationship is made clear by first considering a possible test outcome, as in Table 2.4.

TABLE 2.4. Possible test outcomes

	True Result	
Test Outcome	Positive	Negative
Positive	a	b
Negative	c	d

Sensitivity is then defined as $a/(a + c)$, where a is all true positives detected by a test and $a + c$ represents all true positives.

Specificity is then equal to $d/(b + d)$. Predictive value can now be defined as $a/(a + b)$—that is, the percent of cases identified as positive that are actually positive [2]. An increase in sensitivity must bring with them some degree of cost in terms of type II error—that is, an increase in the number of false positives. The statistical characteristics of test performance have been discussed elsewhere [5].

If the operating technology or basis of interpretation of a test system is changed, of course, this redefines the basic operating characteristics. Thus, for example, cultured cell systems directly incorporating some form of metabolic activation into their operations (say, by being based on coculture with hepatocytes) have potentially more favorable values for a, b, c, and d as a starting place.

2.4 PROBLEMS IN INTERPRETATION AND EXTRAPOLATION

Perhaps the principal barrier against more widespread use of existing *in vitro* tests (and against gaining support for development of new ones) is the difficulty in interpreting the outcome of tests (especially when one considers the issues of sensitivity and predictive value presented earlier) and in extrapolating these results to potential effects in people. What changes are looked at in a model (say, cell culture) system to provide prediction of a specific intact animal endpoint? Simple lethality to a cell-based system does not imply target organ toxicity simply because the cells in question are those that constitute the organ in question. It may simply be cytotoxicity, which is meaningful only if the cells in question are selectively more sensitive than are other cell types. Alterations in functionality that are specific to the cell type (or organ/tissue in question) is more likely to be predictive of a selective toxicity. It is also important that the concentrations of toxicant that yield a positive outcome in the media of an *in vitro* system be relevant to (and hopefully in a known manner, likely related to) those tissue or plasma concentrations causing effects in intact animals. If it takes higher levels of *in vitro* to produce a toxicity than it does of *in vivo*, the relevance of the model should be questioned, which also means, by the way, that pharmacokinetic data are of significant value in designing and interpreting *in vitro* test systems.

At the same time, *in vitro* systems will tend to respond differently than does the intact animal in some ways, which will concern traditional toxicologists. For example, concentration responses of cultured cell systems tend to be fairly sharp (somewhat "all or none") even though for the same target organ endpoint, one will see a graded dose response in animals or humans, because the cells in a culture system are much more homogeneous than are those in a group of animals (or even the cells comparing the target organs of a single animal), because they are near-clonal. That is, they have been derived from a small number of parent cells.

Extrapolation to outcome in humans requires knowledge of the (at least projected) pharmacokinetics of the compound in question and an appreciation of the limitation (time course or limits on cell-to-cell or organ-to-organ interactions) of the *in vitro* system in question. The most likely reasons for failure in such extrapolation were presented earlier in Table 2.3.

Both interpretation and extrapolation can be made less error-prone if proper controls are incorporated into a test system. During development, it is optimal if agents for which data exist in intact animals (the donor species for the *in vitro* system) and humans are available and if a human cell or tissue-based system can also be evaluated. Using this approach, standards (i.e., known positive and negative response compounds) should be established. Subsequent use of the new test system must incorporate regular reference back to the response of these compounds. Likewise, simple osmolarity and cell preparation viability controls should also be included.

2.5 VALIDATION

Validation is a somewhat ill-defined concept that currently is the principal stumbling block impeding use of *in vitro* tests for many toxicologists. It is ill-defined because no fixed process exists and many people mean "regulatory (or peer) acceptance" when they say validation. The issues and considerations involved in conducting a validation have been most recently (and extensively) covered by Frazier [3], and the interested reader is referred to that source for a detailed discussion. In general, the major points to consider are as follows:

1. Is the method reproducible (will it give the same results to all who use it)? This reproducibility must be established both intralab and interlab.

2. Is the method predictive of the outcome in the species of concern? It is not essential that it gives the same results as those of established animal tests, but they should be close.

3. Are the ways in which the method fails known?

Historically, new test systems in the biomedical sciences were proposed in the literature. If they withstood the tests of peer review and being reproduced by others, they were used by more and more people until they became the "accepted method" and were eventually picked up in guidelines and regulations. This was the traditional scientific process, but it is now viewed as not being defined, rigorous, and timely enough. Under the Interagency Coordinating Committee on the Validation of Alternative Methods (ICCVAM) initiative, starting in 1994, a formal process was developed for validations and regulatory acceptance of toxicologic testing methods. This initiative has provided the vital missing links for the process to move forward [10].

References

1. Bennenuto AJ, Cohen R. A realistic role for non-animal test. *Pharmacol Exec* 1990;June.

2. Cooper JA, Saracci R, Cole P. Describing the validity of carcinogen screening test. *Br J Cancer* 1979;39:87–89.

3. Frazier JM. *Scientific Criteria for Validation of In Vitro Toxicity Tests*. Brussels: Organization for Economic Co-Operation and Development, 1990.

4. Gad SC, ed. *Handbook of Product Safety Evaluation*, 2nd ed. New York: Marcel Dekker, 1999.

5. Gad SC. Statistical analysis of screening studies in toxicology: with special emphasis on neurotoxicity. *J Am Coll Toxicol* 1989;8(1):171–183.

6. Gad SC. A tier testing strategy incorporating in vitro testing methods for pharmaceutical safety assessment. *Hum Innov Alter Anim Exp* 1989;3:75–79.

7. Gad SC. Industrial application of in vitro toxicity testing methods: a tier testing strategy for product safety assessment. In: Frazier J, ed. *In Vitro Toxicity Testing*. New York: Marcel Dekker, 1991.

8. Gad SC, Chengelis CP. *Acute Toxicology: Principles and Methods*, 2nd ed. San Diego, CA: Academic Press, 1997.

9. Gad SC. Current status and unmet model/assay needs in the use of alternatives in biologic safety testing in the United States. *Toxicol Methodol* 1998;7:311–318.

10. *Validation and Regulatory Acceptance of Toxicological Test Methods*. NIH Publication 97-3981. March 1997.

11. Gad SC. Strategies of application of in vitro methods to the development of pharmaceuticals and devices. In: Salem H, Katz SA, eds. *Advances in Animal Alternatives for Safety and Efficacy Testing*. Philadelphia: Taylor & Francis, 1998:293–302.

3

Ocular Toxicity Assessment *In Vitro*

J. F. Sina[1] and P. D. Gautheron[2]

[1]Merck Research Laboratories, West Point, Pennsylvania
[2]Laboratoires Merck Sharp and Dohme-Chibret, Riom, France

Given the number of chemicals workers as well as the general public are exposed to each day, a real need exists to identify potential hazards associated with this exposure. For predicting ocular irritation potential, the Draize method [1] has been, and continues to be, a standard procedure, despite a number of criticisms. These drawbacks include a substantial intralaboratory and interlaboratory variability [2,3], subjectivity of the scoring, questions of extrapolation to humans, and animal welfare concerns. Because of these issues, much work has been done in recent years to find modifications or alternatives for the Draize test. This research initially centered on both modifications to the *in vivo* test and the search for *in vitro* or *ex vivo* techniques. At this point, however, the emphasis seems to be shifting from development of new tests to attempts to validate existing tests and to apply data derived from these methods as part of a hazard assessment process. In this chapter, we will explore the types of information available (exclusive of animal data), development and selection of *in vitro* models, incorporation of multiple pieces of data into a decision-making paradigm, efforts at and barriers to moving these paradigms into more general use within and across industries, and regulatory efforts to deal with acceptance of alternative data in lieu of an *in vivo* test.

3.1 IN VIVO IRRITATION TESTING

As will be discussed more fully below, to successfully develop and apply an *in vitro* method, one needs an understanding of the technical basis, underlying mechanism, and limitations of the *in vivo* test. Some of the key points of the technical performance of the Draize test will be discussed here, but the reader is referred to the review of Chan and Hayes [4] for a more detailed examination of the standard Draize methodology as well as modifications. Basically, the Draize assay is a subjective test in which 0.1 ml of a liquid or 0.1 g of a solid test material is placed into the conjunctival sac of one eye of a rabbit, the other eye serving as the control. At various times after dosing, observations are made and a numerical score is assigned based on the extent and severity of corneal opacity, redness of the iris, chemosis of the conjunctiva, and discharge. The bulk of the score (80 of a possible 110) comes from observations of corneal opacity (and, thus, one would expect that any measurement of opacity or related parameters should identify those compounds that would induce the most dramatic changes *in vivo*). The maximal score for conjunctival changes is 20, and 10 for effects on the iris. Based on the total score, chemicals are classified as nonirritating, mild, moderate, severe, or extreme.

Many different scoring systems have been developed to try to more precisely reproducibly describe and reproduce the irritation potential of chemicals (cf., Kay and Calandra [5]; cosmetic, AFNOR, and Organization for Economic Cooperation and Development (OECD), as cited in ref. 6). And the number of distinct irritation categories can vary, for example, from a 1 to 10 scale [7] in a four-category (nonirritating, mild, moderate, severe) classification by Green et al. [8] to an FHSA scale [9], in which a material is either irritating or nonirritating. This diversity in scoring methods and categorization of chemicals has contributed to sometimes significant discrepancies in comparing irritation potential among chemicals, and it has highlighted the subjective character of the test. In addition, as will be illustrated below, the diversity of scoring makes it difficult to establish a consistent "gold standard" against which to compare an *in vitro* result.

3.2 DEVELOPMENT OF ALTERNATIVES

Given these issues with the Draize, if one could develop a method for testing ocular irritation potential using a nonwhole-animal approach with objective measurement(s), it would be a major improvement over the current rabbit eye test. One strategy for approaching this problem would be to determine what information is already available, on which a prediction of ocular irritation may be made (i.e., assessment based on no further biologic testing.) If the data are insufficient, and further testing is required, one must determine which biologic parameters will best define the *in vivo* response to irritants, and then develop a model adequately measuring these endpoints. For the purposes of this chapter, we will first discuss some of the types of information that might be used to predict ocular irritation without further testing, and then we will discuss *in vitro* models.

3.2.1 Prediction from Preexisting Data

Because of the diversity of chemicals to be tested and the number of different ways in which the results will be used (worker safety, transportation regulation, consumer product safety, possible litigation), it is important to find out as much as possible about any test substance. And because of the large number of chemicals needing to be tested, one needs to proceed as quickly as practical without sacrificing accuracy. As a starting point, then, it would be practical to ask whether any information is already available that can be used for making decisions?

Databases. Using ocular irritation data previously collected by other scientists and published in the literature or contained in computer databases seems like a reasonable place to start investigating a new chemical entity. The problem, however, is that such data are limited. Because the Draize test is a routine assay, not contributing new expertise, irritation data are not often published. Some effort is being made to provide a forum for this type of data, for instance, in the Acute Toxicity Data section of *The Journal of the American College of Toxicology*, but to date a limited amount of information is available. When data exist, for instance, in the publications of Carpenter and Smyth [7], in which they are reduced to an arbitrary 1 to 10 scale, or in Grant [10], because the data were collected over a number of years and from a number of sources, discrepancies in scoring tend to increase (because of variations in individuals performing the test, animals, test protocols, etc.) and interpretation becomes confusing. Contributing to the discrepancies is the variability and lack of strict reproducibility in the first place (as shown by the study of Weil and Scala [2]). Further, in view of the number of different scoring methods (cited above), making comparisons from laboratory to laboratory is difficult. And when the data are reduced to irritating or nonirritating, or to a broad category, an accurate appraisal is difficult to make because one cannot make an independent evaluation and reconcile any apparent discrepancies.

Computer databases are constructed from the published experience and thus perpetuate the same sorts of problems. In order to be useful, such databases need to incorporate as much information as possible, which means that even more laboratories are involved, with the attendant variability in performance of the assay and in scoring. And the personnel establishing the database generally have little means of making an assessment as to the quality of the data going in. So, one is left with a much more extensive, more easily accessible body of information, but without any assurance of increased quality.

In-house databases may provide the most appropriate information available. Usually, the raw data are available, and generally, the methodology is more consistent so that test results and scoring change less with time or personnel performing the test. Another potential advantage is that within an industry group, the compounds comprising the database are more likely to have similarities with the unknowns that need to be examined. However, this result can also be a drawback in that the materials tested by, for instance, a cosmetics company, may have no practical relevance to those tested by a pharmaceutical company, and therefore, some in-house data may be too specific for general application. Thus, though it is a good idea to examine databases, it is likely that the data will be useful only for flagging a potential problem. The quality of the data, or lack of data on related structures, may not allow a judgment to be made with confidence.

Computer modeling. Computer modeling is an alternative that has been proposed by many people. Such models are based generally on computer databases, and although modeling may be of some use, it suffers from the same drawbacks as do databases; i.e., a computer simulation is only as reliable as the data used to generate the mathematical equations. Health Designs Inc. (Rochester, NY) has developed such a model (TOPKAT), which is available commercially. The model was developed by quantitating various parameters, and then comparing these measurements with the irritation potential of known materials to determine which endpoints correlate best. Although this process allows one to weight each measurement as to its importance for irritation, and to develop a mathematical model, it highlights the dependence on the quality of the underlying data.

According to their own literature ([11]; TOPKAT manual), Health Designs Inc. had difficulty constructing this model because of the inherent variability in compound classification. The model that was generated has two sets of equations. In a first step, nonirritating compounds are separated from all other materials. Then, a second set of equations is used to distinguish severe compounds from the rest of the materials. The result is a three-category estimate, nonirritating, severe, or other. A further problem is that the developers predict that approximately 30% of materials cannot be handled by this model. In fact, in a study evaluating various alternatives to the Draize test [12], TOPKAT was amenable to evaluating only approximately 50% of compounds. And although the method gave no false negatives (presumably because a strong representation of compounds of similar structure in the database gave assurances of a potential for induction of irritation), nonirritating compounds were only correctly identified approximately 55% of the time.

Some investigators are, however, using models based on their own in-house data with more success than was found with commercially available packages. It would stand to reason that within a particular company many of the materials comprising the database might be generally related to the unknowns being tested. Thus, although these individual models might not be useful for as broad a range of materials as one might like (i.e., for screening across different industries), they may be practical for use within an individual company or within an industry group (pharmaceuticals, soaps and detergents, toiletries, specialty chemicals, etc.)

Physical/chemical data. As alluded to above, in constructing a model, one looks at various measurements that might correlate with irritation potential and then attempts to determine which are truly important (causal) and which are secondary. Various parameters have been examined. For instance, many investigators look first at pH, assuming that compounds at the pH extremes (for instance, <3 or >12) are severe irritants not needing animal testing. Support for this assumption can be found in Walz [13] and Guillot et al. [6], in which materials at pH extremes were generally very irritant (although with exceptions). Extending this idea, a study sponsored by the Soap and Detergent Association (SDA) has found that the alkalinity of a test material (i.e., the strength of the acid or base) may be the key point rather than a simple measurement of pH [14]. Other physical/chemical parameters that have been proposed as possibly predictive of irritation potential include whether a compound possesses surfactant properties (because many surfactants are severe irritants) or octanol/water partition coefficients (because a material tending to partition out of the aqueous phase may penetrate the eye more deeply [15,16]).

The general problem with using physical/chemical parameters for prediction of irritation is that although some correlations and hypotheses exist, the data to adequately support the use of the various parameters are lacking. Thus, although certain characteristics of a chemical may be used to raise a warning flag, or to set testing priorities, at this time, prediction of irritation potential cannot be based solely on physical or chemical data. In many cases, this issue likely relates to the concept, which will be further discussed below, that good correlations do not necessarily imply a relationship to the underlying mechanism of irritation.

Data from other tests. Because multiple toxicity tests (acute and chronic) are generally performed on most chemicals, the question develops as to whether the results from these tests could predict ocular irritation and eliminate the need for a Draize test. Comparisons have been made between dermal and ocular irritation data under the hypothesis that if a material is irritating to the skin, it will also be irritating to the eye.

For instance, Gad et al. [17] found some correlation between ocular and dermal irritation, but the degree of correlation was dependent on the *in vivo* classification scale used. When chemicals were classified as either irritating or nonirritating, the correlation between ocular and dermal data was better than when a categorical ranking (nonirritating, mild, moderate, severe) was used. Gilman et al. [18], on the other hand, found that a reliable correlation between ocular and dermal irritation could not be established for a series of petrochemicals and consumer products. Guillot et al. [6,19] examined both ocular and dermal irritation scored by different protocols. Correlating the data in the two manuscripts, one finds that all dermal irritants (11 chemicals) were ocular irritants, but the degree of ocular irritation was not predictable from the dermal data. Only 18 of 45 nonirritating-to-slightly irritating materials dermally were nonirritating to slightly irritating in the eye. The remaining 27 compounds showed ocular irritation ranging from mild/moderate to extreme. And Williams [20,21] reported similar results. He found that 65% (39 of 60) of severe dermal irritants tested were severe ocular irritants, whereas 10% were moderate and 25% were mild or nonirritating.

These data suggest that if a compound is a severe dermal irritant, it is likely that it will be an ocular irritant as well. However, a significant number of exceptions (false positives in the ocular test) exist. And, the fact that a compound is nonirritating or mild on the skin does not appear to correlate with lack of ocular irritation. Thus, although dermal irritation data may be of some utility, they cannot be relied on to adequately predict ocular irritation potential.

3.2.2   In Vitro Approach

If one wishes to avoid using an animal test, and a decision cannot be made based on preexisting data, the next step would be to evaluate in vitro models for testing. Given this subjectivity and variability in the in vivo test, how does one approach the development of an in vitro model system that would more objectively quantitate specific parameters of the irritation index. Obviously, one important question is what is the Draize test really measuring, or put another way, what is the mechanism(s) of irritation? The answer is complex because, in all likelihood, multiple mechanisms of action lead to the clinical signs identified with varying degrees of ocular irritation.

Mechanistically, Igarashi [22] has suggested that opacity may be caused by precipitation of proteins in the cornea. This hypothesis is based on observations with surfactants that coagulated egg-white solutions, in which the extent of coagulation paralleled the ability of the compounds to cause corneal opacity. The author noted, however, that some anionic surfactants produced entirely opposite results, suggesting that opacity cannot be fully explained on the basis of protein precipitation. Basu [23] has presented data suggesting that damage to the peripheral cells of the cornea results in altered fluid permeability, which leads to distortion of the corneal layers and altered transparency. Additionally, Burstein and Klyce [24] studied the effects of some components of ophthalmic preparations (benzalkonium chloride, thimerosal, amphotericin B, etc.) on morphology and electrophysiologic parameters of isolated corneas. These authors found that many of these test materials (which can be irritants in high enough concentrations) can cause destruction of the epithelial cell layers followed by alteration in transport properties of the cornea. These data support the idea that parameters, such as altered cell morphology or viability and ion or water transport, may be early indicators of irritation potential.

Although corneal damage is important, other responses to irritants exist that need to be considered in modeling in vivo irritation. For instance, to provide a basis for developing alternative methods, Parish [25] attempted to define the histologic features associated with irritants. He found that mild-to-moderate chemicals caused a thinning (through exfoliation) of the corneal epithelium, but no irreversible damage. Further, the conjunctiva showed edema, leukocyte infiltration, and congestion of the blood vessels, suggesting that damage to this tissue may also need to be examined.

In addition, inflammation is a major clinical sign recorded in scoring the Draize assay, and it may be either a cause of or a response to irritation. If necrosis of the corneal epithelium or conjunctiva occurs, autolysis of the cells may release factors initiating an acute inflammatory response to "clean up" the area of damage, resulting in a secondary response. Alternatively, it has been demonstrated [26] that neutrophil or macrophage infiltration of either the endothelial or epithelial corneal surface can cause substantial damage, resulting in cell loss as well as separation of the cell layers and consequent opacification.

From the discussion above, it should be clear that ocular irritation is most likely a result of multiple mechanisms causing distinct clinical signs (opacity, inflammation, congestion, necrosis, etc.). Opacification could be caused by altered ion fluxes, protein precipitation, or cross-linking, etc., whereas inflammation may involve chemotactic factors or other components of the arachidonic acid cascade. And the cells involved include the epithelial cells of the cornea, stromal cells, or conjunctiva. The questions that develop, then, are how do we model these events and how many of these events need to be isolated and analyzed to give an accurate prediction of irritation potential? In the next section, we will review some of the approaches being taken, as well as their advantages and disadvantages.

3.2.3 *In Vitro* Models

An extensive list of models that have been proposed as alternatives to the Draize test was compiled by Frazier et al. [27], and although many of these have received limited attention, substantial effort has been invested in evaluating a number of the assays. For example, in November 1993, the Interagency Regulatory Alternatives Group (IRAG) held a workshop to evaluate the information then available on alternatives, with the goal of establishing a basis for future efforts. A call for data resulted in the submission of over 74 data sets from 59 different laboratories on approximately 26 different test methods [28]. The IRAG program grouped the assays as organotypic, chorioallantoic membrane (CAM)-based, cell function assays, cytotoxicity tests, and other. Because we believe that *in vitro* tests should measure specific components of the Draize when possible, we will discuss the more commonly used assays within the context of the *in vivo* measurement that they appear to be most closely addressing.

Opacity. Because corneal opacity is the most heavily weighted component of the Draize score, it is important to account for this response in an *in vitro* system. At present, three organotypic tests and one commercial assay based on coagulation of a protein matrix have been used extensively to measure endpoints related to opacity.

Muir [29,30] reported the development of an opacity assay, using isolated bovine cornea, in which decreases in light transmission through the cornea were monitored as the endpoint. His work with surfactants and some industrial chemicals indicated a good correlation between the *in vivo* and *in vitro* data. Igarashi et al. [31] developed a variation on this method, in which isolated porcine corneas (with endothelial cells removed) are exposed to test compounds. Changes in voltage across the corneal epithelium are used as a measure of opacification.

In our own laboratory, we extended the technique of Muir and developed a bovine corneal opacity assay coupled with a fluorescein permeability test [32]. This assay has been described in detail [33] and has been included in a number of evaluation/validation studies [12,34,35]. The bovine corneal opacity and permeability (BCOP) assay uses corneas obtained from an abattoir as the target tissue, and it measures both opacification and the penetration of fluorescein dye through the cornea as endpoints. Initial work focused simply on the measurement of opacity based on decreased transmission of light, but we found that some compounds cause sloughing of epithelial cells from the cornea, resulting in false-negative readings. (Because the damaged corneas had fewer cell layers, a greater transmission of light was detected than might be expected.) To measure this type of damage, we adapted the fluorescein dye penetration concept of Tchao [36]. Because the epithelial layers of the cornea form a barrier resistance to chemical penetration, the amount of dye penetrating through to the posterior chamber should be proportional to the degree of damage to the epithelium.

We have found this two-endpoint assay (BCOP) to be reliable for assessing irritation potential of manufacturing intermediates and raw materials. Over a number of years, we have tested approximately 250 commercially available chemicals as well as in-house materials representative of a broad range of chemical classes, and we have found that the assay results correctly predict *in vivo* irritation potential approximately 80–85% of the time. A similar correlation has been found by Casterton et al. [37], with a modification of this test.

We have found relatively few limitations with this method. For instance, working with insoluble materials is difficult in most *in vitro* assays. However, with the BCOP assay, we are able to test most materials because the corneas can be exposed with the holders in a horizontal position, allowing material in suspension to settle out onto the cornea and interact with the cells. The exception to this rule is hydrophobic, insoluble compounds, because the material would float on the medium and never come into contact with the target cells. Another potential problem would occur if a compound caused minimal irritation *in vivo* for some period of time (24 to 36 hours), but then induced increasing irritation over time. This type of delayed reaction would be difficult to detect with most *in vitro* assays, and the BCOP is no exception.

Two other organotypic methods focus on corneal measurements. One method developed by Burton et al. [38], and described further by Price and Andrews [39], uses whole, isolated rabbit eyes, with measurement of corneal thickness, opacity, and fluorescein penetration as the endpoints. The authors found this test reliable, showing a good correlation with *in vivo* irritation potential for the 60 compounds tested. The isolated rabbit eye test has been evaluated in a number of studies [35,40,41] with varying results. A comparison of BCOP and the isolated rabbit eye test has recently been published as part of the results of a workshop on the BCOP [33].

Prinsen and Koeter [42] describe a similar method using an enucleated chicken eye (CEET) as an alternative to making measurements in a laboratory animal. They also measure corneal swelling, opacity, and fluorescein retention as endpoints, and Prinsen reports [43] that the correlation between the CEET and *in vivo* Draize scores is excellent.

Another assay designed as an alternative measure of corneal opacification is the Eytex assay. This method stems from the observations that transparency of the cornea depends on the hydration and organization of proteins and that the presence of high molecular weight aggregates of protein caused opacity [44]. Basically, the test measures the reduction in light transmission resulting from precipitate caused by interaction of the test material with a proprietary protein matrix [45]. In certain applications, this method has proven useful [46,47], giving good correlations with *in vivo* results. On the other hand, Sanders et al. [48] and Bruner et al. [49] found that the correlation of Eytex data with *in vivo* data was unsatisfactory and that other *in vitro* tests gave better correlations. Although this test has been included as part of recent validation studies [35,41,50], the correlation with *in vivo* data has not been as good as might be expected if this endpoint fully relates to opacification of the cornea.

Cytotoxicity. Most assays proposed as alternatives to the Draize test are cytotoxicity tests. Usually, these methods are simple, straightforward, and relatively rapid, with a defined endpoint that can be accurately measured and reproduced. Many of the tests make use of immortalized cell lines as the target tissue, so no use of animals is required. A primary disadvantage is the fact that the assays are not usually mechanistically based and therefore may not provide information specifically related to why and how a chemical causes irritation. Despite this disadvantage, some of these tools [neutral red uptake, fluorescein leakage, 3-(4, 5-dimethylthiazol-2-yl)-2,5 diphenyl tetrazolium bromide (MTT), RBC lysis] have given good correlations with certain types of materials, notably, surfactants (cf., ref. 50). Although in studies in which a variety of materials are evaluated, the overall predictive value of cytotoxicity assays is generally not as good.

This point can be illustrated by examining a specific cytotoxicity assay, the neutral red assay developed by Borenfreund and Puerner [51,52]. This test has been widely used and adapted to a commercial test kit [53]. The basis of the test is the sequestration of the vital dye neutral red into the lysosomes of viable cells. Nonviable cells are unable to retain the dye during harvesting, and thus, the amount of dye per test culture (determined spectrophotometrically) is proportional to the number of viable cells per culture. Either the uptake of the dye or the release of the dye from preloaded cells can be measured as an endpoint.

This assay has been used by a number of laboratories, and it reportedly gives good correlations with *in vivo* ocular irritation data [49,54,55] for cosmetic ingredients, alcohols, or surfactant-based products. However, the correlations between neutral red data and *in vivo* irritation tend to break down when comparisons are made across chemical classes. For instance, Thomson et al. [56] reported that correlations within certain chemical groupings were substantially better than for other chemical types and that the overall correlation of the assay with 94 compounds was relatively poor. They concluded that for their purposes, this assay had limited usefulness. Similarly, in testing a number of diverse materials in V79 and rabbit corneal epithelial cells, we found the overall correlation between the neutral red endpoint and ocular irritation *in vivo* was relatively poor ($r = -0.33$) [57]. However, if the rank correlation was limited to alcohols, the correlation improved greatly, giving a correlation coefficient (r) of -0.67.

MTT [58] has also been used extensively as an endpoint with a variety of cell targets, most notably, in a tissue equivalent model [59]. Based on the conversion of a yellow soluble tetrazolium dye into an insoluble, purple formazan product in mitochondria of viable cells, this test is scored as either the dose of compound necessary to reduce dye conversion (directly proportional to cell number) by 50% or as the time necessary for a given dose of test material to cause a 50% reduction in conversion. In certain systems, the latter scoring method seems to give more accurate correlation with *in vivo* data [59].

Dye penetration assays monitor the integrity of a cell sheet as a measure of penetration of a test material into the tissue of the cornea, the hypothesis being that a compound destroying cell-to-cell connections will penetrate the cornea and potentially cause extensive damage. For example, Brooks and Maurice [60] have developed a proptosis mouse eye/permeability test, in which eyes of freshly sacrificed mice are exposed to test agent and sulforhodamine B penetration is measured. An assay more often used, however, is the fluorescein penetration assay developed by Tchao [36]. In this model, MDCK cells, a dog kidney cell line forming tight junctions at confluence, are grown on a filter separating two compartments, so that polarity of the monolayer is established and the sides can be separated. Fluorescein is placed in the inner chamber (analogous to the external surface of the cornea), and the passage of the dye through the cell sheet to the other chamber is monitored. When an intact monolayer covers the filter, the passage of fluorescein through the cell sheet is restricted. If a chemical damages the cells, extensive leakage of fluorescein occurs. In a large validation study sponsored by COLIPA (a European fragrance, cosmetic, and toiletries industry group), the fluorescein leakage test was one of the most predictive with cosmetic ingredients and formulations [50]. However, when a broad range of pharmaceutical intermediates was tested in this assay, the results were much less impressive [61].

These data seem to suggest that cytotoxicity assays may measure specific mechanisms of tissue destruction, e.g., membrane damage, which would be reflected in general toxicity to the eye. Generalized toxicity, obviously, is one mechanism associated with ocular irritation, but the data also suggest that this mechanism is applicable only to specific chemicals, for instance, surfactants. Thus, although it is clear that cytotoxicity assays are applicable in some situations [62–64], it is equally clear that in many instances the ability of *in vitro* cytotoxicity assays to predict *in vivo* irritation is inadequate [54,65,66]. Thus, cytotoxicity endpoints may represent a complementary component of a multiple endpoint model (as in the BCOP assay) or as part of a battery of tests.

Inflammation. The inflammation component is another important aspect of the Draize score that needs to be addressed in any attempt to develop *in vitro* alternatives. And the mechanistic rationale for the involvement of inflammation in ocular irritation and damage has been amply demonstrated. For instance, Elgebaly et al. [26,67] have shown that leukocytes, attracted to an inflammation site by the release of chemotactic factors, can cause significant damage to the corneal epithelium or endothelium. Bazan et al. [68] have demonstrated that after cryogenic injury to rabbit cornea, both prostaglandins and HETEs are released from epithelium, stroma, and endothelium. Srinivasan and Kulkarni [69] have demonstrated that prostaglandins and prostacyclin are involved in mediating the polymorphonuclear leukocyte response to different types of corneal injury, and Bhattacherjee et al. [70] have reported a study in rabbits that showed that both prostaglandins and leukotrienes are important in the development of ocular inflammation, inducing both vascular and cellular changes in ocular tissue.

But one of the problems in trying to develop an inflammation model *in vitro* is our limited understanding of the interactions among cells and molecules (prostaglandins, leukotrienes, histamine, serotonin, thromboxane, etc.) involved in inflammation. That is to say, it is clear that certain cells and molecules are present at the site of inflammation, and so they are likely to be involved early in the response, but which might be causative and which are secondary is less well understood.

One type of assay proposed as an indicator of inflammation (although obviously measuring other components of the irritation process, such as toxicity) is a test using the CAM as a target, e.g., CAM vascular assay (CAMVA) [71–73], bovine epithelial chorioallantoic membrane (BECAM) [74], and hen's egg test/CAM (HET/CAM) [75,76]. Basically, the assay scores morphologic alterations in the CAM of chicken eggs on exposure to test compounds, and it is therefore somewhat subjective. The three modifications of this test can be differentiated as follows. The HET/CAM assay concentrates the scoring of effects on blood vessels of the CAM and the egg albumen, is performed after an acute (0.5 to 5 minute) exposure to test material, and uses 10-day-old eggs. This latter point is key in that testing on the CAM of an egg greater than 10 days beyond fertilization may be considered to be use of a live animal in the United Kingdom [77]. The BECAM assay scores alterations primarily to the blood vessels of the CAM, with scoring and timing of exposure similar to the HET/CAM technique. In addition, this assay combines the CAM observations with observations of opacity and fluorescein permeability made in isolated bovine eye. The CAMVA assay is a modification developed at Colgate-Palmolive in which 14-day-old eggs are exposed to compound for 30 min and scoring is focused on changes in vascularity (hemorrhaging, capillary injection, ghost vessels). Using this modification, Bagley et al. [73] have found a good correlation with irritation data from a traditional Draize score for surfactant-based materials.

As with many of the other assays discussed here, the CAM assay has proven useful in some applications, but inadequate in others. The reason(s) for this discrepancy may be because of the compound classes tested or the various methodological differences between laboratories. For instance, Lawrence et al. [77,78] found the CAM assay inadequate for their purposes. When they tested nine materials (surfactants, alcohols, miscellaneous), they found that in only four cases were they able to predict the *in vivo* irritation potential from the CAM data. They hypothesize that this result may be because of observations that indicate that inflammation in the rabbit conjunctiva differs mechanistically from inflammation in the CAM, with the former being associated with infiltration of neutrophils and macrophages, whereas the latter is basically chemically induced necrosis. Price et al. [79] also reported that the CAM assay was of limited value for their applications. Using a protocol with short exposure times (similar to HET-CAM), they found that although the assay was capable of accurately predicting 23 of 30 materials as to whether they were irritating or nonirritating, the degree of irritation could not be assessed based on the CAM response. Yet, in other validation studies with the HET-CAM test, the results were satisfactory [80,81]. Similarly with the CAMVA, evaluations such as one performed with pharmaceutical intermediates [12] showed the CAMVA to give one of the better correlations among the tests evaluated. Results like these serve to reinforce the idea that different compounds cause ocular irritation by different mechanisms and, thus, measurement of a defined endpoint, even in organotypic tissue, will not suffice for every test material. The key to applying these tests, as will be discussed further below, is to understand the limitations as well as the strengths of various assays and use them together in a rational approach to risk assessment.

3.3   USING ALTERNATIVES

Given the array of methods available, how does one approach reducing or replacing *in vivo* ocular irritation testing in a practical situation? First, it needs to be remembered that although an ultimate result of these efforts will be to reduce the number of animals used in testing, the primary consideration is obtaining information that can accurately predict a hazard. In deciding how to structure an overall approach, then, two questions that impact the integration of alternatives come immediately to mind: (1) what types of materials will be tested in the alternative assay(s), and (2) how will the test results be applied in a decision-making scheme?

With regard to the first question, as discussed above with reference to strengths and weaknesses of individual assays, some test methods seem to provide good *in vivo/in vitro* correlations with a specific chemical class (e.g., surfactants) but perform poorly with mixed groups of materials. Presumably, this reaction is caused by the endpoint measurement closely tracking a mechanism of action (irritation) specific to the chemical class, but one not as critical in the action of other chemicals (perhaps cell lysis resulting in necrosis for surfactants versus protein coagulation leading to opacity for other chemicals). In establishing how an assay can be applied, it is as important to objectively determine when an assay cannot be used, as to establish its utility.

For instance, Bruner et al. [49] evaluated seven proposed ocular testing alternatives (silicon microphysiometer, luminescent bacteria, neutral red, total protein, *Tetrahymena* species motility, BECAM, and Eytex) with test materials relevant to their own situation (emphasis on surfactants). The assays selected covered a range of endpoints and potential mechanisms of action. They concluded that for their consumer products all of the tests, except BECAM and Eytex, had some utility as screens before limited confirmation studies *in vivo*. The Cosmetics, Toiletries, and Fragrance Association (CTFA) has taken a similar approach [82], looking at approximately 12 types of endpoints and proceeding to evaluate their predictive value with a series of related compounds (hydro-alcoholic formulations in Phase I [83], oil/water emulsions in Phase II [84], and surfactant-based formulations in Phase III [41]), with the object being to determine which assays would be useful for which chemicals. In their work [14,85] the SDA has not only evaluated a number of assays with a view toward choosing the one(s) most useful for this product category, but it has also identified areas in which further understanding is necessary (e.g., pH versus alkalinity).

In our own laboratory, we have assessed cytotoxicity endpoints (neutral red, leucine incorporation, MTT) and opacity evaluation (BCOP) as predictors for evaluating pharmaceutical agents, compounds spanning a broad range of chemical classes and physical forms. Our data indicate that measuring cytotoxicity alone is insufficient for predicting irritation potential [57], but that determination of chemically induced opacity in combination with a measurement of disruption of the epithelial cell sheet (fluorescein permeability) is accurate [32]. An additional example is a study in which a number of pharmaceutical companies evaluated pharmaceutical intermediates in seven alternative tests (BCOP, CAMVA, Eytex, TOPKAT structure-activity database, neutral red [53], microtox [86], and MTT in the Living Dermal Equivalent [87]). The data showed that the organotypic assays (BCOP and CAMVA) were the most predictive of the Draize result with a correlation of approximately 80–85%. In trying to maximize the information obtained from the *in vitro* assay, we asked whether a combination of two or more of these particular tests would increase the reliability of the prediction, but unfortunately found that this was not the case.

In establishing how data from *in vitro* methods can be applied in risk assessment, the second question is how precisely should the irritation potential be defined? Three scenarios come to mind: (1) a simple irritating versus nonirritating determination, (2) a ranking of greater or lesser irritation potential relative to a known irritant(s), or (3) a continuous scale ranking from nonirritating through severe. For determining whether a compound is irritating, one needs an endpoint measurement providing clear-cut, objective data in which a distinct "threshold" of effect versus no effect can be defined. For continuous scale ranking, this measurement might be totally inadequate, with an assay providing a more subjective endpoint proving necessary. In both of these cases, two or more complementary assays may need to be performed to provide confirmation of the results. On the other hand, for a measurement of irritation potential relative to a known irritant, a single assay may suffice, because any variation within the test would likely affect both known and unknown similarly.

For instance, in our laboratory, we test process intermediates for worker safety purposes. We have chosen to first test all materials in the BCOP assay [32]. If the material is moderate or severe based on the *in vitro* assay criteria established in our validation studies, we label it as such and take appropriate precautions in the workplace. Because the question is whether to use additional protection in handling a material, no attempt is made to determine exactly how irritating a material might be; the question is simply irritating/nonirritating. To be conservative, however, we have decided to confirm any mild or nonirritating *in vitro* results in a limited *in vivo* study, because we know that the BCOP has given 19.4% false negatives over the last seven years. As we gain more familiarity and comfort with the *in vitro* test, we will hopefully be able to eliminate this *in vivo* confirmation step for most test chemicals.

It is clear from the literature that laboratories are, in fact, incorporating *in vitro* methods when practical, even as they recognize the limitations of the individual assays [37,88,89]. Batteries of methods are being used to overcome some of the problems associated with individual tests [90–92]. Yet, it is also clear from large-scale validation studies (cf., ref. 35) that *in vitro* methods are not presently valid alternatives to the Draize test. How does one reconcile this discrepancy?

3.4 VALIDATION

Although the word validation evokes different meanings for different investigators, in a practical situation, one can simplify the process to a single question: What information must be obtained so that one is comfortable using the method for decision making? That is to say, one has to explicitly define criteria under which the assay(s) will be deemed acceptable for routine usage and measure how well a particular *in vitro* method or battery meets these criteria. Although this definition lends itself well to in-house validation, it does not fully address the question of broader validation leading to acceptance by other industry groups and regulatory authorities. The following section will explore the difference between in-house and extramural validation.

As a starting point, good overall reviews of points to be considered in approaching formal validation work have been presented by various organizations: OECD [93], Center for Alternatives to Animal Testing (CAAT)/European Research Group for Alternative Toxicity Tests (ERGATT) [94], CAAT [95], and ECVAM [96]. In the simplest sense, in-house development work entails evaluation of different assays and endpoints until an approach that predicts the *in vivo* data is found. This evaluation comprises limited validation because before being put into routine use, the target tissue/endpoint combination(s) would have been found to be sensitive, robust, and predictive, and the limitations as well as the advantages would be appreciated. This appreciation would come about after enough compounds with known irritation profiles and mechanisms of action appropriate for the need had been tested. But what does this mean? Critically speaking, it only means that if someone uses the test in the way that the developers had in mind (classes/types of chemicals, scoring, prediction model, etc.), they would be likely to obtain an answer from the *in vitro* test predicting with some accuracy the results that would be obtained in an animal study. What it does not mean is that the irritation potential of any compound could be predicted or that the test can be used in ways unrelated to the scope addressed during development.

A number of examples of assays in use in various companies have been cited above, but what about moving the most successful, i.e., predictive, assays from an individual laboratory into broader use? Interlaboratory transferability of methods has generally not been a problem technically. For instance, fairly good laboratory-to-laboratory correlation has been found with the BCOP [34], the rabbit enucleated eye test [40], the neutral red and the HET-CAM assays [97], and a variety of endpoints in a CTFA study [41]. However, the data from the IRAG workshop [28] lead to the conclusion that although alternative methods were being used in industry, the data were "insufficient to support the total replacement of *in vivo* ocular irritancy testing with *in vitro* methods." This conclusion has been echoed after large-scale validation studies. For instance, the EC/HO study [35] evaluated 60 chemicals in nine different tests, each performed in at least four different laboratories. The result was that none of the tests adequately predicted the modified maximum average Draize score obtained *in vivo* for the diverse set of test compounds, with the possible exception of surfactants. A similar conclusion was reached in a study sponsored by COLIPA [50], in which 10 alternative methods were applied to a set of 55 cosmetic ingredients and formulations. Again, none of the assays could predict eye irritation reliably enough to be considered a valid replacement for the Draize test. The question is what are the barriers to validation and acceptance in a broad setting?

A good general overview of the challenges surrounding replacement of animals in toxicity testing has been published [98], and specific issues impeding "global" validation have also been described [99]. Clearly, the quality of the animal data is one of those issues. Data from Weil and Scala [2], recently confirmed by Earl et al. [3], suggest a coefficient of variation of >50% when scores are compared for the same material tested in different laboratories. Further, a recent study showed that when the animal test was conducted 15 times on a single material by the same laboratory, the coefficient of variation was still 24% [100]. Given the variability in both the *in vivo* and *in vitro* scores, perhaps expecting a prediction of MMAS by *in vitro* tests is inappropriate; perhaps the correlation would be better if one asked whether the categorical classification was predicted by the *in vitro* test. After all, should it be important to hazard prediction if the numerical score is 60 or 110? Additionally, some of the discrepancy in assessing the predictive value of *in vitro* tests comes from the different scoring schemes in place in different industries because of regulation by various agencies. In analyzing the results of a multiple test evaluation study with pharmaceutical intermediates, Sina et al. [12] found that the extent of correlation shifted dramatically depending on whether one compared the *in vitro* data to *in vivo* data obtained by the scoring of Kay and Calandra [5], IRAG [101], the French cosmetic directive [102], or various other agencies [103]. Although attempts are being made to harmonize these scoring systems [104], the current variability presents a significant obstacle in that what is predictive for one industry group is not necessarily adequate for a different industry.

An issue that, perhaps, has not received enough attention is the choice of rabbit for irritation testing and the consequent influence the existing rabbit data have in the development and validation of alternatives. The rabbit has been accepted for so long as the species of choice in predicting human eye irritation that the assumption is naturally made that the experimental database with rabbit represents the gold standard for mechanistic as well as predictive comparisons. However, if one looks into the older literature, one finds caveats. For instance, in 1984, Bito [105] wrote:

> The acceptance of the rabbit eye as a suitable model for the mammalian eye has apparently been based on the assumption that its sensitivity to irritation constitutes only a quantitative difference in the expression of mechanisms that are identical to those of other species. If this were the case, the extreme sensitivity of the rabbit eye might even offer an experimental advantage over the use of the much less sensitive and much more costly primate eye. However, there is no experimental evidence to support this assumption.

A scan of the literature suggests that in the last 15 years we have not provided substantial experimental data to test the hypothesis. So, although the rabbit data form the most extensive, available *in vivo* database, it may not always be suitable for the risk assessment questions we would like to address with alternative methods.

It is clear in the literature that some endpoints work well for certain chemical classes, but are totally inadequate for others, and that a single test does not seem to address all questions (the hazard from acute versus chronic exposure, reversibility, recovery, etc.). This limitation may stem from what is understood, or not understood, of the underlying biology of ocular irritation. Conceptually, as Flint [106] suggests, we are using *in vitro* models analogous to the animal situation and, in so doing, are isolating some important components of the underlying mechanism, but not incorporating all components. Is it, then, so surprising that no test taken in isolation has been found to be predictive of the wide variety of test materials used in large-scale validation studies, or that within a company or industry specific tests are reported to be predictive for specific products? Perhaps, as has been suggested in a number of publications, we simply do not understand enough of the underlying biology of ocular irritation, or we do not understand which components are most important and should be analyzed together (either in a single assay or multiple tests) to provide a more predictive result.

Still another consideration is the importance of having a robust prediction model. As Bruner et al. [107,108] have pointed out, for each *in vitro* assay, a prediction model must exist to unambiguously link the *in vitro* data with their meaning in hazard assessment. This model sets the framework of expectation for performance of *in vitro* assays, and it makes possible a judgment as to utility or adequacy of prediction in a validation study. A validation study sponsored by COLIPA [50] used this concept, establishing prediction models (and thus, to some degree, expectations) before the start of the study. Although the results showed that none of the methods tested met all criteria for validation, the data are being reviewed to establish new prediction models that might be applied in future studies.

In fact, establishing prediction models is a key point in another important consideration for validation, the process of prevalidation. Prevalidation [109,110] encompasses, and would extend and formalize, a number of phases normally occurring during assay development and assessment. In the first phase, a protocol is refined with a Good Laboratory Practices (GLP)-compliant standard operating procedure and intralaboratory reproducibility being established. Next, the laboratory-to-laboratory transferability would be assessed and appropriate adjustments made to the protocol. In the third phase, the performance of the test would be assessed against the prediction model. Completion of these three phases would give objective evidence as to whether testing in a formal validation study with a view toward incorporation into regulatory guidelines is appropriate for that assay. Given the substantial costs in materials and effort to conduct formal validation studies, this procedure should identify those alternative tests with the best chance of success and thus maximize the return on the investment for all involved.

Two points are clear from the effort that has been described above: Progress has been made in using alternatives at a company or an organization level, and we are clearly at a point where regulatory bodies around the world are involved with scientists from academia and industry to establish a framework for acceptance of alternatives in hazard assessment. The European Centre for the Validation of Alternative Methods (EVCAM) was created with the goal of promoting acceptance of valid alternatives, and it has played a role through workshops (the reports from which can be found on the Web at http://altweb.jhsph.edu) and other activities in attempts to respond to the EC Directive that sought to ban the Draize test for cosmetics by January 1, 1998, if possible [111]. In the United States, IRAG served to evaluate the state-of-the-art with respect to ocular testing alternatives and to seek a way forward [28]. This goal has lead to formalization of the process through the Interagency Coordinating Committee for Validation of Alternative

Methods (ICCVAM), which has developed a framework for validation and acceptance of alternative methods into routine testing [112]. More information can be found regarding the mission and current activities of ICCVAM at its website: http://iccvam.niehs.nih.gov/home.htm.

3.5 CONCLUSIONS

We have tried to briefly outline the range of approaches being used to reduce or eliminate animal testing for ocular irritation. A number of alternative assays has been reported in the literature, each with its uses and limitations. But given the complexity of the *in vivo* response being modeled, and the fact that not all mechanisms of irritation are completely understood, it is probably not reasonable to depend on a single test, measuring a single endpoint, to be predictive of *in vivo* ocular irritation potential in all cases. The broad approach of using a tiered testing system, gathering as much data as practical before making predictions, seems likely to be the most fruitful in the long term.

What is clear is that over the last few years the field seems from be moving from a focus on development of testing methods to a critical evaluation of the most promising assays. Large-scale validation studies have been disappointing if one only looks at the bottom line results. But a closer reading of the literature indicates that we have learned a great deal about the process of validation and understand better where the value and weaknesses have been in our approach. This understanding has brought us significantly closer to being able to incorporate alternative methods into an overall decision-making paradigm.

References

1. Draize JH, Woodard G, Calvery HO. Methods for the study of irritation and toxicity of substances applied topically to the skin and mucous membranes. *J Pharmacol Exp Ther* 1944;82:377–390.

2. Weil CS, Scala RA. Study of intra- and interlaboratory variability in the results of rabbit eye and skin irritation tests. *Toxicol Appl Pharmacol* 1971;19:276–360.

3. Earl LK, Dickens AD, Rowson MJ. A critical analysis of the rabbit eye irritation test variability and its impact on the validation of alternative methods. *Toxicol In Vitro* 1997;11:295–304.

4. Chan P-K, Hayes AW. Assessment of chemically induced ocular toxicity: a survey of methods. In: Hayes AW, ed. *Toxicology of the Eye, Ear, and Other Special Senses*. New York: Raven Press, 1985:103–143.

5. Kay J, Calandra J. Interpretation of eye irritation tests. *J Soc Cosmet Chem* 1962;13:281–289.

6. Guillot JP, Gonnet JF, Clement C, Caillard L, Truhaut R. Evaluation of the ocular-irritation potential of 56 compounds. *Food Chem Toxicol* 1982;20:573–582.

7. Carpenter CP, Smyth HF. *J Indust Hyg (Arch Indust Hyg)* 1944–1974.

8. Green WR. *A Systematic Comparison of Chemically Induced Eye Injury in the Albino Rabbit and Rhesus Monkey.* New York: The Soap and Detergent Association, 1978.

9. FHSA. Code of federal regulations, Title 16: subchapter C—Federal Hazardous Substances Act, Part 1500.42, revised January 1, 1981 (test for eye irritants).

10. Grant WM. *Toxicology of the Eye.* Springfield, IL: Charles C. Thomas, 1986.

11. *HDI Toxicology Newsletter.* Rochester, NY: Health Designs Inc., 1987:No. 6.

12. Sina JF, Galer DM, Sussman RG, Gautheron PD, Sargent EV, Leong B, Shah PV, Curren RD, Miller K. A collaborative evaluation of seven alternatives to the Draize eye irritation test using pharmaceutical intermediates. *Fundam Appl Toxicol* 1995;26:20–31.

13. Walz D. Irritant action due to physico-chemical parameters of test solutions. *Food Chem Toxicol* 1985;23:299–302.

14. Booman KA, De Prospo J, Demetrulias J, Driedger A, Griffith JF, Grochoski G, Kong B, McCormick, WC, North-Root H, Rozen MG, Sedlak RI. The SDA alternatives program: comparison of *in vitro* data with Draize test data. *J Toxicol—Cutan Ocul Toxicol* 1989;8:35–49.

15. Halle W, Baeger I, Ekwall B, Spielmann H. Correlation between *in vitro* cytotoxicity and octanol/water partition coefficient of 29 substances from the MEIC programme. *ATLA* 1991;19:338–343.

16. Babich H, Borenfreund E. Structure-activity relationship (SAR) models established *in vitro* with the neutral red cytotoxicity assay. *Toxicol In Vitro* 1987;1:3–9.

17. Gad SC, Walsh RD, Dunn BJ. Correlation of ocular and dermal irritancy of industrial chemicals. *J Toxicol—Cutan Ocul Toxicol* 1986;5:195–213.

18. Gilman MR, Jackson EM, Cerven DR, Moreno MT. Relationship between primary dermal irritation index and ocular irritation. *J Toxicol—Cutan Ocul Toxicol* 1983;2:107–117.

19. Guillot JP, Gonnet JF, Clement C, Caillard L, Truhaut R. Evaluation of the cutaneous-irritation potential of 56 compounds. *Food Chem Toxicol* 1982;20:563–572.

20. Williams SJ. Prediction of ocular irritancy potential from dermal irritation test results. *Food Chem Toxicol* 1984;22:157–161.

21. Williams SJ. Changing concepts of ocular irritation evaluation: pitfalls and progress. *Food Chem Toxicol* 1985;23:189–193.

22. Igarashi H. The opacification of the bovine isolated cornea by surfactants and other chemicals—a process of protein denaturation? *ATLA* 1987;15:8–19.

23. Basu PK. Toxic effects of drugs on the corneal epithelium: a review. *J Toxicol—Cutan Ocul Toxicol* 1983;2:205–227.

24. Burstein NL, Klyce SD. Electrophysiologic and morphologic effects of ophthalmic preparations on rabbit cornea epithelium. *Invest Ophthal Vis Sci* 1977;16(10):899–911.

25. Parish WE. Ability of *in vitro* (corneal injury—eye organ—and chorioallantoic membrane) tests to represent histopathological features of acute eye inflammation. *Food Chem Toxicol* 1985;23:215–227.

26. Elgebaly SA, Gillies C, Forouhar F, Hahem M, Baddour M, O'Rourke J, Kreutzer DL.

An *in vitro* model of leukocyte mediated injury to the corneal epithelium. *Curr Eye Res* 1985;4:31–41.

27. Frazier JM, Gad SC, Goldberg AM, McCulley JP. A critical evaluation of alternatives to acute ocular irritation testing. In: Goldberg AM, ed., *Alternative Methods in Toxicology*, Vol. 4. New York: Mary Ann Liebert, 1987.

28. Bradlaw JA, Wilcox NL. Workshop on eye irritation testing: practical applications of non-whole animal alternatives. *Food Chem Toxicol* 1997;35:1–11.

29. Muir CK. A simple method to assess surfactant-induced bovine corneal opacity *in vitro*: preliminary findings. *Toxicol Lett* 1984;22:199–203.

30. Muir CK. Opacity of bovine cornea *in vitro* induce by surfactants and industrial chemicals compared with ocular irritancy *in vivo*. *Toxicol Lett* 1985;24:157–162.

31. Igarashi H, Katsuta Y, Nakazato Y, Kawasaki T. The use of an opacitometer to compare the *in vitro* cornea opacifying effects of timolol with and without benzalkonium chloride. *ATLA* 1991;19:263–270.

32. Gautheron P, Dukic M, Aliz D, Sina JF. The bovine corneal opacity and permeability test: an *in vitro* assay of ocular irritancy. *Fundam Appl Toxicol* 1992;18:442–449.

33. Sina JF, Gautheron PD. The bovine corneal opacity and permeability assay: an historical perspective. *In Vitro Mol Toxicol*. In press.

34. Gautheron P, Giroux J, Cottin M, Audegond L, Morilla A, Mayordoma-Blanco L, Tortajada A, Haynes G, Vericat JA, Pirovano R, Tos EG, Hagemann C, Vanparys P, Deknudt G, Jacobs G, Prinsen M, Kalweit S, Spielmann H. Interlaboratory assessment of the bovine corneal opacity and permeability (BCOP) assay. *Toxicol In Vitro* 1994;8:381–392.

35. Balls M, Bothan PA, Bruner LH, Spielmann H. The EC/HO international validation study on alternatives to the Draize eye irritation test. *Toxicol In Vitro* 1995;9:871–929.

36. Tchao R. Trans-epithelial permeability of fluorescein *in vitro* as an assay to determine eye irritants. In: Goldberg AM, ed., *Alternative Methods in Toxicology*, Vol. 6. New York: Mary Ann Liebert, 1988; 271–283.

37. Casterton PL, Potts LF, Klein BD. A novel approach to assessing eye irritation potential using the bovine corneal opacity and permeability assay. *J Toxicol—Cutan Ocul Toxicol* 1996;15:147–163.

38. Burton ABG, York M, Lawrence RS. The *in vitro* assessment of severe eye irritants. *Food Cosmet Toxicol* 1981;19:471–480.

39. Price JB, Andrews IJ. The *in vitro* assessment eye irritancy using isolated eyes. *Food Chem Toxicol* 1985;23:313–315.

40. Whittle E, Basketter D, York M, Kelly L, Hall T, McCall J, Botham P, Esdaile D, Gardner J. Findings of an interlaboratory trial of the enucleated eye method as an alternative eye irritation test. *Toxicol Methods* 1992;2:30–41.

41. Gettings SD, Lordo RA, Hintze KL, Bagley DM, Casterton PL, Chudkowski M, Curren RD, Demetrulias JL, Dipasquale LC, Earl LK, Feder PI, Galli CL, Glaza SM, Gordon VC, Janus J, Kurtz PJ, Marenus KD, Moral J, Pape WJW, Renskers KJ, Rheins LA,

Roddy MT, Rozen MG, Tedeschi JP, Zyracki J. The CTFA evaluation of alternatives program: an evaluation of *in vitro* alternatives to the Draize primary eye irritation test. (Phase III) Surfactant-based formulations. *Food Chem Toxicol* 1996;34:79–117.

42. Prinsen MK, Koeter HBWM. Justification of the enucleated eye test with eyes of slaughterhouse animals as an alternative to the Draize eye irritation test with rabbits. *Food Chem Toxicol* 1993;31:69–76.

43. Prinsen MK. The chicken enucleated eye test (CEET): a practical (pre)screen for the assessment of eye irritation/corrosion potential of test materials. *Food Chem Toxicol* 1996;34:291–296.

44. Kelly C. EYTEX: an in-vitro method of predicting ocular safety. *Pharmacopeial Forum* 1989;Jan–Feb:4,815–4,824.

45. Gordon VC, Kelly CP. An *in vitro* method for determining ocular irritation. *Cosmet Toiletries* 1989;104:69–73.

46. Lawrence-Beckett EM, James JT. Initial experience with the Eytex *in vitro* irritation test system. *Toxicologist* 1991;10:259.

47. Soto RJ, Servi MJ, Gordon VC. Evaluation of an alternative method for ocular irritation. In: Goldberg AM, ed. *Alternative Methods in Toxicology*, Vol. 7. New York: Mary Ann Liebert, 1989:289–296.

48. Sanders C, Swedlund TD, Stephens TJ, Silber PM. Evaluation of six *in vitro* toxicity assays: comparison with *in vivo* ocular and dermal irritation potential of prototype cosmetic formulations. *Toxicologist* 1991;11:282.

49. Bruner LH, Kain DJ, Roberts DA, Parker RD. Evaluation of seven *in vitro* alternatives for ocular safety testing. *Fundam Appl Toxicol* 1991;17:136–149.

50. Brantom PG, Bruner LH, Chamberlain M, DeSilva O, Dupuis J, Earl LK, Lovell DP, Pape WJW, Uttley M, *for the management team*, Bagley DM, Baker FW, Bracher M, Courtellemont P, Declercq L, Freeman S, Steiling W, Walker AP, *for the task force*, Carr GJ, Dami N, Thomas G, *for the statistics subgroup*, Harbell J, Jones PA, Pfannenbecker U, Southee JA, Tcheng M, *for the lead laboratories*, Argembeaux H, Castelli D, Clothier R, Esdaile DJ, Itigki H, Jung K, Kasai Y, Kojima H, Kristen U, Larnicol M, Lewis RW, Marenus K, Moreno O, Peterson A, Rasmussen ES, Robles C, Stern M, *for the participating laboratories*. A summary report of the COLIPA international validation study on alternatives to the Draize rabbit eye irritation test. *Toxicol In Vitro* 1997;11:141–179.

51. Borenfreund E, Puerner JA. A simple quantitative procedure using monolayer cultures for cytotoxicity assays. *J Tissue Cult Methods* 1984;9:7–9.

52. Borenfreund E, Puerner JA. Short-term quantitative *in vitro* cytotoxicity assay involving an S-9 activating system. *Cancer Lett* 1987;34:243–248.

53. Triglia D, Wegener PT, Harbell J, Wallace K, Matheson D, Shopsis C. Interlaboratory validation study of the keratinocyte neutral red bioassay from Clonetics Corporation. In: Goldberg AM, ed. *Alternative Methods in Toxicology*, Vol. 7. New York: Mary Ann Liebert 1989:357–365.

54. Bracher M, Faller C, Spengler J, Reinhardt CA. Comparison of *in vitro* cell toxicity with *in vivo* eye irritation. *Mol Toxicol* 1987;1:561–570.

55. Shopsis C. Validation study: ocular irritancy prediction with the total cell protein, uridine uptake, and neutral red assays applied to human epidermal keratinocytes and mouse 3T3 cells. In: Goldberg AM, ed. *Alternative Methods in Toxicology*, Vol. 7. New York: Mary Ann Liebert, 1989:273–287.

56. Thomson MA, Hearn LA, Smith KT, Teal JJ, Dickens MS. Evaluation of the neutral red cytotoxicity assay as a predictive test for the ocular irritancy potential of cosmetic products. In: Goldberg AM, ed. *Alternative Methods in Toxicology*, Vol. 7. New York: Mary Ann Liebert, 1989:297–305.

57. Sina JF, Ward GJ, Laszek MA, Gautheron PD. Assessment of cytotoxicity assays as predictors of ocular irritation of pharmaceuticals. *Fundam Appl Toxicol* 1992;18:515–521.

58. Mossman T. Rapid colorimetric assay for cellular growth and survival: application to proliferation and cytotoxicity assays. *J Immunol Methods* 1983;65:55–63.

59. Osborne R, Perkins MA, Roberts DA. Development and intralaboratory evaluation of an *in vitro* human cell-based test to aid ocular irritancy assessments. *Fundam Appl Toxicol* 1995;28:139–153.

60. Brooks D, Maurice D. A simple fluorometer for use with a permeability screen. In: Goldberg AM, ed. *Alternative Methods in Toxicology*, Vol. 5. New York: Mary Ann Liebert, 1987:173–177.

61. Gautheron P, Duprat P, Hollander CF. Investigations of the MDCK permeability assay as an *in vitro* test of ocular irritancy. *In Vitro Toxicol* 1994;7:33–43.

62. Shopsis C, Borenfreund E, Walberg J, Stark DM. *In vitro* cytotoxicity assays as potential alternatives to the Draize ocular irritancy test. In: Goldberg AM, ed. *Alternative Methods in Toxicology*, Vol. 2. New York: Mary Ann Liebert, 1984:103–114.

63. North-Root H, Yackovich F, Demetrulias J, Gacula M, Heinze JE. Evaluation of an *in vitro* cell toxicity test using rabbit corneal cells to predict the eye irritation potential of surfactants. *Toxicol Lett* 1982;14:207–212.

64. Tachon P, Cotovio J, Dossou KG, Prunieras M. Assessment of surfactant cytotoxicity: comparison with the Draize eye test. *Int J Cosmet Sci* 1989;11:233–243.

65. Flower C. Some problems in validating cytotoxicity as a correlate of ocular irritancy. In: Goldberg AM, ed. *Alternative Methods in Toxicology*, Vol. 5. New York: Mary Ann Liebert, 1987:269–274.

66. Kennah HE, Albulescu D, Hignet S, Barrow CS. A critical evaluation of predicting ocular irritancy potential from an *in vitro* cytotoxicity assay. *Fundam Appl Toxicol* 1989;12:281–290.

67. Elgebaly SA, Forouhar F, Gillies C, Williams S, O'Rourke J, Kreutzer DL. Leukocyte-mediated injury to corneal endothelial cells: a model of tissue injury. *Am J Pathol* 1984;116:407–416.

68. Bazan HEP, Birkle DL, Beuerman RW, Bazan NG. Inflammation-induced stimulation of the synthesis of prostaglandins and lipoxygenase-reaction products in rabbit cornea. *Curr Eye Res* 1985;4:175–179.

69. Srinivasan BD, Kulkarni PS. The role of arachidonic acid metabolites in the mediation

of the polymorphonuclear leukocyte response following corneal injury. *Invest Ophthalmol Vis Sci* 1980;19:1,087–1,093.

70. Bhattacherjee P, Hammond B, Salmon JA, Stepney R, Eakins KE. Chemotactic response to some arachidonic acid lipoxygenase products in the rabbit eye. *Eur J Pharmacol* 1981;73:21–28.

71. Kong BM, Viau CJ, Rizvi PY, De Salva SJ. The development and evaluation of the chorioallantoic membrane (CAM) assay. In: Goldberg AM, ed. *Alternative Methods in Toxicology*, Vol. 5. New York: Mary Ann Liebert, 1987:59–73.

72. Bagley DM, Rizvi PY, Kong BM, De Salva SJ. An improved CAM assay for predicting ocular irritation potential. In: Goldberg AM, ed. *Alternative Methods in Toxicology*, Vol. 6. New York: Mary Ann Liebert, 1988:131–138.

73. Bagley DM, Kong BM, De Salva SJ. Assessing the eye irritation potential of surfactant-based materials using the chorioallantoic membrane vascular assay (CAMVA). In: Goldberg AM, ed. *Alternative Methods in Toxicology*, Vol. 7. New York: Mary Ann Liebert, 1989:265–272.

74. Weterings PJJM, Van Erp YHM. Validation of the BECAM assay—an eye irritancy screening test. In: Goldberg AM, ed. *Alternative Methods in Toxicology*, Vol. 5. New York: Mary Ann Liebert, 1987:515–521.

75. Luepke NP. Hen's egg chorioallantoic membrane test for irritation potential. *Food Chem Toxicol* 1985;23:287–291.

76. Luepke NP. HET-chorioallantois test: an alternative to the Draize rabbit eye test. In: Goldberg AM, ed. *Alternative Methods in Toxicology*, Vol. 3. New York: Mary Ann Liebert, 1985:592–605.

77. Lawrence RS. The chorioallantoic membrane in irritancy testing. In: Atterwill CK, Steele CE, eds. *In Vitro Methods in Toxicology*. New York: Cambridge University Press, 1987:263–278.

78. Lawrence RS, Groom MH, Ackroyd DM, Parish WE. The chorioallantoic membrane in irritation testing. *Food Chem Toxicol* 1986;24:497–502.

79. Price JB, Barry MP, Andrews IJ. The use of the chick chorioallantoic membrane to predict eye irritants. *Food Chem Toxicol* 1986;24:503–505.

80. de Silva O, Rougier A, Dossou KG. The HET-CAM test: a study of the irritation potential of chemicals and formulations. *ATLA* 1992;20:432–437.

81. Gilleron L, Coecke S, Sysmans M, Hansen E, Van Oproy S, Marzin D, Van Cauteren H, Vanparys, Ph. Evaluation of a modified HET-CAM assay as a screening test for eye irritancy. *Toxicol In Vitro* 1996;10:431–446.

82. Gettings SD, McEwen GN. Development of potential alternatives to the Draize eye test: the CTFA evaluation of alternatives program. *ATLA* 1990;17:317–324.

83. Gettings SD, Di Pasquale LC, Bagley DM, Chudkowski M, Demetrulias JL, Feder PI, Hintze KL, Marenus KD, Pape W, Roddy M, Schnetzinger R, Silber P, Teal JJ, Weise SL. The CTFA evaluation of alternatives program: an evaluation of *in vitro* alternatives to the Draize primary eye irritation test (Phase I) hydro-alcoholic formulations; a preliminary communication. *In Vitro Toxicol* 1990;3:293–302.

84. Gettings SD, Di Pasquale LC, Bagley DM, Casterton PL, Chudkowski M, Curren RD, Demetrulias JL, Feder PI, Galli CL, Gay R, Glaza SM, Hintze KL, Janus J, Kurtz PJ, Lordo RA, Marenus KD, Moral J, Muscatiello MJ, Pape W, Renskers KJ, Roddy MT, Rozen MG. The CTFA evaluation of alternatives program: an evaluation of *in vitro* alternatives to the Draize primary eye irritation test (Phase II) oil/water emulsions. *Food Chem Toxicol* 1994;32:943–976.

85. Booman KA, Cascieri TM, Demetrulias J, Driedger A, Griffith JF, Grochoski GT, Kong B, McCormick WC, North-Root H, Rozen MG, Sedlak RI. *In vitro* methods for estimating eye irritancy of cleaning products phase I: preliminary assessment. *J Toxicol—Cutan Ocul Toxicol* 1988;7:173–185.

86. Bulich AA, Tung KK, Scheibner G. The luminescent bacteria toxicity test: its potential as an *in vitro* alternative. *J Biolumin Chemilumin* 1990;5:71–77.

87. Bell E, Parenteau N, Gay R, Nolte C, Kemp P, Bilbo P, Ekstein B, Johnson E. The living skin equivalent: its manufacture, its organotypic properties and its responses to irritants. *Toxicol In Vitro* 1991;5:591–596.

88. Curren RD, Harbell JW, Raabe HA, Sussman RG, Kimmel TA. Comparison of the BCOP and CAMVA *in vitro* assays to predict the ocular irritancy of drug intermediates. *Toxicologist* 1997;36:43.

89. Swanson JE, Lake LK, Donnelly TA, Harbell JW, Huggins J. Prediction of ocular irritancy of full-strength cleaners and strippers by tissue equivalent and bovine corneal assays. *J Toxicol—Cutan Ocul Toxicol* 1995;14:179–195.

90. Rougier A, Cottin M, de Silva O, Catroux P, Roguet R, Dossou KG. The use of *in vitro* methods in the ocular irritation assessment of cosmetic products. *Toxicol In Vitro* 1994;8:893–905.

91. Spielmann H, Liebsch M, Moldenhauer F, Holzhutter H-G, de Silva O. Modern biostatistical methods for assessing *in vitro/in vivo* correlation of severely eye irritating chemicals in a validation study of *in vitro* alternatives to the Draize eye test. *Toxicol In Vitro* 1995;9:549–556.

92. de Silva O, Cottin M, Dami N, Roguet R, Catroux P, Toufic A, Sicard C, Dossou KG, Gerner I, Schlede E, Spielmann H, Gupta KC, Hill RN. Evaluation of eye irritation potential: statistical analysis and tier testing strategies. *Food Chem Toxicol* 1997;35:159–164.

93. Frazier JM. Scientific criteria for validation of *in vitro* toxicity tests. *OECD Monogr* 1990:36.

94. Balls M, Blaauboer B, Brusick D, Frazier J, Lamb D, Pemberton M, Reinhardt C, Roberfroid M, Rosenkranz H, Schmid B, Spielmann H, Stammati A, Walum E. Report and recommendations of the CAAT/ERGATT workshop on the validation of toxicity test procedures. *ATLA* 1990;18:313–337.

95. Goldberg AM, Frazier JM, Brusick D, Dickens MS, Flint O, Gettings SD, Hill RN, Lipnick RL, Resnskers KJ, Bradlaw JA, Scala RA, Veronesi B, Green S, Wilcox NL, Curren RD. Framework for validation and implementation of *in vitro* toxicity tests: report of the validation and technology transfer committee of the Johns Hopkins Center for Alternatives to Animal Testing. *In Vitro Toxicol* 1993;6:47–55.

96. Balls M, Blaauboer BJ, Fentem JH, Bruner L, Combes RD, Ekwall B, Fielder RJ, Guillouzo A, Lewis RW, Lovell DP, Reinhardt CA, Repetto G, Sladowski D, Spielmann H, Zucco F. Practical aspects of the validation of toxicity test procedures. *ATLA* 1995;23:129–147.

97. Spielmann H, Liebsch M, Kalweit S, Moldenhauer F, Wirnsberger T, Holzhutter H-G, Schneider B, Glaser S, Gerner I, Pape WJW, Kreiling R, Krauser K, Miltenburger HG, Steiling W, Luepke NP, Muler N, Kreuzer H, Murmann P, Spengler J, Bertram-Neis E, Siegemund B, Wiebel FJ. Results of a validation study in Germany on two *in vitro* alternatives to the Draize eye irritation test, the HET-CAM test and the 3T3 NRU cytotoxicity test. *ATLA* 1996;24:741–858.

98. Purchase IFH, Botham PA, Bruner LH, Flint OP, Frazier JM, Stokes WS. Workshop overview: scientific and regulatory challenges for the reduction, refinement, and replacement of animals in toxicity testing. *Toxicol Sci* 1998;43:86–101.

99. Curren R, Bruner L, Goldberg A, Walum E. 13th meeting of the scientific group on methodologies for the safety evaluation of chemicals (SGOMSEC): validation and acute toxicity testing. *Environ Health Perspect* 1998;106(suppl 2):419–425.

100. Doucet O, Lanvin M, Zastrow L. Comparison of three *in vitro* methods for the assessment of the eye irritating potential of formulated cosmetic products. *In Vitro Mol Toxicol.* In press.

101. Anon. Workshop on Updating Eye Irritation Test Methods, September 26–27, 1991.

102. Journal Officiel de la République Française, Arrêté du 09 Juillet 1992, J. O. du 10 Juillet 1992: Test d'irritation oculaire: evaluation de l'irritation oculaire des produits cosmétiques ou d'hygiène corporelle.

103. Eye irritation testing. *ECETOC Monogr* 1988;11:1–65.

104. OECD Environment Directorate, Step 2: proposal for a harmonized system for the classification of chemicals which cause eye irritation/corrosion, 1997.

105. Bito LZ. Species differences in the responses of the eye to irritation and trauma: a hypothesis of divergence in ocular defense mechanisms, and the choice of experimental animals for eye research. *Exp Eye Res* 1984;39:807–829.

106. Flint OP. What is an alternative? *In Vitro Toxicol* 1997;10:165–168.

107. Bruner LH, Carr GJ, Chamberlain M, Curren RD. Validation of alternative methods for toxicity testing. *Toxicol In Vitro* 1996;10:479–501.

108. Bruner LH, Carr GJ, Chamberlain M, Curren RD. No prediction model, no validation study. *ATLA* 1996;24:139–142.

109. Curren RD, Southee JA, Spielmann H, Liebsch M, Fentem JH, Balls M. The role of prevalidation in the development, validation, and acceptance of alternative methods. *ATLA* 1995;23:211–217.

110. Balls M, Fentem JH. Progress toward the validation of alternative tests. *ATLA* 1997;25:33–43.

111. Balls M, De Klerck W, Baker F, van Beek M, Bouillon C, Bruner L, Carstensen J, Chamberlain M, Cottin M, Curren R, Dupuis J, Fairweather F, Faure U, Fentem J,

Fisher C, Galli C, Kemper F, Knaap A, Langley G, Loprieno G, Loprieno N, Pape W, Pechovitch G, Spielmann H, Ungar K, White I, Zuang V. Development and validation of non-animal tests and testing strategies: the identification of a coordinated response to the challenge and the opportunity presented by the sixth amendment to the cosmetics directive (76/768/EEC). *ATLA* 1995;23:398–409.

112. NIEHS. *Validation and Regulatory Acceptance of Toxicological Test Methods*. Bethesda, MD: National Institutes of Health, 1997. NIH Publication 97-3981.

4

In Vitro Methods to Predict Skin Irritation

Ai-Lean Chew and Howard I. Maibach

Department of Dermatology, University of California, San Francisco, California

A complicated series of chemical and physiologic responses result in skin irritation. When skin is exposed to toxic substances, the Draize rabbit skin test, first outlined by Draize et al. in 1944, remains an important source of safety information for government and industry [1]. In this test, the cutaneous irritation caused by a substance is investigated by observing changes ranging from erythema and edema to ulceration produced in rabbit skin when irritants are applied. These skin reactions are produced by diverse physiologic mechanisms, although they are easily observed visually and by palpation.

The applicability of irritation or sensitization evaluation based on the visual assessment of reactions in animals has been a source of controversy for years [2,3]. Levels of skin damage are judged by observation, a procedure that has long been noted as highly subjective and unreliable, leading to problems of interlaboratory variability and calling the accuracy of the data into question [3]. Also, the differing skin reactions exhibited by varying species have cast doubt on the applicability of the results derived from animal studies as they pertain to human irritation [2]. Furthermore, the fact that the guinea pig and rabbit *in vivo* systems yield little information about the physiologic mechanisms underlying skin irritation has contributed to the search for objective *in vitro* investigational methods. Recent ethical concerns about the humane treatment of animals have also increased efforts to develop improved methods of *in vitro* toxicology evaluation.

Thus, in response to scientific and sociologic issues, research on *in vitro* skin irritation methods has recently been active. Many investigators are developing *in vitro* irritation systems that elicit more specific information about actual mechanisms involved in the complex cascade of events causing irritation.

4.1 CURRENT *IN VITRO* METHODS

Proposed *in vitro* methods are based on cell cytotoxicity, inflammatory or immune system response, alterations of cellular, bacterial, or fungal physiology, cell morphology, biochemical endpoints, macromolecular targets, and structure activity analysis [4–9]. With a decrease in animal testing, additional *in vitro* testing has been more often used in a comprehensive toxicology program. These methods can be broadly placed into six categories (see Table 4.1).

TABLE 4.1. Current *in vitro* methods

	Methods	Examples
1.	Physicochemical analysis	pH
		Absorption spectra
		Partition coefficient
		SKINTEX
		SOLATEX-PI
2.	Cell culture techniques	Conventional keratinocyte/fibroblast cultures
		Skin explants or organ cultures
3.	Microorganism studies	Microtox (*Photobacterium phosphorium*)
		Tetrahymena thermophila
		Daniels
4.	Human skin recombinants	EPISKIN
		EpiDerm
		SKINETHIC
		TESTSKIN
		Skin²
5.	Embryonic testing	HET/CAM
6.	Computer modeling	Quantitative structure-activity relationships

4.1.1 Physicochemical Test Methods

Analysis of the physicochemical properties of test substances, including the pH, absorption spectra, and partition coefficients, often indicates potential cutaneous toxicity. The potential corrosivity or irritancy of strong acids and bases has been well established. According to previous Organization for Economic Cooperation and Development (OECD) guidelines, substances with a pH of less than 2 or greater than 11.5 are regarded as corrosive and do not require testing for irritancy *in vivo* [10]. However, the single-parameter pH may not always be an accurate predictor, because not all corrosive or irritant chemicals have a mechanism of action directly related to pH. The OECD guidelines have recently been revised to recognize the importance of the buffering capacity of acid/alkali over the single-parameter pH [11]. Accordingly, Young and How [12] have formulated an equation to express the relationship between pH–acid/alkali reserve and classification of irritancy:

If pH + 1/6 alkali reserve ≥13 or pH – 1/6 acid reserve ≤1, the preparation is irritant.

Physicochemical analysis has evaluated the particular chemical properties of test substances identified as key structural components contributing to penetration, irritation, or sensitization. Absence of absorption in the ultraviolet (UV) range also has been used to suggest a lack of photoirritant potential [13]. Physicochemical tests are rapid, cost-effective, easily standardized, and reproducible. For penetration, a partition coefficient of the test sample provides a useful guide. The size of a chemical is also indicative of potential penetration. Many of the physicochemical properties of surfactants are potential indicators of their action on skin [14].

Target macromolecular systems. Test methods using analysis of biochemical reactions or changes in organized macromolecules evaluate toxicity at a subcellular level. Because of their simplicity, they can be readily standardized and transferred to outside laboratories to provide yardstick measurements for varying degrees of cutaneous toxicity.

One *in vitro* irritation prediction method that uses nonhuman substrates can be described as a biomembrane-barrier–macromolecular-matrix system. This method, known as the SKINTEX (In Vitro International, Irvine, CA) system, uses a two-compartment physicochemical model incorporating a keratin/collagen membrane barrier and an ordered macromolecular matrix [15]. The effect of irritants on this membrane is detected by changes in the intact barrier membrane through the use of an indicator dye attached to the membrane. The dye is released after membrane alteration or disruption, which can occur when the synthetic membrane barrier is exposed to an irritant. A specific amount of dye corresponding to the degree of irritation can be liberated and quantified spectrophotometrically. The second compartment within the system is a reagent macromolecular matrix responding to toxic substances by producing turbidity. This second response provides an internal detection for materials disrupting organized protein conformation after passing through the membrane barrier [15].

Test samples can be applied directly to the barrier membrane as liquids, solids, or emulsions and inserted into the liquid reagent. The results are directly compared with the Draize cutaneous irritation results.

More than 5,300 test samples have been studied in the SKINTEX system, including petrochemicals, agrochemicals, household products, and cosmetics. The reproducibility with standard deviations of 5–8% is excellent. New protocols applicable to low irritation test samples and alkaline products have increased the applicability of this method. SKINTEX validation studies resulting in an 80–89%

correlation to the Draize scoring have been reported by Yves Rocher, S.C., Johnson & Johnson, and the Food and Drug Safety Center [16–18].

Thus far, most *in vitro* irritation methods, including most SKINTEX protocols (such as the upright membrane assay, the standard labeling protocol, and the high sensitivity assay) have relied heavily on the vast Draize rabbit skin database for validation. As previously discussed, the discrepancies in the information generated by the Draize system raise questions about the applicability of this information to irritation reactions in humans. A new SKINTEX protocol called the "human response assay" optimizes the model to predict human irritation. Good correlations to human response have been demonstrated for pure chemicals, surfactants, vehicles, and fatty acids [19–21].

The SKINTEX test is a rapid, standardized approach with well-refined protocols and an extensive database. The results produced are contiguous with the historical *in vivo* database. However, the method cannot predict immune response, penetration, or recovery after the toxic response.

SOLATEX-PI (In Vitro International, Irvine, CA) uses the two-compartment physicochemical model of SKINTEX to predict the interactive effects of specific chemicals and UV radiation. SOLATEX-PI has demonstrated the capability to predict the potential for photoirritation of certain materials [22]. SOLATEX-PI is being validated by FRAME and the BGA (Zebet) as an *in vitro* test to predict photoirritants.

4.1.2 Cell Culture Techniques

Cell culture models developed to study the cutaneous irritation potential of chemicals include *in vitro* monolayer cell cultures comprising keratinocytes, fibroblasts, or melanocytes, immortalized cell lines, and skin explants or organ cultures.

Conventional cell cultures. Typically, only fibroblasts and keratinocytes are used in skin irritation investigations. Cells of the inflammatory response, such as polymorphonuclear leukocytes, tend to be absent. Further, monolayer cell cultures lack a stratum corneum to convey barrier protection. They are, thus, inaccurate models of irritation prediction, often resulting in overestimation of the toxicity of a compound [23]. A major limitation of cell culture systems is that only water-soluble substances may be tested. To address these concerns, recent developments have been directed toward human skin equivalents (HSEs) (*see below*).

Organ cultures or skin explants. The effects of chemical irritants in human and animal skin organ cultures have been investigated [24]. Skin organ culture models are two-dimensional, containing all dermal and epidermal cell types (including stratum corneum) involved in the irritation response [25]. Skin explants involve excision of skin from animals or humans, which are then maintained on cell culture media, epidermis side up at the air interface and the dermal component immersed in media.

Good correlations to *in vivo* models have been obtained with dilute chemicals, but not with high concentrations [24]. However, disadvantages exist to this model. These methods are difficult to implement in routine testing because of the short survival of the tissue. The technique is unsuitable for assessing mild irritants because the damage induced by excising and culturing the skin stimulates the release of mediators [23]. The limited availability of viable human skin also restricts this predictive method. Animal skin is an alternative; however, it is largely recognized that the barrier function of most animal skin is less than that of human skin. Thus, animal skin models tend to overestimate irritation [24].

Endpoint measurements for cytotoxicity tests (colorimetric bioassays). As the process of cutaneous irritation is complex, no parameter has emerged as the ideal predictor of irritation potential. *In vitro* cytotoxicity tests indicating basic cell toxicity by measuring parameters, such as cell viability, cell proliferation, membrane integrity, DNA synthesis, or cellular metabolism, have been used as indicators of cutaneous toxicity [26–29]. Cytotoxicity tests use various assays to assess these biologic endpoints. The most commonly used endpoint measurements use colorimetry, namely, the neutral red uptake assay (for cell viability), the Lowry (labeled proline) Coomassie blue and Kenacid blue assays (for measuring total cell protein and hence cell proliferation), the 3-(4, 5-dimethylthiazol-2-yl)-2,5 diphenyl tetrazolium bromide (MTT) or tetrazolium assay (for assessing mitochondrial function and hence cellular metabolism), and the intracellular lactate dehydrogenase (LDH) activity test (for assessing cell lysis).

In the neutral red uptake (cell viability) and total protein (cell proliferation) assays, cells are treated with various concentrations of a test substance in Petri or multiwell dishes; after a period of exposure, the substance is washed out of the medium. (An analytical reagent is added in the case of protein measurements.) Neutral red is a supravital dye accumulating in the lysosomes of viable, uninjured cells, and it can be washed out of cells that have been damaged. In the protein test, Kenacid blue is added and reacts with cellular protein. Controlled cells are dark blue; killed cells are lighter colored. In both tests, the cellular dye uptake may be quantified spectrophotometrically. The IC_{50} (the concentration inhibiting by 50%) is determined; the test can be rapidly performed with automation. However, materials must be solubilized into the aqueous media for analysis. For many test materials, this process will require large dilutions eliminating properties of the materials causing irritation.

The MTT test assays mitochondrial function by measuring reduction of the yellow MTT tetrazolium salt to an insoluble blue formazan product. It has been compared with the neutral red technique for testing the cytotoxicity of 28 test substances, including drugs, pesticides, caffeine, and ascorbic acid. With the mouse

BALB/c 3T3 fibroblast cell line, for any given cell density, the two assays ranked the test substances with a correlation coefficient of 0.939, on the basis of IC_{50} concentrations. The two assays did differ in sensitivity for a few test agents, suggesting that a combination of the two might be most effective [27].

Enzyme leakage may detect sublethal cell injury that might not be observed histologically. Skin in organ culture has been analyzed to determine quantifiable parameters to assess injury, such as cellular enzyme leakage, glucose metabolism, DNA synthesis, water loss, and changes in electrolyte concentration [36]. Rat skin *in vivo* exposed to toxicants causes release of acid phosphatase, LDH, and N-acetylglucosaminidase, which is associated histologically with epidermal edema and an increase in dermal leukocytes [35]. The activity of these enzymes may be analyzed using a colorimetric method.

Evaluation of cutaneous toxicity (noncolorimetric methods). *In vitro* methods are based on years of laboratory and clinical research determining the basic features of skin penetration, irritation, and sensitization. The targets are so complex that the effect of toxic substances on the structure of the skin is poorly understood. Studies have elucidated considerable information about the mechanisms of damage and repair that occur in skin. Typical events identified in the cutaneous irritation process include protein denaturation, epidermal cell lysis, cytotoxicity, enzyme leakage, and production of epidermal antigens and cytokines [30–33]. Noncolorimetric means of evaluating the evidence of cell damage include examining morphology, signs of the inflammatory reaction initiation, cellular toxicity, and electrical properties [34]. Also, synthetic models of epithelium have been designed to mimic irritant damage characteristics [35]. Some investigators have combined two or more of these modalities and compared them to assess the differences.

Helman et al. [36] compared the morphologic responses of *in vitro* and *in vivo* skin exposed to chemicals with light microscopy. They found that the absence of an intact vascular system in *in vitro* skin specimens did not interfere significantly with the ability to detect graded microscopic epidermal lesions and concluded that the morphologic response of skin maintained in an organ culture is an accurate indicator of skin toxicity. In addition to the altered histology seen with light microscopy, electron-microscopic analysis of irritant-damaged skin reveals characteristic changes, including spongiosis of epidermis, disappearance of tonofilament–desmosome complexes, and dissolution of horny cells [37,38].

Irritation has been evaluated by analyzing epidermal edema with other techniques. Sodium lauryl sulfate produced swelling in *in vitro* skin disks prepared from excised human skin and dermal calf collagen [39,40]. In an *in vitro* system without skin, tritiated water uptake (i.e., swelling) of a collagen film was proportional to the degree of *in vivo* irritation in a series of surfactants [39].

A device using cellular metabolic activity as an endpoint is the microphysiometer. This device employs a silicon-based electrode, known as a light-activated potentiometric sensor (LAPS), which can detect subtle changes in the pH of cell culture media by determining the rate at which cells excrete acidic metabolic byproducts, such as lactic acid and carbon dioxide [41]. These metabolic changes can be observed dynamically, on a time scale of seconds to minutes, and thereby, they can assess recovery of the cell monolayer after toxicologic insults.

Inflammatory mediator release. More recently, studies have been published on measurements of inflammatory mediator release, such as interleukins (IL1α, IL6, IL8), tumor necrosis factor (TNF-α), and arachidonic acid metabolites (e.g., prostaglandins, leukotrienes) [23]. These inflammatory mediators are synthesized by viable cells and released into the extracellular matrix as part of the cells' response to irritation.

A variety of analytic methods exist for quantification of inflammatory mediators. Bioassays are available; a typical endpoint of a bioassay for measuring inflammatory mediators is cellular proliferation, as measured by ^{3}H-thymidine uptake by dividing cells [42]. The use of bioassays has now declined, with the availability of more reliable quantitative methods, such as enzyme-linked immunosorbent assay (ELISA) and polymerase chain reaction (PCR) analysis. A recent review of cytokines in dermatotoxicology by Gerberick et al. details methods of cytokine analysis and elucidates current knowledge on the cytokine profile in cutaneous irritation [43].

4.1.3 Microorganism Studies

The chemical processes of microorganisms as a measure of toxic effect are employed by some *in vitro* assay systems. The Microtox system uses reduction in fluorescence normally emitted by luminescent bacteria (*Photobacterium phosphorium*) after exposure to irritants [44]. Another system uses a ciliated protozoan *Tetrahymena thermophila* [45,46]. Normal motility of these organisms is impaired after irritant exposure, and it can be compared with motility in untreated organisms.

Phototoxicity studies also use microorganism assays. The Daniels's test for phototoxicity uses the yeast *Candida albicans* as the test organism. A 1988 study compared favorably the results of this test with the results of photopatch testing in volunteers for samples from six furocoumarin-containing plants [29]. Many test materials producing an erythematous response in the photoirritant test are not analyzed as positive in this test.

4.1.4 Human Skin Equivalents

Limitations of the conventional cell culture models have resulted in development of three-dimensional, reconstituted human skin models, which closely mimic human skin. These skin equivalents, originally developed as engineered grafts for burns patients, were subsequently used for testing potential dermatotoxic effects of substances.

One of the first HSEs commercially available was TESTSKIN (Organogenesis, Inc., Cambridge, MA), which consisted of human keratinocytes seeded onto a bovine collagen base or collagen–glycosaminoglycan matrix containing human fibroblasts. The production of TESTSKIN was discontinued in 1993. Another skin equivalent, developed by Marrow-Tech, Inc. (Elmsford, NY), consists of a dermal layer of fibroblasts and naturally secreted collagen and an epidermal layer of keratinocytes separated by a dermal–epidermal junction. Whereas TESTSKIN uses bovine collagen, Marrow-Tech's skin model consists solely of human tissue. Another early HSE was Skin^2 (Advanced Tissue Sciences, La Jolla, CA). This three-dimensional skin equivalent comprised neonatal skin cells cultured on a nylon mesh. Although validation studies showed promising results, production of Skin^2 ceased in 1996.

Currently, the three main commercial HSEs used for skin irritancy testing are EPISKIN (Imedex, Chaponost, France), EpiDerm (MatTek Corporation, Ashland, MA), and SKINETHIC (SkinEthic Laboratories, Nice, France). The skin recombinant, differentiated keratinocyte cultures are grown at the air–liquid interface on various substrates, thus resulting in a stratified, differentiated epithelium. The EPISKIN cultures consist of seeded adult human keratinocytes on a dermal support of collagens I and III, covered with a thin film of collagen IV. The EpiDerm model comprises normal human epidermal keratinocytes grown on permeable membranes to form a multilayered, differentiated epidermis. The SKINETHICcultures consist of normal human adult keratinocytes on an inert polycarbonate filter at the air–liquid interface in modified and supplemented chemically defined medium.

In general, the same endpoints used for the monolayer cell culture systems are used in these multilayer skin equivalents. The use of HSEs represents a major advance in *in vitro* irritation testing. HSEs use human cells instead of animal cells, thus eliminating any discrepancies in results caused by species variation. HSEs are grown at the air–liquid interface, which generates a stratified layer, similar to the *in vivo* human stratum corneum (SC). This functional SC confers barrier properties to the HSE, analogous to the *in vivo* situation, and allows topical products (both water-soluble and water-insoluble) to be applied directly to the surface [47]. The major disadvantage, clearly, is that these HSEs still lack intact vascular systems and inflammatory cell components.

4.1.5   Embryonic Testing

The hen's egg test/chorioallantoic membrane system (HET/CAM) uses fertilized chicken eggs, the vascular network (CAM) of which is exposed by cutting a small opening into the eggshell [7]. Test substances are applied directly to the CAM, and their effects are assessed by scoring visual changes in the blood vessel network (such as hemorrhage and coagulation), at 0.5, 2, and 5 minutes after treatment. The basis of this model is that the inflammatory processes involved in irritation (e.g., erythema, edema) depend on vascular changes, which may be monitored via the CAM. This method is mainly used in Europe for ocular irritancy testing; however, its basic principles may be employed for skin irritancy.

4.1.6 Computer Modeling/QSAR

Quantitative structure-activity relationship (QSAR) models are used to predict the extent of an anticipated toxic effect by relating physicochemical and structural properties of a closely related group of chemicals to a given toxicologic endpoint (e.g., irritation, sensitization). The biologic properties of structurally similar compounds can then be predicted. Although QSARs are more established for predicting allergenic potential, models for predicting skin irritation potential of chemicals are currently being investigated. Several expert systems for predicting toxicity incorporate skin irritation as one of the endpoints, for example, DEREK, TOPKAT (Health Designs Inc., Rochester, NY), and Hazardexpert (Case Western Reserve, Clevelsand, OH), although these have yet to be validated [24].

4.2 HUMAN VOLUNTEER STUDIES

Human volunteer studies are *in vivo* studies, but they will be discussed briefly here, because they are widely used to assess skin irritation, penetration, and sensitization, and they are regarded as an important alternative method to *in vivo* animal studies. A review by Patil et al. is recommended for a broader coverage of human predictive assays of irritation [48].

Single-application patch tests are often used to assess the irritation potential of products. The 24-hour acute irritation assay originally described by Draize et al. [1] is the most commonly used in its various modified forms, although other exposure periods are also used, such as 4 hours, 6 hours, and 48 hours.

Many industries regularly conduct repeat insult patch tests or cumulative irritation assays on human volunteers to evaluate topical irritancy. Groups of human volunteers are patched with the test substance. One to five concentrations can be tested simultaneously, which is a wide enough range to yield results relevant to the usage. Cumulative skin irritancy is measured by applying patch applications each day for 3 weeks [21].

Skin irritation is usually assessed visually—erythema, edema, and vesiculation are scored on a visual scale. Nowadays, skin bioengineering data are often used as quantitative adjuncts, such as transepidermal water loss (as a measure of skin barrier function), laser Doppler flow (to measure skin blood flow), and colorimetry (to quantify erythema). In these noninvasive tests, dose-response curves can be obtained.

Human volunteers are also used in many industries in tests for allergic sensitization by cosmetic substances and formulations. The repeat insult patch test includes an induction phase (repeat applications during 3 weeks) and a 2-week rest period (incubation phase), followed by a challenge to see if sensitization has occurred. A pilot study of 20 human volunteers can be followed by more extensive testing (80–100 subjects). Positive results at more than the 10% level in the human volunteers would suggest a major problem with the formulation. Using tests with the sensitized individuals and nonreactive matched control subjects can often determine the importance of these results, i.e., determine whether the sensitivity is significant under normal conditions of product use. Broader tests can be carried out with 250–500 subjects [21].

4.3   CONCLUSIONS

Whole-animal tests represent true physiologic and metabolic relationships of macromolecules, cells, tissues, and organs evaluating the reversibility of toxic effects. However, these tests are costly, time consuming, insensitive, and difficult to standardize, and they are sometimes poorly predictive of human *in vivo* response.

A wide range of *in vitro* methods based on diverse endpoints have been developed to provide information on the complex series of chemical and physiologic responses of the skin to toxic substances. This series of responses concentrates on dermal toxicity, which has been studied *in vivo* using the Draize rabbit skin irritation test, the guinea-pig sensitization test, and the skin penetration test.

New *in vitro* test methods target the behavior of macromolecules, cells, tissues, and organs in well-defined methods controlling experimental conditions and standardizing experimentation. These tests provide more reproducible, rapid, and cost-effective results. In addition, more information at a basic mechanistic level can be obtained from these tests.

The challenge of the new millennium will be to understand the capabilities and limitations of the existing methods, to refine these methods, and to develop newer methods and assays. Combining test methods can provide a greater understanding of the mechanisms of toxic molecules. Test batteries evaluating cell cytotoxic responses at high dilutions and changes in macromolecules at low dilutions will be more informative than is visual scoring of complex events *in vivo*.

References

1.   Draize JH, Woodard G, Calvery HO. Methods for the study of irritation and toxicity of substances applied topically to the skin and mucous membranes. *J Pharmacol Exp Ther* 1944;82:377–389.

2.   Patrick E, Maibach HI. Comparison of the time course, dose response and mediators of chemically induced skin irritation in three species. In: Frosch PJ, et al., eds. *Current Topics in Contact Dermatitis*. New York: Springer-Verlag, 1989:399–403.

3.   Kastner W. Irritancy potential of cosmetic ingredients. *J Soc Cosmet Chem* 1977;28:741–754.

4. Borenfreund E, Puerner JA. Toxicity determined *in vitro* by morphological alterations and Neutral Red absorption. *Toxicol Lett* 1985;24:119–124.

5. Borenfreund E, Puerner JA. A simple quantitative procedure using monolayer cultures for cytotoxicity assays. *J Tissue Cult Methods* 1984;9:7–9.

6. Bulich AA, Greene MW, Isenberg DL. Reliability of bacterial compounds and complex effluents. In: Branson DR, Dickson KL, eds. *Aquatic Toxicology and Hazard Assessment*. Philadelphia: American Society for Testing and Materials, 1981:338–347. ATSM Publication 737.

7. Luepke NP, Kemper FH. The HET-CAM test: an alternative to the Draize eye test. *Fundam Chem Toxicol* 1986;24:495–496.

8. Parce JW, Owicki JC, Kercso KM, et al. Detection of cell-affecting agents with silicon biosensor. *Science* 1989;246:243–247.

9. Silverman J. Preliminary findings on the use of protozoa (*Tetrahymena thermophila*) as models for ocular irritation testing rabbits. *Lab Anim Sci* 1983;33:56–58.

10. Acute dermal irritation/corrosion. *Guidelines for Testing Chemicals*, Section 404. Paris: OECD, 1981.

11. Acute dermal irritation/corrosion. *Guidelines for Testing of Chemicals*, Section 404. Paris: OECD, July 17, 1992.

12. Young JR, How MJ. Product classification as corrosive or irritant by measuring pH and acid/alkali reserve. In: Rougier A, Goldberg AM, Maibach HI, eds. *In Vitro Skin Toxicology: Irritation, Phototoxicity, Sensitization*. New York: Mary Ann Liebert, 1994.

13. Morrison WL, McAuliffe DJ, Parrish JA, Bloch JJ. *In vitro* assay for phototoxic chemicals. *J Invest Dermatol* 1982;78:460–463.

14. Serban GP, Henry SM, Cotty VF, et al. *In vivo* evaluation of skin lotions by electrical capacitance: I. The effect of several lotions on the progression of damage and healing after repeated insult with sodium laurel sulfate. *J Soc Cosmet Chem* 1981;32:407–419.

15. Gordon VC, Kelly CP, Bergman HC. Evaluation of SKINTEX, an *in vitro* method for determining dermal irritation. *Toxicologist* 1990;10(1):78.

16. Soto RJ, Servi MJ, Gordon VC. Evaluation of an alternative method for ocular irritation. In: Goldberg AM, ed. *In Vitro Toxicology*. New York: Mary Ann Liebert, 1989.

17. Khaiat A. Evaluation of SKINTEX at Yves Rocher. Presented at 1st European *In Vitro* Symposium, Paris, France, June 1990.

18. Food and Drug Safety Center Report. Presented at JSAAE, Tokyo, Japan, November 1990.

19. Bason M, Gordon VC, Maibach H. Skin irritation *in vitro* assays. *Int J Dermatol* 1996;30:623–626.

20. Bason M, Harvell J, Gordon V, Maibach H. Evaluation of the SKINTEX system. Presented at the Irritant Contact Dermatitis Symposium, Groningen, The Netherlands, 1991.

21. Harvell J, Bason M, Gordon V, Maibach H. Evaluation of dermal irritation of pure chemicals. Presented at the *In Vitro* Workshop, FDA/AAPS Symposium, Washington,

DC, December 2, 1991.

22. Gordon VC, Acevedo J. SOLATEX-PI, An *in vitro* method to predict photoirritation. Presented at JSAAE, Tokyo, Japan, November 13, 1991.

23. Ponec M. *In vitro* models to predict skin irritation. In: van der Valk PGM, Maibach HI, eds. *The Irritant Dermatitis Syndrome.* Boca Raton, FL: CRC Press, 1996.

24. Botham PA, Earl LK, Fentem JH, et al. Alternative methods for skin irritation testing: the current status. ECVAM Skin Irritation Task Force report 1. *ATLA* 1998;26:195–211.

25. Van der Sandt JJM, van Schoonhoven J, Maas WJM, et al. Skin organ culture as an alternative to *in vivo* dermatotoxicity testing. *ATLA* 1993;21:443–449.

26. Clothier RH, Hulme L, Ahmed AB, Reeves HL, Smith M, Balls M. *In vitro* cytotoxicity of 150 chemicals to 3T3-L1 cells, assessed by the FRAME Kenacid Blue method. *ATLA* 1988;16:84–95.

27. Borenfreund E, Babich H, Martin-Alguacil N. Comparisons of two *in vitro* cytotoxicity assays—the Neutral Red (NR) and tetrazolium MTT test. *Toxicol In Vitro* 1988;2:1–6.

28. Chan KY. Chemical injury to an *in vitro* ocular system: differential release of plasminogen activator. *Clin Eye Res* 1986;5:357–365.

29. Jackson EM, Hume RD, Wallin RF. The agarose diffusion method for ocular irritancy screening: cosmetic products, Part II. *J Toxicol—Cutan Ocular Toxicol* 1988;7:187–194.

30. Nago S, Stroud JD, Hamada T, et al. The effect of sodium hydroxide and hydrochloric acid on human epidermis: an EM study. *Acta Derm Venereol (Stockh)* 1972;52:11–23.

31. Kanerva L, Lauharanta J. Variable effects of irritants (methylmethacrylate, terphenyls, dithranol and methylglyoxal-bisguanylhadrazone) on the fine structure of the epidermis. *Arch Toxicol* 1986;9:455.

32. Gibson WT, Teall MR. Interactions of C12 surfactants with the skin: changes in enzymes and visible and histological features of rat skin treated with sodium laurel sulfate. *Fundam Chem Toxicol* 1983;21(5):587–593.

33. SOT Position paper. Comments on the LD 50 and acute eye and skin irritation test. *Fundam Appl Toxicol* 1989;13:621–623.

34. Oliver GJA, Pemberton MA, Rhodes C, et al. An *in vitro* model for identifying skin-corrosive chemicals. 1. Initial validation. *Toxicol In Vitro* 1988;2:7–17.

35. Bell E, Gay R, Swiderek M, et al. Use of fabricated living tissue and organ equivalents as defined higher order systems for the study of pharmacologic responses to test substances. Presented at the NATO Advanced Research Workshop, Pharmaceutical Application of Cell and Tissue Culture to Drug Transport, Bandol, France, September 4–9, 1989.

36. Helman RG, Hall JW, Kao JV. Acute dermal toxicity: *in vivo* and *in vitro* comparisons in mice. *Fundam Appl Toxicol* 1986;7:94–100.

37. Geller W, Kobel W, Seifert G. Overview of animal test methods for skin irritation. *Fundam Chem Toxicol* 1985;23(2):165–168.

38. Bloom E, Maibach HI, Tammi R. *In vitro* models for cutaneous effects of glucocorticoids using human skin organ and cell culture. In: Maibach HI, Lowe NJ, eds. *Models in*

Dermatology, Vol. 4. New York: Karger, 1989:12–19.

39. Blake-Haskins JC, Scala D, Rhein LD, et al. Predicting surfactant irritation from the swelling response of a collagen film. *J Soc Cosmet Chem* 1986;37:199–210.

40. Choman BR. Determination of the response of skin to chemical agents by an *in vitro* procedure. *J Invest Dermatol* 1963;44:177–182.

41. Parce JW, Owicki JC, Kercso KM, et al. Detection of cell-affecting agents with a silicon biosensor. *Science* 1989;246:243–247.

42. Schreiber S, Kilgus O, Payer E, et al. Cytokine pattern of Langerhans cells isolated from murine epidermal cell cultures. *J Immunol* 1992;149:3,524–3,534.

43. Gerberick GF, Sikorski EE, Ryan CA, Limardi LC. Use of cytokines in dermatotoxicology. In: Marzulli FN, Maibach HI, eds. *Dermatotoxicology Methods: The Laboratory Worker's Vade Mecum*. Washington, DC: Taylor & Francis, 1997:187–206.

44. Bulich AA, Greene MW, Isenberg DL. Reliability of bacterial luminescence assay for determination of the toxicity of pure compounds and complex effluents. Presented at Aquatic Toxicology and Hazard Assessment, 4th Conference. Branson DR, Dickson KL, eds., American Society for Testing and Materials (ASTM 737). Philadelphia, PA, 1981:338–437.

45. Silverman J. Preliminary findings on the use of protozoa (*Tetrahymena thermophila*) as models for ocular irritation testing rabbits. *Lab Anim Sci* 1983;33:56–58.

46. Silverman J, Pennisi S. Evaluation of *Tetrahymena thermophila* as an *in vitro* alternative to ocular irritation studies in rabbits. *J Toxicol—Cutan Ocul Toxicol* 1987;6:33–42.

47. Prunieras M. Skin and epidermal equivalents: a review. In: Rougier A, Goldberg AM, Maibach HI, eds. *In Vitro Skin Toxicology: Irritation, Phototoxicity, Sensitization*. New York: Mary Ann Liebert, 1994:97–105.

48. Patil SM, Patrick E, Maibach HI. Animal, human and *in vitro* test methods for predicting skin irritation. In: Marzulli FN, Maibach HI, eds. *Dermatotoxicology*, 5th ed. Washington, DC: Taylor & Francis, 1996:411–436.

5

Lethality Testing

Shayne Cox Gad

Gad Consulting Services, Raleigh, North Carolina

Several assays have been developed since 1980 to evaluate the acute toxicity and lethality of chemicals in an *in vitro* assay system. These assays, by their nature, are simple, rapid indicators of relative toxicity with the intent of predicting the toxic effects of chemicals in animals and humans. *In vitro* assays are attractive alternatives to traditional animal tests not only because they limit the use of animals, but also because they provide a cost-effective approach for the rapid screening of a large number of environmental xenobiotics and new drugs in development. Advantages also exist to using *in vitro* systems for mechanistic studies. *In vitro* studies are simplistic in that they can be limited to the target cell or tissue of interest. In addition, culture conditions can be easily controlled without the influence of exogenous factors, including diet, drug usage, and environmental chemical exposures, or endogenous systemic factors, such as metabolism, immunologic status, and hormonal variation.

Many factors exist that must be considered when using *in vitro* methods to extrapolate *in vivo* toxicity. The major differences existing between growth of many cell types in culture and their growth in an intact organ originate from displacement and changes in spatial rearrangement. Normal cells *in situ* grow in a complex three-dimensional arrangement of different cell types, whereas *in vitro* they are forced to grow on a two-dimensional substrate. Not only is the normal spatial architecture of the tissue lost, but also as cells spread out they begin to lose their capability for normal cell-to-cell interactions and communications. *In vitro* cultures also lack the hormonal and neural control mechanisms required for the normal homeostatic regulations inherent *in vivo*. Although *in vitro* systems have their limitations, it should not be the intention to totally replace the need for animal testing with *in vitro* tests, but rather to augment primary quantitative *in vivo* tests and to provide supportive mechanistic information.

Many of the same principles that govern *in vivo* toxicity assessment are also applicable to the study of *in vitro* cytotoxicity. The cytotoxicity of a xenobiotic in an *in vitro* system is dependent on the quantity of a compound reaching the cell and the duration of exposure, as is true for *in vivo* toxicity. However, in an *in vitro* test system, chemicals come into direct contact with cell membranes and diffuse into cells without the protection afforded by barriers such as skin and limitations of absorption. In addition, *in vitro* systems have limited metabolic and detoxification mechanisms. *In vitro* systems are therefore highly sensitive and can often exaggerate the magnitude of the response. The *in vitro* cytotoxic properties of a given xenobiotic are also dependent on the biologic system itself and the endpoints that are evaluated. Thus, it is imperative that the investigator use caution in the experimental design and interpretation of *in vitro* tests. In addition, both the cell system and the endpoints should be well characterized and validated as a basis for a meaningful interpretation of the results.

5.1 CELL SYSTEMS

A variety of cell culture systems have been developed and used to assess *in vitro* toxicity and lethality. Before choosing a particular cell system for *in vitro* testing, it is important to understand the fundamental differences between the various sources and types of cultures available. Once a cell system is selected, it is then necessary to define the optimal growth conditions for that system.

5.1.1 Primary Cultures

Primary cultures are prepared from tissues or organs taken directly from an organism immediately after necropsy. A single-cell suspension is obtained by mechanical or chemical dispersion of the cells with the use of enzymes such as collagenase. The cells are suspended into growth medium and incubated for at least 24 hours in culture, during which time they may attach to the surface of the vessel or remain in suspension, depending on the growth characteristics of that particular tissue. With the exception of mature hematopoietic cells, most normal cells attach to a substrate, spread, and grow as a monolayer in culture. Transformed cell lines and lines established from tumor cells can proliferate in suspension because they have lost the need for attachment. Viability of the culture is affected by mechanical or enzymatic damage to the cell membrane during cell isolation and by depletion of essential nutrients or hormones needed for cell survival. Primary cultures are difficult to reproduce because the viability and growth kinetics of the cells will vary between cultures as a result of individual genetic, hormonal, and age differences.

Even replicate cultures derived from the same tissue contain varying proportions of different cell types.

Primary cell cultures are initially heterogeneous and well differentiated, and thus, they retain many of the complex biochemical functions of the animal tissue from which they are derived. However, primary cultures have limited life spans in culture. Within days, in culture, faster growing and more rigorous subpopulations begin to take over and predominate through the selective pressures of the culture conditions in which they are maintained. With time, the cultures become more homogeneous, less differentiated, and begin losing specific cell functions and metabolic capabilities. Thus, most *in vitro* systems are established with permanent cell lines that will grow consistently for relatively long periods of time in culture.

5.1.2 Permanent Cell Lines

Permanent cell lines are usually derived by subculture from a primary culture to form a diploid, continuous (established), or clonal line that may either grow in suspension or attach as a monolayer on the surface of the culture dish. At least 75% of the population of cells in a diploid cell line contains the same normal complement of genetic material as the organism from which it was derived [20] and will usually undergo approximately 40–50 divisions in culture. The typical life cycle of human fibroblast cultures has been characterized into three phases on the basis of growth characteristics in culture [42]. The initial phase I primary culture has a long population doubling time. After the first subculture, the primary culture is referred to as a cell line. Selection pressures and phenotypic drift gradually change the cell population and growth characteristic of the cell line with each subsequent subculture. By phase II, the culture has become more stable and hardy and the growth kinetics have changed, resulting in a period of short, consistent population doubling times. Phase III is characterized by the onset of senescence, with progressively longer doubling times followed by degeneration and subsequent death.

Continuous or established cell lines are permanent cell lines derived from tumor cells by either a spontaneous or induced transformation of a primary culture or diploid cell line resulting in immortalization of the line. Transformation can be induced by viruses, chemical mutagens, or irradiation to result in cells with acquired, inheritable morphologic growth characteristics stably transmitted from generation to generation in all of the progeny cells. The genome of continuous cell lines deviates genetically from that of the normal cells from which they originated. Most continuous cell lines are aneuploid and heteroploid compared with the normal, consistently diploid karyotype. Transformed cells are similar to tumor cells in that they have altered morphology and lack contact inhibition, anchorage dependence, and density-dependent inhibition of cell multiplication, which results in unlimited cell division in culture.

5.1.3 Clonal Cell Lines

Clonal cell lines are derived from the mitosis of a single cell of a primary, diploid, or continuous line to form a genetically homogeneous subline termed a *cell strain*. Frequent reisolation and cloning is required to minimize heterogeneity over time produced by genetic drift and, thus, to maintain a genetically homogeneous population of cells. Cells with specific phenotypic traits or markers can be selected by growing the cells in selection medium containing a drug or chemical that will allow only the clones exhibiting the trait to survive. Alternatively, subculturing at a relatively low density, which allows single cells to attach and form colonies that can be selected for their desired phenotypes, can clone cells.

5.1.4 Growth Conditions

To maximize viability and reproducibility, the culture medium and growth conditions must be optimized and standardized for any given cell system. Tissues and cell lines differ in their nutritional requirements and preference for growth in suspension, as a monolayer, or in more specialized three-dimensional matrices, such as collagen gel. The cell concentration in a culture can also be critical for adequate cell viability. Normal (untransformed) cells suspend division by culture by a mechanism known as contact inhibition once the cells begin to touch and the culture becomes confluent. Therefore, most cultures actively dividing must be regularly subcultured to reduce the cell density to an optimal level for growth and cell division. However, the process of subculture itself can adversely affect the viability of the cells because of mechanical or chemical (trypsin) damage to the cell membranes from the methods and agents required to detach cells growing in monolayer. In addition, cell densities that are too low can also adversely affect the viability of the cultures. Cells actively growing secrete essential growth factors and chemical messengers into the culture medium that can be taken up by adjacent cells by a mechanism known as cross feeding. If adequate cell densities cannot be maintained to facilitate cross feeding, it is necessary to supplement the culture medium with the necessary growth factors. Cell density can also affect the uptake of compound: As cell density increases with time in culture, tendency toward decreased uptake exists. Decreased uptake may be a function of reduced growth rate and a decreased cell surface area exposure as the result of increased cell-to-cell contacts [77].

In vitro systems, by their nature, experience nonphysiologic conditions and periods of hypoxia, regardless of how well the system is defined and controlled. Because *in vitro* systems do not have inherent methods of clearance, they are dependent on regular changes of growth medium, which results in dramatic fluctuations in levels of nutrients and metabolites over the culture period. Buffering the culture medium and providing CO_2 is necessary to minimize extreme fluctuations in pH that can occur in metabolically active cultures. A humidified atmosphere in the incubator is also necessary to minimize evaporation of culture medium that can result in hypertonicity.

5.1.5 Metabolic Activating Systems

Several chemicals are metabolically activated *in vivo* to a more toxic metabolite. Primary cultures of hepatocytes and several hepatocellular tumor cell lines are metabolically competent. However, most cell cultures contain little, if any, cytochrome P-450 mixed function oxidase (MFO) metabolic capability. Cocultivation of these cells with primary hepatocytes is one means of providing P-450 activity. The addition of exogenous metabolic activation systems to the culture medium, similar to those routinely used in established *in vitro* tests for genotoxicity, can also be used to provide some metabolic functions in *in vitro* toxicity assays. These systems can be prepared with either an S9 (9,000-g postmitochondrial supernatant) fraction of liver homogenate, which provides both microsomal and cytosolic enzymes, or with a more purified microsomal fraction, prepared by first centrifuging at 15,000 g and subsequently centrifuging the supernatant at 105,000 g. Either fraction must be mixed with a nicotine adenine dinucleotide phosphate (NADPH)-generating cofactor system to be metabolically active. Approximately 5 days before collecting hepatocytes, the animals can be pretreated with chemicals, such as Aroclor-1254 (a mixture of polychlorinated biphenyls), phenobarbitol, 3-methylcholanthrene, or ß-naphthoflavone, which will induce the synthesis of various P-450 isozymes, thus increasing the metabolic activity of the liver fraction.

The use of induced S9 systems has been shown to increase the cytotoxicity of various chemicals, such as cyclophosphamide, an antineoplastic agent metabolically activated to a cytotoxic and mutagenic metabolite [45]. Both S9 fractions and purified microsomal fractions are cytotoxic in themselves. However, the S9 fraction is much more cytotoxic than is the microsomal fraction [4], which limits its usage to no more than 2–4 hours. In tests with cyclophosphamide, the microsomal fraction was shown to be approximately twice as active as the S9 fraction in producing the cytotoxic metabolite, with 50% inhibitory concentration (IC_{50}) values of 35 µg/ml and 70 µg/ml, respectively [4].

Although these systems have improved the overall correlations with *in vivo* toxicity, they are highly artificial and do not always reflect the level or type of reactive metabolites produced *in vivo*. Because phase II conjugation and detoxification enzymes, such as glutathione S-transferase, sulphotransferases, and glucuronosyltransferases, are not available at sufficient levels in S9 or microsomal fractions, these exogenous activation systems may overestimate toxicity with some chemicals. An apparent decrease in cytotoxicity with S9 is not necessarily indicative of metabolism to a less toxic species, but it may be a consequence of nonspecific binding to S9 proteins. Toxicity may be either underestimated or overestimated with some chemicals, depending on species differences in the types and levels of isozymes and differences between induction systems. Discordance with *in vivo* results may still occur even if the appropriate metabolite is generated *in vitro*. For example, paraoxon, the toxic metabolite of parathion *in vivo*, is produced *in vitro* by exogenous activation systems. However, in contrast to what is observed *in vivo*, paraoxon exerts its toxicity by inhibiting cholinesterase in the nervous system, a mechanism irrelevant *in vitro*.

5.2 ENDPOINTS

The goal of most endpoints selected for evaluation is to objectively measure either a cytotoxic or a cytostatic effect. Cytotoxic effects ultimately reduce cell viability, whereas cytostatic effects inhibit normal cell division without necessarily affecting the viability of the affected cell. However, either effect can result in toxicity or lethality in an organism if the function of the affected cell is critical and if a significant number of cells is affected.

Cytotoxicity, as perceived *in vitro*, is dependent on the endpoints and methods used to define it in a given test system. The endpoint measured must be both specific and well defined. Qualitative and quantitative endpoints have been used to assess cytotoxicity. Although quantitative endpoints are advantageous in that they can be objectively measured, qualitative data can also provide a wealth of information if the investigator strictly adheres to defined evaluation criteria. Although no *in vitro* assay can totally replace *in vivo* testing, a battery of well-designed complementary *in vitro* assays can augment *in vivo* results and help prioritize compounds for *in vivo* testing and further development [32].

The most-valuable endpoints are indicative of a wide variety of chemical damage [70]. Cell death is an endpoint that is easy to define and may result from a diverse array of cellular insults. Xenobiotics can produce cellular lethality by either direct damage to the structural components of the cell or indirect interference with the normal physiology and metabolism of the cell through impaired protein synthesis, respiration, ion exchange, and DNA synthesis capabilities. Thus, cell death, as an endpoint, is nonspecific and all-or-nothing, providing no opportunity to establish mechanisms or the reversibility of the damage. Various other endpoints have been used to measure cellular toxicity *in vitro*, including those characterized by altered cell morphology, abnormal cell behavior, and reductions in growth. More specific endpoints, such as those affecting specific metabolic or enzymatic systems may detect compounds that are toxic without necessarily resulting in acute lethality. However, these systems may be overly sensitive at detecting endpoints of little *in vivo* consequence. Thus, the ideal endpoint is one that is sensitive only to serious insults and relevant by any mode of insult [77].

5.2.1 Basal Cytotoxicity

The term *basal cytotoxicity* has been used to describe the toxic effects that a chemical may have on the basic cellular functions and structures common and critical to all eukaryotic cells [21]. Target organ toxicity has been correlated with the chemical distribution and basal cytotoxicity to that organ *in vivo* [20,21]. *In vitro* models may be developed in conjunction with *in vivo* pharmacokinetic data to simulate the distribution and plasma levels of a xenobiotic that may be achieved *in vivo*. In theory, the lowest concentration of a toxic chemical to the basal function of all cells *in vitro* can be compared with the concentrations toxic to various target tissues to determine a critical concentration for producing toxicity to a particular target organ *in vivo*. Once plasma concentrations have reached this level in a vital target organ *in vivo*, ensuing toxicity would be expected. Thus, the relevant endpoints to basal cytotoxicity may be assessed *in vitro* as a useful method for studying and predicting the toxic effects responsible for animal lethality.

5.3 VIABILITY

Cell viability has been evaluated using a variety of techniques. Reduced cell numbers can be quantified microscopically using a hemocytometer or by using electronic cell counters. In either case, only a reduction in the number of intact cells produced by cell degeneration or lysis can be accurately obtained. Intact dead and damaged cells are difficult to differentiate from living cells on the basis of cell counts alone. Other methods for assessing the cell viability are based on changes in membrane permeability assessed as dye exclusion of viable cells and membrane leakage of dead or damaged cells.

5.3.1 Dye Exclusion/Uptake

Dyes, such a trypan blue and nigrosin, can be used as vital stains to detect a large portion of dean- and membrane-damaged cells that have lost their ability for dye exclusion. Cells with damaged membranes allow the stain to pass into the cytoplasm, whereas undamaged cells are capable of dye exclusion. Conversely, supravital stains, such as neutral red, can diffuse through the plasma membrane and concentrate in the lysosomes of living cells. Neutral red uptake, measured by extraction and spectrophotometric absorption, has been used as a reliable, reproducible, and inexpensive *in vitro* assay for viability [8,9]. Damage to the cell surface or lysosomal membranes leads to lysosomal fragility and, ultimately, decreased uptake and binding of neutral red. A dual florescence technique combining fluorescein diacetate with diethidium bromide can be used to simultaneously stain living and dead cells using a fluorescence microscope with an epifluorescence condenser [50,79]. With this procedure, fluorescein diacetate is converted to fluorescein by cellular esterases in living cells, resulting in green fluorescence, whereas ethidium bromide, excluded by living cells, can penetrate damaged cell membranes and stain the nuclear component of dead and damaged cells red. Each of these stains can be used in conjunction with manual microscopic counts with a hemacytometer or electronic counts with more sophisticated methods involving flow cytometry or colorimetric analysis using spectrophotometers. With flow cytometry, several thousands of cells per second pass through a laminar flow system, then one by one through a laser beam scattered by the cells, and the red incident light is absorbed by the trypan blue stain.

5.3.2 Membrane Leakage

Leakage of soluble cellular cytosolic enzymes, such as lactate dehydrogenase (LDH), into the cytoplasm has also been used to quantitate lethality. The advantage of this endpoint is that it quantitates enzyme leakage from cells that have been lysed, in addition to those that are dean and damaged with leaky membranes. The amount of enzymes that the dead and damaged cells release into the culture medium can be assayed using sodium pyruvate as substrate and nicotine adenine dinucleotide (NADH) as a cofactor. In the presence of pyruvate, LDH is assayed by conversion of NADH to NAD^+. The rate of change in NADH absorbance at 340 nm can be measured with a spectrophotometer or by using a microcentrifugal analyzer. Lysing the remaining cells and measuring the levels of enzymes subsequently released can also assess the LDH content of the surviving cells. Using this technique, enzyme leakage can be presented as the percentage released relative to the total amount of enzyme in the culture. Alternatively, if the number of cells in each culture is variable, enzyme leakage can be presented on the basis of the amount of enzyme per million cells by determining the cell number, the total amount of DNA, or the protein content of each culture.

Total cellular protein can be assayed after treatment using the classical methodology of Lowry [55] or by more recent methodology. With the Lowry procedure, cells are incubated with a solution containing NaOH to lyse the membranes and an alkaline copper sulfate and potassium tartrate solution to produce a colorimetric reaction. Phenol is then added 30 minutes before assaying absorbance at 660 nm with a spectrophotometer. An alternative procedure involves first lysing the cells at 37°C for 60 minutes with NaOH and then mixing the suspension with Coomassie brilliant blue G-250 dye [53]. Absorbance with procedure is measured at 630 nm with a spectrophotometer or microplate reader. Protein content can also be determined by binding of Kenacid blue dye, as described by Knox et al. [52]. This method was shown to be faster and the results less variable than they were with the Lowry method.

Similar methods have been based on the release of a radiolabeled compound, proteins, or DNA into the medium to provide a sensitive and objective measure at relatively low levels of cell damage. Typically, cells are cultured in the presence of the radioactive labeled marker compound for a period of time to allow uptake of the label. The medium containing the label is then removed, and the cultures are rinsed to remove the surface label. Cells are then treated with the test article, and the supernatant medium is then harvested and counted in a gamma counter or liquid scintillation counter. If desired, the cells can then be lysed and counted to estimate the counts present in the intact cells. Chromium-51 (^{51}Cr) is a common radiolabel used for the study of cell damage because it can be used to assess cell damage before lysis, whereas the [^3H]thymidine- or [^{125}I]deoxyuridine-labeled DNA are only released after nuclear and cellular lysis. Radiochromium-labeled sodium chromate does not covalently bind to cell proteins and other cell constituents as does ^3H- or ^{14}C-labeled thymidine, which was readily incorporated during the synthesis of DNA, RNA, and protein. As a consequence, dead cells subsequently release at least 70% of the radioactivity taken up as ^{51}Cr. Hexavalent sodium chromate is reduced to the trivalent form once incorporated into cells and attaches to proteins and other cellular components. Once the chromate is reduced to the trivalent form, it cannot be reused by other cells [11,44].

A disadvantage of several of these techniques is that they will only detect the most severely and irreversibly damaged cells, not cells otherwise functionally impaired or unable to divide. For example, line 1 carcinoma cells treated with *Vibrio cholerae* neuraminidase showed no evidence of cytotoxicity using dye-exclusion (eosin or trypan value) techniques, but they showed a 73–84% reduction in viability when assessed for cell growth by colony formation or ^{3}H-thymidine incorporation [87].

5.4 CELL GROWTH

5.4.1 Cloning Efficiency

Cell growth and reproduction are widely used endpoints for assessing the viability of cells in culture. Cloning efficiency can be determined for cells growing in monolayers by dispersing a dilute suspension of a known number of cells (100–200) into culture plates. After incubating the cells undisturbed for several days in culture, reproductively competent cells form clones that can be visually counted. Changes in cell number can also assess cell-growth kinetics in culture. Cells cultured at low densities in a buffered, pH-controlled medium will divide logarithmically as long as sufficient nutrients are available and room for growth exists. By counting the cells at various intervals in culture, one can assess reductions in growth relative to untreated control cultures through differences in cell number. A reduction in cloning efficiency can result from cell death or the impaired ability to reproduce.

5.4.2 DNA Synthesis

Reductions in DNA synthesis can be assessed directly by measuring the uptake of tritium-labeled thymidine (^{3}HTdR). Cells actively involved in DNA synthesis incorporate thymidine as a normal constituent of DNA. To assess effects on DNA synthesis, cells treated with the test article are cultured in a medium containing ^{3}HTdR. Cells having retained the ability to synthesize DNA can incorporate the radiolabeled thymidine in place of thymidine. Because the cells can reuse released thymidine, it is necessary to add an excess of cold thymidine before harvest to block the reuse of the labeled thymidine. The DNA is then extracted from the cells, and the extracts are counted in a liquid scintillation spectrometer.

An indirect method [6] for measuring the DNA content of the cells can be performed using DNA-specific fluorescence staining procedures (feulgen hydrolysis followed by Schiff-type acriflavine staining). Photometric readings are then obtained for each cell using a microscope photometer. DNA synthesis curves can then be derived for each culture using the distribution of DNA fluorescence readings, as described by Walker [81].

5.4.3 Mitogenicity

The mitotic index is another useful endpoint for assessing the reproductive competence of cells. Not only must the cells be capable of DNA synthesis, but they must also be able to progress through G2 and the chromatin must be able to condense to form chromosomes. Various methods exist for assessing the mitotic index. Cells can be grown in suspension, as monolayers in culture dishes, or directly attached to coverslips. Cells are treated with the test article 6–48 hours after treatment, and colchicine (or the synthetic, colcemid) is added during the final hours of treatment to arrest the cells in metaphase. Cells grown directly on coverslips can be fixed and stained with Giemsa. Alternatively, cells grown attached to plates are removed by trypsinization, and the cell suspensions are fixed and stained. Before fixation, cells can be treated briefly with a hypotonic solution of KCl or sodium citrate if optimal chromosome morphology is also desired. At least 1,000 cells are then scored for the proportion in mitosis, and the mitotic index is calculated as the number of metaphases per 100 cells.

5.4.4 Cell-Cycle Kinetics

Many cells capable of mitosis may experience treatment-related delays in cell-cycle kinetics at one or more stages of the cell cycle. Various methods are available to assess cell-cycle kinetics, a few of which are based on incorporation of 5-bromo, 2-deoxyuridine (BUdR) during DNA synthesis. BUdR is an analogue of thymidine that can be taken up by the cell and incorporated into DNA in place of thymidine. Cells are treated and grown in the presence of BUdR for approximately two cell cycles. Through semiconservative replication, one strand of DNA will incorporate BUdR with each cell cycle. Thus, after one division, both chromatids of each chromosome will contain one strand that has incorporated BUdR. After a second round of replication in the presence of BUdR, one chromatid will contain one strand of DNA with BUdR and the other chromatid will have BUdR in both strands, resulting in differential staining between the chromatids on staining with Hoechst fluorescent stain. Cells arrested in metaphase are then scored on the basis of their differential staining patterns for the relative frequency of cells in the first (M_1), second (M_2), or third (M_3) division after treatment. The replicative index (RI) of each culture is first calculated as follows: $RI = 1(M_1) + 2(M_2) + 3(M_3)$. The average generation time (AGT) can then be calculated by dividing the total culture time in BUdR by the RI for each culture [46].

More recently, with the advent of monoclonal antibodies, an antibody to

BUdR (anti-BUdR) has become available commercially [39]. This antibody can be used as a primary antibody for the sensitive detection of BUdR incorporation. A secondary antibody selective for the primary antibody and conjugated with a fluorescein isothiocyanate (FITC) stain is then used to attach a fluorescent label to the anti-BUdR/BUdR complex. The amount of fluorescence can then be quantitated using a photometric plate reader or by flow cytofluorometry. With flow cytometry, it is possible to use an improved double-staining method combining anti-BUdR staining with conventional DNA stains, such as propidium iodine or 6-di-amidino-2-phenylindol dihydrochloride (DAPI). This method permits quantitation of the proportions of cells in G1, S, and G2 + M and establishes the distribution of cells in each phase of the cell cycle [51].

Another method useful for assessing cell-cycle kinetics involves treatment of the cells with cytochalasin B. Cytochalasin B interferes with the polymerization of actin, which blocks the cytokinesis of the cell without affecting DNA synthesis or mitosis. Cells appearing mononucleate have not yet undergone replication, whereas binucleate and polynucleate cells have undergone one or more replications, respectively. By assessing the proportions of mononucleated, binucleated, and polynucleated cells relative to a normal, untreated control population of cells, one can estimate the amount of treatment-induced cell cycle and delay.

5.5 CELL AND CULTURE MORPHOLOGY

5.5.1 Cell Morphology

Morphologic and cytologic evaluation can be performed on cells treated in culture to evaluate cellular toxicity using light and electron microscopy. Techniques for the study of whole-cell preparations *in situ* have steadily evolved and have been refined in recent years to include several improvements for fixation, preparation, and viewing using high-voltage electron microscopes. The use of glutaraldehyde as a fixative by Buckley and Porter in 1975 was a major improvement followed by the combination of rapid freezing and freeze-substitution technology more recently [59]. To minimize artifacts during dehydration and drying, aldehyde-fixed cells can be adhered to plastic substrates or electron-transparent melamine foil support mediums. These techniques have been shown useful for maintaining cellular topography and improving the study of fine cellular details [83].

5.5.2 Morphology Indicators of Cytotoxicity

Morphologic changes in the cell membrane associated with toxicity include changes in size (pycnotic or gigantism), shape (blebbing), and integrity of the cell membrane. Morphologic changes in the attachment, spreading, and growth patterns of cells growing in monolayers display a regular polar orientation that may be disrupted under toxic conditions. With fluorescence staining, the nucleolar borders of normal cells fluoresce brightly, and nuclei rich in uncondensed euchromatin fluoresce with a ground-glass–like appearance. Conversely, damaged cells lack nucleolar fluorescence and contain more heterochromatic chromatin in the nuclei progressively becoming more heteropycnotic (condensed) as the cell dies. Toxic effects on the nucleus of the cell and its membrane may appear as blebbing, distribution of the nuclear membrane, reductions in the number of mitotic figures, chromosome damage or stickiness, and multinuclearity. Cytoplasmic changes may also be observed, such as vacuolization, condensation or swelling of mitochondria, precipitation or changes in the distribution of ribosomes, and blebbing of the cytoplasmic membrane. Blebs appear as protrusions of the membrane containing cytosol and are thought to be a symptom of membrane damage [21]. Ultimately, dead cells lyse on release of lysosomes or degrade by necrosis. Necrosis is pathologically characterized *in vivo* by cell swelling, membrane rupture, and disorganized disruption of chromatin. Necrosis can also be observed *in vitro* as an increase in cell detachment, nuclear debris, and chromatin clumping. Apoptosis, programmed cell death, can also be induced by toxins and appears as reduced cell volume, dilation of the endoplasmic reticulum, compaction of organelles, loss of junctional complexes, membrane blebbing, and condensation and margination of chromatin.

5.5.3 Quantification of Morphologic Changes

Although morphologic observations are mostly subjective and descriptive, various methods have been employed to make these data more objective and quantifiable. One of the first attempts to categorize and rank cytotoxicity by degree was proposed by Toplin [80] as a method for standardizing the evaluation of cytotoxicity of various chemotherapeutic drugs. The Toplin scale ranges from 0 to 4, with a score of 0 indicating no cellular damage and 4 indicating complete cell degeneration or cytolysis. Although semiquantitative, this method is based on subjective, qualitative endpoints.

Morphometric methods employing image analysis coupled with computer digitalization have allowed some morphologic endpoints to become quantifiable. One such procedure considers the volume densities of cells relative to the total number of cells as opposed to the area of the section being evaluated to convert two-dimensional morphologic information into three-dimensional quantitative data that can be statistically analyzed [76]. This procedure was validated using primary hepatocyte cultures treated with cadmium, erythromycin, benoxaprofen, and indomethacin and was demonstrated to be a sensitive index of cell injury, even at levels causing detachment of a significant percentage of the cells.

Planimetric methods have also been employed as simple and rapid methods for quantifying gross cytotoxicity by measuring the area occupied by cells on a substratum. To assess viability, cells are cultured as monolayers and evaluated for plaque formation using a double-agar overlay procedure, in which the second agar overlay medium contains neutral red. Living cells take up neutral red, whereas dead cells remain unstained. Planigraphs are then prepared by tracing the areas of cytotoxicity (plaques) onto paper, and cytotoxicity is quantitated by area measurements that can be automated by using a compensating polar planimeter [71]. Computer-assisted planimetry can be used to further assess changes in shape, orientation, or polarization of the cells.

Cell spreading has been shown to be necessary for the survival, growth, and movement of many cells typically growing as monolayers on a substratum [78]. Cell attachment, flattening, and a subsequent increase in surface area characterize cell spreading. Using image analysis, the degree of cell spreading on a substratum can be quantified by measuring the cell perimeter as a function of time [10]. Because cell margins are difficult to contrast, the cells are stained with acridine orange (2,8 bis-dimethylaminoacridine), a fluorescent stain, to produce a bright fluorescence appearing dark on a photonegative. The basic methodology uses a drum scanner as an input device to scan the photonegatives to construct a computer image defined by gray levels.

A sophisticated automated system for assessing more complex cytotoxic endpoints has been developed using Allen video-enhanced contrast-differential interference contrast (AVEC-DIC) microscopy. DIC microscopy was developed to study phase objects with a bright field by providing shadowcast details at interfaces between membrane and organelles, resulting in gradients of optical paths based on differences in the refractive indices. The AVEC method incorporates a video camera especially designed to reject stray light, which can limit contrast and resolution [1]. This system has been used to measure microtubule-related motility, changes in the movement of cellular organelles, and the fine structure of organelles [56]. AVEC-DIC methodology is particularly well suited for assessing sublethal effects in living cells by providing a quantitative analysis of organelle dynamics at earlier time points and at lower concentrations than those required to produce gross effects. This system has been used to evaluate cytoplasm consistency, the appearance of vacuoles, spikes, blebs, and changes in the number, length, and shape of mitochondria vital-stained with a florescent rhodamine-123 dye. Several effects, including cell retraction (which precedes detachment), the appearance of blebs, and changes in the number and fine morphology of mitochondria, were detected with this system at concentrations of a toxin that did not increase LDH release or produce cytotoxic changes using conventional morphometric analysis or viability assessments.

5.6 CELL FUNCTIONS

Although indicative of lethality, morphologic changes are often subjective and time consuming to quantify. Biochemical assays based on vital cell functions are usually easier to quantify, more objective, and readily automated. In addition, these assays can often detect subtle impairments in functions that may occur long before cell death.

5.6.1 Thermodynamic and Metabolic Function

Adenosine triphosphate (ATP) can be used as an indicator of cytotoxicity because it is the primary energy source at the cellular level. In order for the cell to function optimally, it must maintain an intricate balance between energy production and consumption. Several assays for ATP have been developed. A specific and highly sensitive assay for ATP has been developed on the basis of the firefly luciferase bi-oluminiscent reaction, which requires ATP as a source of energy to drive the reaction [15]. This assay is technically complex, requiring an acid extraction of ATP with perchloric acid, followed by neutralization with potassium hydroxide (KOH) and dilution at various concentrations using a luciferin–luciferase test reagent. The intensity of luminescence, which correlates with ATP content, can be measured with a luminometer. To quantitative levels of ATP, a standard curve with known concentrations of ATP must be constructed.

A simpler technique using ^{14}C-labeled ATP has been shown to correlate well with ATP levels as measured by the luciferin–luciferase method [73]. A few hours before treatment, ATP pools are labeled by adding [^{14}C] adenine to the culture medium. Lysing a sample of cells with Triton X-100 and determining the amount of radioactivity in the lysate can verify the amount of uptake. Immediately before treatment, the labeled medium is removed. Cells are then washed and treated with the test article in unlabeled medium. At various time intervals after treatment, an aliquot of medium can be removed and counted for radioactivity using a liquid scintillation counter. With this method, Shirhatti and Krishna demonstrated that approximately 65–70% of the incorporated ^{14}C label was in the ATP pool; the remaining label was incorporated into the adenosine dinucleotide phosphate (ADP) and 5'-adenosine monophosphate (AMP) pools [73]. In the presence of toxic drugs, a marked decrease in cellular ATP levels with a concomitant increase in ^{14}C leakage into the medium was observed to correlate well with LDH leakage. Because the total [^{14}C]adenine uptake can be estimated from the disappearance of [^{14}C]adenine from the medium without the need to lyse the cells, this method is noninvasive. Thus, the advantage is that multiple samples of a medium can be taken from the same cultures at various time points after treatment.

5.6.2 MTT Assay

The primary function of mitochondria is to produce and maintain sufficient levels of ATP for the cell to carry out its energy-requiring activities. Therefore, the more active the cell, the more ATP is required, and subsequently, mitochondria are more active and abundant. To produce ATP, mitochondria actively metabolize pyruvate, a product of glycolysis, that is coupled with coenzyme A to produce acetyl CoA in the matrix of the mitochondria. Acetyl CoA is also provided as a substrate to the mitochondria through amino-acid metabolism and oxidation of fatty acids. Once in the mitochondria, regardless of its source, acetyl CoA is circulated through the tricarboxylic acid cycle (TCA cycle), in which it reacts with various enzymes to produce a chain of substrate and release of electrons, resulting in the production of ATP. One of these enzymes, succinate dehydrogenase, is responsible for converting succinate to fumerate, which provides a pair of electrons (from the hydrogen atoms of the substrate) to be used in ATP production. Tetrazolium (3-(4,5-dimethylthiazol-2-yl)-$_2$ diphenyl tetrazolium bromide) (MTT) is a pale yellow substrate that can also be cleaved by succinate dehydrogenase. Formosan, the product of this reaction, is a dark blue pigment that can be used as an indicator of mitochondrial function. Because the conversion takes place only in living cells, the amount of formazan produced correlates with the number of viable cells present. Mosmann used this reaction as a basis for the development of a rapid colorimetric assay for cell viability [58]. Absorbance at 540 nm can be measured using a spectrophotometer.

5.6.3   Calcium Ions

Changes in homeostasis of Ca^{2+} can also occur as a result of cellular injury. Ca^{2+} ions are essential second messengers and regulators of critical enzyme functions (i.e., DNA endonucleases) and cell division. Calcium ions normally entering cell down an electrochemical gradient through voltage-dependent Ca^{2+} channels are taken up by the endoplasmic reticulum and mitochondria, and packaged. Ca^{2+} stores are then released from the cell by plasma membrane Ca^{2+} ATPase after mediation by hormones, growth factors, and neurotransmitters. Although a sustained increase in calcium influx occurs, the efficiency and sensitivity of the calcium ion pump are enhanced, enabling the calcium efflux to compensate for the increased influx [69]. Thus, a rapid but transient increase in intracellular Ca^{2+} may be caused by receptor-mediated physiologic agonists, such as bradykinin, acting on the cell surfaces [47,60]. In contrast, slow, sustained increases are usually associated with irreversible cell damage [68]. Cell damage results in an increase in intracellular-free Ca^{2+} concentration from normal levels of approximately 10^{-7} M to micromolar levels through various mechanisms, including increased permeability of the cell plasma membrane to external Ca^{2+}, the release of internal stores of Ca^{2+}, damage to the calcium pump, or effects on critical cellular transport proteins.

Intracellular Ca^{2+} concentrations can be measured using aequorin, a Ca^{2+}-sensitive photoprotein. Cells are loaded with aequorin using a low Ca^{2+} centrifugation technique [60], and then cultured in dishes placed over a sensitive photomultiplier tube. At various time points after treatment, the calcium ion concentration is measured by quantitating the light emitted from the aequorin-loaded cells. Alternative methodology involves the use of fluorescent imaging with fluorescent calcium chelators, such as Quin-2, Fura-2, and Indo-1, which can localize and quantify intracellular calcium reserves. This methodology was used to associate a sustained increase of calcium ions with subsequent cell blebbing and loss of membrane permeability [18].

5.7 ASSAY VALIDATION

Validation of assays is necessary to demonstrate the relevance, reliability, and predictability of new methodology, before gaining acceptance and usage as replacements for traditional *in vivo* methods. In order to accurately assess the specificity and sensitivity of an assay, a wide variety of compounds with different mechanisms of action must be tested. Because lethality may be the end result of a variety of toxic mechanisms, a battery of *in vitro* tests with different endpoints should be more predictive than is a single assay. Several factors should be considered when designing a validation study. First, the goals of the validation must be defined as a basis for the selection of appropriate *in vitro* and *in vivo* endpoints, treatment protocols, and test compounds. For example, if the goal is to validate the predictivity of an *in vitro* assay for determining acute *in vivo* lethality, it would be appropriate to select *in vitro* endpoints measuring basal cell cytotoxicity and reflective of the *in vivo* mechanisms responsible for lethality (i.e., impaired basal cell functions or structures). Before defining the treatment protocol, it is advantageous to know the *in vivo* pharmacokinetics and metabolism of each test compound selected for validation. A continuous *in vitro* exposure has been shown to more closely approximate *in vivo* conditions when a slow rate of elimination occurs [30]. If the parent compound is metabolized to a more or less toxic species, an exogenous metabolic activation system should be provided for cell systems other than hepatocytes.

The results of the *in vitro* tests should be compared with an appropriate parameter of *in vivo* lethality, such as LD_{50} values from an acute *in vivo* toxicity test. If an appropriate animal model is used that is not a good predictor of human toxicity, the *in vitro* method developed against that model may be of limited value [12]. When available, human data should be used from an appropriate database. Correlation of *in vivo* LD_{50} does-response data to an *in vitro* IC_{50} response is generally done by linear regression analysis using the actual concentrations of the test agents or the rank order of their toxicity in each test system. The rank order of toxicity is preferable if the absolute values of the *in vitro* and *in vivo* endpoints differ by more than a factor of 1,000 [77]. The correlation coefficient (r) of the line provides a measure of the strength of the relationship of the two test systems, in which an r of 1 is indicative of a perfect positive correlation.

Several factors should be considered when evaluating the predictivity of the assay. Because the compound comes into direct contact with target cells in an *in vitro* assay, intravenous (IV) or intraperitoneal (IP) LD_{50} data, when available, should be more comparable than are data from oral studies, in which the compound may have limited absorption and systemic bioavailability [31,64]. Total cellular protein ID_{50} data from tests with 27 compounds believed to interfere with critical basal cell functions were compared with LD_{50} values [31]. For 21 compounds having both rat oral and mouse IP data available, a weak but significant correlation ($r = 0.49$) was obtained using log oral LD_{50} values. The *in vivo/in vitro* correlation for these same compounds was improved ($r = 0.68$) by using mouse IP data. In addition, by removing three compounds metabolized to more or less toxic metabolites, the correlation was further improved to 0.82. The best correlations ($r = 0.94$ and 0.95) were obtained when compounds with similar mechanisms of action were compared using *in vivo* data from the more sensitive species. Thus, when a human database is not available, a better correlation may be obtained when LD_{50} data from the more sensitive species and route are used, and when compounds are grouped on the basis of their mechanism of toxicity.

5.7.1 Early Validation Studies

As early as 1954, Pomerat and Leake published a cytotoxicity study with 110 compounds tested on primary explants of fetal chicken cells [66]. However, few studies before 1976 compared cytotoxicity *in vitro* with systemic *in vivo* toxicity using standardized procedures. For the most part, good correlation was found with small groups of related compounds [38].

In 1981, Ekwall presented a large-scale study with 205 drugs of various classes tested in HeLa cells using standardized procedures [20]. Cytotoxicity was first assessed microscopically on the basis of the absence or scarcity of spindle-shaped cells at 24 hours, and metabolic inhibition was assessed after 7 days by evaluating pH changes in the growth medium apparent as color change of phenol red. Preliminary tests by the same author with a subsample of the 205 compounds were first performed to (1) compare the toxicity of a sample of 25 drugs in HeLa cells versus other cell systems [26] and (2) compare the toxicity of a random sample of 52 drugs with systemic LD_{50} toxicity values of mice and humans [20]. The first study indicated that the relative levels of inhibitory toxicity showed similar differential sensitivities regardless of the cell system, indicating a qualitatively similar mechanism of action characteristic of vasal cytotoxicity. In the second study, seven compounds were more toxic in humans as a result of target organ toxicity to specialized neuroreceptors not found *in vitro*, whereas the remaining compounds expressed similar *in vitro* and *in vivo* toxicities indicative of basal cytotoxicity. These studies showed that a tiered approach to assessing two endpoints improves the likelihood of detecting toxicity with at least one of the endpoints and provides useful mechanistic information for some compounds.

This same standardized test system was used to test a group of 29 pasticizers *in vitro* in HeLa cells [27]. Seven of the compounds were also tested in other test systems with other cell types, including chick embryo cells, mouse L-cells, human diploid WI-38 cells, and mouse cerebellar explants. All tests had a similar rank order of toxicity, in which cytotoxicity was correlated with increasing chain length of the alcohol groups until the point where solubility was inhibited. Data for 20 of the compounds were also shown to correlate well with rodent IP LD_{50} values, suggesting similar mechanisms of lethality caused by basal cytotoxicity. Data from this test system and another cell line were also compared with cytotoxicity (LDH release) data obtained with primary hepatocyte cultures to determine the value of using metabolically competent hepatocytes in general cytotoxicity screening [22]. Of the 14 hepatotoxic compounds, most expressed similar toxicity with cell lines and liver cells, indicative of acute cytotoxicity not mediated by reactive metabolites. However, four of the hepatoxic compounds, known to cause more specific metabolism-mediated hepatocellular damage, were shown to be more selectively cytotoxic to the hepatocytes.

When possible, a large number of compounds of many different chemical classes should be used to validate a test system. A cytotoxicity study of 114 compounds was conducted [16] by measuring the protein content in inorganic acids, and salts, hepatoma (HEP) G2 cells, a human-hepatoma–derived established cell line. A diverse array of compounds of various chemical classes were tested, including surfactants, alcohols, organic acids, and some of their salts, and heavy metal compounds; and miscellaneous compounds and solvents. Twenty-four hours after treatment, the cell membranes were lysed, and the amount of protein was assayed by the Lowry method. The relative toxicity of each compound was determined as the concentration required to produce a 50% reduction in cell protein content (PI_{50}). Consistently low values were observed with toxic compounds, such as the heavy metals and organic amines. Subsets of the data showed good correlation when compared with data from other *in vitro* assays; however, the lack of adequate *in vivo* data precluded meaningful comparisons.

5.7.2 Multicenter Studies

Over the last decade, with the support groups like Fund for the Replacement of Animals in Medical Experiments (FRAME), more thorough validations have been performed as multicenter programs involving several laboratories testing the same set of compounds, usually in a blind fashion. Multicenter testing not only facilitates the testing of a larger number of compounds in a variety of assays, but it also enhances the credibility of the results. A multicenter approach allows for the evaluation of interlaboratory variability and circumvents the inherent bias of validations performed by the laboratory that first developed the technique. In addition, with more laboratories involved in selecting the compounds, a more diverse sample of compounds is generally tested.

The Fund for the Replacement of Animals in Medical Experiments. In 1982, FRAME established a multicenter research project aimed at developing, standardizing, and validating a cell culture method for assessing cytotoxicity. A large set of compounds was selected to include chemicals of different mechanisms of activity and with different degrees of toxicity, stability, volatility, and solubility. In addition, some of the compounds were metabolically changed to more or less toxic metabolites [3]. The group limited their testing to human BCL-D1 fibroblast-like cells, because results obtained in general cytotoxicity tests have been shown not to depend on the choice of cell type [70]. After a 72-hour treatment period, total protein was assessed as the endpoint for measuring cytotoxicity using Kenacid blue staining [52]. This method, as previously described, is based on a direct relationship between protein content, cell number, and binding of the stain. Results from testing 50 chemicals, of diverse toxicities and mechanisms of action, were generally in close agreement among the four participating laboratories.

A subset of 30 of the FRAME compounds was also tested for cytotoxicity in a blind trial using the Kenacid blue protein assay, the neutral red uptake assay, and a morphometric assay aimed at determining the highest tolerated dose (HTD) based on minimal morphologic alterations [72]. 3T3-L1 cells, a continuous fibroblast cell line derived from mouse embryos, were treated for 24 hours, and then assessed for cytotoxicity by each of the three methods. When ranked in order of toxicity, a close correlation was observed between the relative cytotoxicities of chemicals tested by all three methods. Because an exogenous metabolic system was not provided, some of the chemicals requiring metabolic activation to a more toxic metabolite were less toxic *in vitro*.

In addition to developing and validating alternative *in vitro* test systems, FRAME manages an alternative test validation scheme, in which sets of compounds are coded and supplied to research groups to validate new test methods blindly at a number of laboratories throughout the world. The ATP assay, as previously described, was validated in a test using 20 of the FRAME compounds [49]. L929 mouse fibroblasts were treated for 4 hours, and then ATP levels were determined using the luciferin–luciferase bioluminescent assay. ATP_{50} values were calculated as the concentrations reducing cellular ATP by 50%. The ranking of the test compounds by ATP_{50} values was similar to that obtained by cell death (CD_{50}). However, animal lethality data were not available at that time for comparison.

In vivo data may be derived from a variety of sources using different species, stains, routes of administration, and treatment protocols. Thus, one of the biggest problems with the FRAME study, as well as other studies, was the unbiased selection and availability of *in vivo* toxicity data of sufficiently high quality and reproducibility. When possible, rat oral and mouse IP lethality data produced at one laboratory (Imperial Chemical Industries; ICI) were used to provide *in vivo* toxicity profiles for the FRAME validation compounds, 50 of which have been published [67]. These data, along with additional data provided from the Registry of Toxic Effects of Chemical Substances (RTECS, compiled by NIOSH), were used to compare rat oral and mouse IP LD_{50} values to *in vitro* ID_{50} data [13] for 59 of the 150 compounds tested in the FRAME *in vitro* screen. Using linear regression analysis to compare log-transformed *in vitro* and *in vivo* data, correlation coefficients of 0.76

and 0.80 were obtained with the rat oral and mouse IP data, respectively [14]. As was suggested by Fry et al. [30,31], the best correlations ($r = 0.81$) were obtained using LD_{50} data form the most sensitive species for each compound.

Multicenter Evaluation of *In Vitro* Cytotoxicity. A similar multicenter program to FRAME was initiated in 1983 by the Scandinavian Society of Cell Toxicology under the title Multicenter Evaluation of *In Vitro* Cytotoxicity (MEIC). Whereas the main emphasis of the FRAME study has been to test the interlaboratory variability of methods, the MEIC study has concentrated on the predicitivity of *in vitro* results compared with the *in vivo* response, and the relevance of the results to various types of human toxicity [7]. The use of a battery of *in vitro* test systems with different cell types and endpoints should provide a higher predictive value. Initially, 50 reference compounds were selected, without bias, on the basis of known human and rodent acute lethality dosages and toxicokinetics [5]. To date, more than 50 international laboratories have tested at least a portion of the compounds in over 50 *in vitro* test systems. The primary findings of a few of these studies will be reviewed.

As a preliminary validation study, Ekwall et al. [19,23,25] published the results of the first ten MEIC compounds tested in a battery of *in vitro* cytotoxicity assays, encompassing four different cell systems (primary rat hepatocytes treated 1 hour or 24 hours, and 3T3 and HepG2 fibroblasts treated 24 hours). Each cell system was evaluated for three different endpoints—intracellular LDH, total protein, and MTT. These results were combined with previously published data from the same compounds tested in the MIT24 assay with HeLa cells and used to derive a multivariate partial least-squares (PLS) model. As a means of predicting human lethality, the model was then compared with rodent LD_{50} values from the RTECS. In general, mouse LD_{50} values were more predictive of human lethality than were rat values. Although the sample of compounds in this preliminary study was small, the collective prediction using the PLS model was shown to be as predictive of human lethality as are the mouse LD_{50} values.

The first ten MEIC compounds were also tested in rat hepatocytes for 24 or 48 hours [61]. Measuring mitochondrial activity assessed cytotoxicity, and the MTT and Coomassie blue dye assays assessed cell number. Good correlations with oral rat LD_{50} data were obtained with both the MTT and Coomassie blue assays after a 24-hour treatment ($r^2 = 0.86$ and 0.83, respectively). Although the values obtained in this study were not tested for correlation with human data, the results from this study correlated better with rat LD_{50} data than did the results of the preliminary study described above. In a more recent study, MTT data with cultured human hepatocytes was compared with data from primary rat hepatocytes and the nonhepatic 3T3 murine line using the first ten MEIC compounds [48]. This study showed that acute toxicity in humans was most accurately predicted with cultured human hepatocytes than it was with either rat hepatocytes or mouse 3T3 cells, suggesting that, when available, human hepatocytes should be included in *in vitro* test batteries.

A few recent studies have been published on the results from testing the first 20 MEIC compounds. In an attempt to estimate the effects of chronic exposure, long-term cytotoxicity was investigated with human embryonic lung (MRC-5) cells [17], which can be maintained in culture for more than 6 weeks without requiring subculture. Cytotoxicity was quantified by assaying total protein. PI_{50} values obtained from cultures treated for 6 weeks were compared with those obtained with the human epithelial HepG2 cell line treated for 24 hours with the same compounds. Although the PI_{50} values at 6 weeks were substantially lower than were the 24-hour values, with the exception of digoxin, good correlation was observed ($r^2 = 0.94$). The first 20 compounds were also tested for total protein and LDH release after a 24-hour exposure to HEP-2 human epithelial cells [88]. LDH release was only slightly less sensitive than was the total protein for measuring cytotoxicity. In general, the data from this study were in good agreement with the other studies with these compounds.

Another study of the first 20 MEIC compounds evaluated metabolic and functional endpoints as well as viability. Primary, cultured rat skeletal muscle cells were treated for 1 hour or 24 hours with each reference compound [40]. Viability was assessed by changes in intracellular creatine kinase and total protein, and decreased energy metabolism was assayed as a reduction in glucose consumption. Effects on function were determined by assaying changes in spontaneous contractility, which reflects not only the structural integrity of the excitable cell membranes, but also the functional abilities of electromechanical coupling and contraction. Decreased contractility after exposure for 1 hour or 24 hours was the most sensitive measure of critical biologic activities or cytotoxicity in this study, and it was reflective of therapeutically intended or acute toxic effects observed *in vivo*.

Cell growth and morphology were assessed for 48 of the MEIC compounds in primary rat hepatocytes, Madin–Darby bovine kidney (MDBK) cells, and McCoy cells, a human epithelial line derived from synovial fluid [74]. Hepatocytes were observed for morphologic changes for 24 hours after a 4–6-hour treatment, whereas the cell lines were observed for growth and morphology at 72 hours posttreatment. Cell viability was determined by trypan blue exclusion and LDH release. For each compound tested, average values from all three parameters were used to determine minimum cytotoxic concentrations (CT_{50} and CT_{100}), defined as follows. CT_{50} was the concentration that induced morphometric changes in 50% of the cells or 50% cell death, or a 50–100% increase in hepatocyte LDH release. CT_{100} was the concentration that induced marked morphometric changes or >50% cell death along with >100% increases in hepatocyte LDH release. The log of these values was compared with oral log LD_{50} values for rats and mice using linear regression analysis. The correlations between LD_{50} values and CT_{50} values were $r = 0.77$, 0.80, and 0.83 for McCoy cells, hepatocytes, and MDBK cells, respectively. These results agreed with those of Ekwall and Johannson, who also demonstrated that cell type had little effect on the overall relative cytotoxicity values [26]. An accurate *in vivo* LD_{50} dose was predicted for at least 75% of the compounds studied. Using an empirical approach, Shrivastava et al. [75] also showed that CT_{50} and CT_{100} values could be used to predict an *in vivo* maximum tolerated dose (MTD). *In vitro* CT_{50} and CT_{100} values for 25 compounds in each of the three cell systems were shown to have a greater than 80% correlation with actual *in vivo* results from MTD studies with dogs and rats.

After completion of the initial test phase with numerous test systems, the MEIC group will evaluate the relevance and effectiveness of each assay for predicting human lethality using toxicokinetic models. Multivariate modeling will be used to select tests and batteries of tests that may be useful as supplements or alternatives to animal testing. In the final phase of validation, the tests best predicting human toxicity will be evaluated for reliability by contract laboratories using coded compounds. This approach should minimize the costs and time of validation by selecting test systems on the basis of relevance to be subsequently tested for reliability [24].

French multicenter study. Over the past few years, the French Ministry for Research and Higher Education has established a multicenter study with the goal of setting up a model of acute *in vivo* toxicity predictive of acute *in vivo* toxicity [28]. Four endpoints—LDH release, neutral red uptake, the MTT assay of mitochondria function, and total cellular protein content—are to be evaluated in primary cultures of rat hepatocytes treated with various compounds. Recently, the validity and predictability of this model were evaluated by comparing the IC_{50} cytotoxicity results for several compounds with IV LD_{50} values using multivariate analysis [64]. The 30 compounds tested were selected from those used in the FRAME project. IC_{50} values were closely correlated for the four endpoints ($r >$ 0.97). Values from the neutral red assay were the most sensitive indicators of *in vitro* cytotoxicity (lowest IC_{50} values), as were the IV or IP LD_{50} values *in vivo*. Linear regression between these two parameters for 25 of the compounds yielded a statistically significant ($p < 0.001$) correlation coefficient ($r = 0.877$). With the limited number of compounds tested in this study, the model had a predictability of 95% with a confidence interval of 75–100% at concentrations up to 1,500 g/ml. Additional compounds must still be tested to reduce the confidence interval to an acceptable limit before the use of this model can be considered.

5.8 THE ROLE OF PHARMACOKINETICS AND STRUCTURE-ACTIVITY RELATIONSHIPS IN MODELING

Preliminary results from numerous *in vitro* assays indicate that these tests are relevant for predicting the human lethality of most chemicals, particularly those that interfere with critical basal functions. As would be expected, compounds not accurately predicted by *in vitro* tests often have specific metabolic or toxicokinetic requirements resulting in variations in the time that the compound is maintained at concentrations high enough to produce a toxic response. Although metabolism can be easily achieved *in vitro* with cultured hepatocytes or an exogenous microsome system, the rate of biotransformation *in vivo* is affected by factors, such as species differences, sex differences, and the route of administration. These factors subsequently influence the degree and rate of uptake and distribution, which are difficult to fully simulate *in vitro*. Even *in vivo*, under controlled conditions using the same sex and species, as much as a 25-fold variation can exist in activity between individuals because of genetic variability, and within individuals over time because of age, diurnal variability, seasonal changes, illness, and nutritional status.

5.8.1 Pharmacokinetics

In classic pharmacokinetic testing, animals are administered the parent compound by the applicable route, and then concentrations of the parent compound and its metabolites are determined at various time points in the blood and excreta. Rate constants for absorption, distribution, and elimination can be calculated to describe the transfer of compound among various tissue compartments, which depends on the rate of blood flow to the tissue and the rate of diffusion across cell membranes. The volume of distribution (V_d) can also be calculated as a measure of the fluid volume in the body (plasma, extracellular fluid, and intracellular fluid) available to contain the entire compound at the same concentration as in the plasma.

If a compound is extensively bound to plasma proteins, the volume of distribution is equivalent to that of the plasma volume alone for acidic agents, but not lipophilic bases, such as propranolol. Many xenobiotics bind reversibly to plasma proteins (albumin, glycoproteins) at nonspecific sites that can become saturated. Bound compound is held to be essentially biologically inactive because it is not free to bind to active sites on tissue membranes or diffuse across membranes into cells. The volume of distribution can increase as the unbound fraction of the compound in the plasma increases by the relationship $V_d = V_p + VTf_p/f_T$, where V_p is the plasma volume, V_T is the volume of other body tissue water, and f_p and f_T are the fractions of unbound compound in the plasma and tissue, respectively [33]. When unbound plasma concentrations in plasma water and cell water concentrations of a nonionizable compound are plotted against time, the concentration of the drug in cell water is in equilibrium with that of the plasma water at the highest concentration of a drug [35]. Based on this relationship, the highest concentration of a nonionizable compound not actively transported across the cell membrane cannot be higher in the cell water than is the highest concentration in the plasma water. In addition, once steady state is achieved, the range of concentrations of the unbound xenobiotic in plasma water would reflect the maximum range of concentrations of the xenobiotic in the cells exposed to the xenobiotic through passive diffusion. Thus, the concentrations of unbound compound in the plasma water, and the clearance rates and distribution volumes measured by concentration in plasma water, are the appropriate parameters for comparisons with *in vitro* concentration-response curves [37,57].

The concentration of unbound (free) xenobiotic in the plasma water can be measured through ultrafiltrations. Ultrafiltration partitions free compound from relatively small plasma volumes by applying a pressure gradient on plasma, thus forcing a protein-free ultrafiltrate through a semipermeable membrane using centrifugal force. The concentration of free compound can then be determined directly from the ultrafiltrate. However, ultrafiltration may be unsuitable for many compounds because of nonspecific binding of the drugs to the devices. In these cases, equilibrium dialysis may be more appropriate.

5.8.2 Physiologic Pharmacokinetic Modeling

A physiologically based model has been described as an approach for comparing the time course of the concentration of a xenobiotic *in vivo* with its effective concentration *in vitro* [37]. These types of models attempt to integrate basic physiologic and biochemical information into a comprehensive model for predicting the distribution and disposition of a compound. Because the concentration of compound at the local site of a target tissue or organ may not necessarily reflect the concentration in the blood, an organ in its simplest form is assumed to consist of two pharmacokinetic compartments, the blood in the organ and the cells in the organ. By repeated dosing with an optimal interval between doses, the concentration of the free compound within the cells of each organ can be maintained in steady state with the concentration in the blood flowing through the capillaries of the organs, minimizing the range of concentrations of free compound in the cells. The total concentration of a xenobiotic in an organ (C_T) can then be used to relate the concentration of compound in the cells at any given time to the concentration in the capillaries at that time. C_T can be calculated using the formula $C_T = R_{conb}$ where the partition ratio (R) is estimated for the organ by measuring the compound and its metabolites in the plasma water obtained from the blood flowing through the organ at various times, and then calculating the areas under the curves (AUC) for the organ and blood.

If the unbound compound exists in only one form and is passed into and out of the cells solely by passive diffusion, and assuming that the compound or its metabolites reach an immediate steady state between the two compartments, the appropriate values can be substituted into rate formulas and physiologically based pharmacokinetic models, as described by Gillette [36,37]. However, most compounds are weak acids or bases passing through cell membranes in their nonionized forms. Thus, the total concentration of unbound compound would depend on the pK_a of the compound and the pH values of the blood and cells. In addition, if the compound is eliminated from the cells by metabolism or active excretion, the calculations must be further modified. These modifications require knowledge of the clearance values of the compound into and out of the cell, as well as measurement of enzyme activities and their effects on the intracellular concentration of the compound. The formulas are further complicated if the compound is actively transported into cells, then metabolized or cleared by passive diffusion. These parameters are best studied using *in vitro* pharmacokinetic experiments with purified enzyme [34,35].

A well-integrated approach for *in vitro* extrapolation should encompass data from a variety of pharmacokinetic and toxicokinetic assays. The validation of physiologically based models requires large amounts of *in vitro* as well as *in vivo* data. Although these models require complex sets of equations, they can often be simplified to include only the target organs or the organs in which the compound has been shown to accumulate [2]. Regardless of the complexity, the integration of *in vitro* data with pharmacokinetic models can greatly enhance the interpretation and the reliability of risk assessment using *in vitro* test systems. For example, combining cytokinetic data with *in vitro* lethality data improved the prediction of acute human toxicity in multivariate analysis with the first ten MEIC compounds [23].

5.8.3 Computer Structure-Activity Relationships Modeling

Over the last few decades, numerous attempts have been made to apply the concepts of quantitative structure-activity relationships (QSAR) to predicting *in vivo* mammalian toxicity on the basis of chemical structure. QSARs are statistical models that have been based on empiricism. These models assume that the total magnitude of a compound's interaction with a biologic receptor can be modeled as an additive combination of each function group's physicochemical interactions [65,41]. Numerous physicochemical parameters, such as molecular weight, quantum mechanical index, and reactivity constants, may be incorporated into the model. In addition, substituent interaction constants may be included, such as the Hammett electrostatic constant, the Taft E_s steric constant, the Hansch hydrophobic constant, and dispersive descriptors based on molar refractivity.

A comprehensive review of the methodology and use of QSAR is beyond the scope of this chapter. The interested reader should refer to Phillips et al. [65] for a thorough and critical review of the use of QSAR for predicting *in vivo* lethality. In general, these authors concluded that QSAR is not yet useful for predicting LD_{50} values of unrelated compounds or when based on electronic or structural descriptors of the substituents. QSAR has had greater predictive success when using small groups of related compounds and with certain physiocochemical properties of compounds, particularly hydrophobicity. The hydrophobicity of a compound may be expressed as the logarithm of the partition coefficient (1-octanol/water partition coefficient), which is determined from the distribution of the compound between two immiscible solvents (water and 1-octanol), one polar and the other nonpolar [40]. Hydrophobicity, estimated by the 1-octanol/water partition coefficient, may be a more useful predictor of biologic activity than are the electronic characteristics of a compound, because hydrophobicity correlates well with the lipophilicity of the compound and, in turn, its ease of transport across biologic membranes. For some classes of compounds, such as intercalators, the use of charged partial surface area (CPSA) descriptors may also be appropriate. CPSAs encode the surface area and partial charge of the compound simultaneously. A more recent approach has been to link the functional group contributions determined by traditional QSAR and the pharmacophore geometry based on the steric and electrostatic fields presented to the active site on an enzyme. X-ray crystallographic data can also be used to predict interactions between enzymes, substrates, and inhibitors.

In general, structure-activity data alone do not appear to be sufficient to successfully predict *in vivo* lethality. However, the integration of descriptors based on the biologic mechanisms of lethality should improve the ability of QSARs to correlate the LD_{50} values using large groups of unrelated compounds. The computer optimized molecular parametric analysis for chemical toxicity (COMPACT) system is one such system developed by Parke et al. [63] using integrated molecular modeling to predict the mechanism of chemical toxicity. Using prior knowledge of receptor interactions, the particular receptor with which a compound interacts, the metabolic fate of the compound, and, in turn, its toxicity can be predicted from the molecular and electronic structures of the chemical. Ultimately, an analysis of a chemical structure using COMPACT could be combined with other QSAR predictions to select a drug candidate with the least toxic potential and with maximal pharmacologic activity, before a compound being synthesized [62].

5.9 APPROACHES FOR EVALUATING HUMAN RISK

Risk can be defined as the probability of a hazard or injury occurring in humans, animals, or the environment under a given set of conditions. Although a compound may be toxic, risk may be minimal if a poor likelihood of exposure exists. Thus, the assessment of risk requires not only toxicity data, but also knowledge of the physical state, concentration, route, and frequency of exposure, physicochemical properties, SARs to similar compounds of known activity, and pharmacokinetic data from a relevant route of exposure. Physicochemical properties, such as the volatility, solubility, and reactivity of a compound, can be used to estimate the general likelihood of exposure. Other parameters, including molecular weight, ionization, steric properties, lipophilicity, and protein binding, can be used to estimate the risk of exposure of the target cells. An optimal approach to modeling risk is to include and integrate as many of these parameters as possible with *in vitro* pharmacokinetic, metabolism, and toxicity date [82].

The development of an integrative approach for predicting human risk from *in vitro* data, as well as physicochemical properties and pharmacokinetics, can be facilitated by computerized model systems that incorporate an assortment of these complex *in vivo* and *in vitro* parameters. An analogy model can be built on the basis that a model system should behave analogously to the system that it predicts. Thus, the more similar the model system is to the predicted system, the more likely it is to contain relevant information. However, because of ethical considerations, we must increasingly rely on *in vitro* cytotoxicity model systems and physicochemical modeling and less on human and animal models.

Hellberg et al. [43] described a reasonable approach to the development of prediction models for human toxicity that combines several different and complementary model systems primarily based on *in vitro* and physicochemical data, supplemented by a limited number of *in vivo* systems. Multivariate data analysis was then applied to analyze the data, because this method can be used with a large number of possible collinear variables. In contrast, multiple regression analysis, typically used in modeling, is limited to applications in which the number of test systems is small compared with the number of chemicals in the study, and the results may not be reliable if test systems are correlated. Multivariate analysis can be employed in analogy modeling by using the projection method PLS with cross validation, as described by Wold et al. [84–86]. The PLS method is based on correlation studies between different data sets (*in vitro*, physicochemical, etc.), with the aim of finding combinations between the parameters of each set, and then correlating between the combinations to predict *in vivo* toxicity [64]. The individual contribution of each assay to the overall prediction can also be evaluated. The PLS method has been shown to handle multivariate data with strong correlations among the *x*-variables, as is seen with related *in vitro* cytotoxicity test systems. Applying this model to the first ten MEIC compounds, Hellberg et al. [43] showed that this model was relevant for predicting human toxicity using cytotoxicity data combined with physiocochemical data.

The toxicity associated with most compounds appears to be related to basal cytotoxicity produced by interference with vital cell structures or functions essential for the survival and reproduction of cells, which should be detected by almost any dividing cell. However, because a variety of mechanisms can lead to systemic toxicity and lethality, no single assay can be expected to be reliably predictive of human lethality for all compounds. In addition, some compounds produce organ-specific toxicity because the compound is selectively concentrated in the target tissue. For many of these compounds, pharmacokinetic information can be integrated with a battery of *in vitro* data to model the *in vivo* response, as previously described. However, other compounds adversely affect organ-specific functions by noncyto-toxic-receptor–mediated mechanisms, which cannot be accurately predicted using poorly differentiated cell systems. Thus, it will be difficult to completely eliminate the need for animal testing, but the number of animals can be effectively reduced by combining results of *in vitro* tests with pharmacokinetic studies in animals and with computer models based on structure activity.

References

1. Allen RD, Allen NC, Travis JL. Video-enhanced contrast, differential interference contrast (AVIC-DIC) microscopy: a new method capable of analyzing microtubule-related

motility in the reticulopodial networks of *Allogromia laticollaris*. *Cell Motil* 1981;1:291–302.

2. Balant LP, Gex-Fabry M. Review: physiological pharmacokinetic modeling. *Xenobiotica* 1990;20(11):1,241–1,257.

3. Balls M, Horner SA. The *FRAME* interlaboratory programme on *in vitro* cytotoxicology. *Food Chem Toxocol* 1985;23(2):209–213.

4. Benford DJ, Reavy HJ, Hubbard SA. Metabolizing systems in cell culture cytotoxicity tests. *Xenobiotica* 1988;18(6):649–656.

5. Bernson V, Bondesson I, Ekwall B, Stenberg K, Walum E. A multicentre evaluation study of *in vitro* cytotoxicity. *ATLA* 1987;14:144–146.

6. Bohm N, Sprenger E. Fluorescence cytophotometry: a valuable method for the quantitative determination of nuclear feulgen-DNA. *Histochemic* 1968;16:100–118.

7. Bondesson I, Ekwall B, Hellberg S, Romert L, Stenberg K, Walum E. MEIC: a new international multicenter project to evaluate the relevance to human toxicity of *in vitro* cytotoxicity tests. *Cell Biol Toxicol* 1989;5:331–347.

8. Borenfreund E, Puerner JA. A simple quantitative procedure using monolayer cultures for cytotoxicity assays (HTD/NR-90). *J Tissue Cult Methods* 1984;65:55–63.

9. Borenfreund E, Puerner JA. Toxicity determined *in vitro* by morphological alterations and neutral red absorption. *Toxicol Lett* 1985;24:119–124.

10. Brugmans N, Cassiman JJ, Van der Heydt L, Oosterlinck AJJ, Vlietinck R, and Van den Berghe H. Quantification of the degree of cell spreading of human fibroblasts by semiautomated analyses of the cell perimeter. *Cytometry* 1982;3:262–268.

11. Bunting WL, Keily JH, Owen CA, Jr. Tadiochromium-labelled lymphocytes in the rate. *Proc Soc Exp Biol* 1963;113:370–374.

12. Chamberlain M, Parish WE. Hazard and risk based on *in vitro* test data. *Toxicol In Vitro* 1990;4(4/5):694–697.

13. Clothier RH, Hulme LM, Ahmed AB, Reeves HL, Smith M, Balls M. *In vitro* cytotoxicity of 150 chemicals to 3T3-L1 cells assessed by the *FRAME* Kenacid blue method. *ATLA* 1988;16:84–95.

14. Clothier RH, Hulme LM, Smith M, Balls M. Comparison of the *in vitro* cytotoxicities and acute *in vivo* toxicities of 59 chemicals. *Mol Toxicol* 1989;1:571–577.

15. DeLuca MAS, McElroy WD. *Bioluminescence and Chemiluminescence*. New York: Academic Press, 1981:122.

16. Dierickx PJ. Cytotoxicity testing of 114 compounds by the determination of the protein content in Hep G2 cell cultures. *Toxicol In Vitro* 1989;3:189–193.

17. Diericks PH, Ekwall B. Long-term cytotoxicity testing of the first twenty MEIC chemicals by the determination of the protein content in human embryonic lung cells. *ATLA* 1992;20:285–289.

18. Duffy PA. Mechanisms of cell toxicity: I. Cell death. *Toxicol In Vitro* 1992;6(1):91–92.

19. Ekwall B. Toxicity to HeLa cells of 205 drugs as determined by the metabolic inhibition test supplemented by microscopy. *Toxicology* 1980;17:273–295.

20. Ekwall B. Preliminary studies on the validity of *in vitro* measurement of drug toxicity using HeLa cells: IV. Therapeutic effects and side effects of 50 drugs related to the HeLa toxicity of the therapeutic concentrations. *Toxicol Lett* 1981;7:359–366.

21. Ekwall B. Screening of toxic compounds in mammalian cell cultures. *Ann NY Acad Sci* 1983;407:64–77.

22. Ekwall B, Acosta D. *In vitro* comparative toxicity of selected drugs and chemicals in HeLa cells, Chang liver cells, and rap hepatocytes. *Drug Chem Toxicol* 1982;5:219–231.

23. Ekwall B, Bondesson I, Castell JV, et al. Cytotoxicity evaluation of the first ten MEIC chemicals: acute lethal toxicity in man predicted by cytotoxicity in five cellular assays and by oral LD50 tests in rodents. *ATLA* 1989;17:83–100.

24. Ekwall B, Bondesson I, Hellberg S, Hogberg J, Romert L, Stenberg K, Walum E. Validation of *in vitro* cytotoxicity tests: past and present strategies. *ATLA* 1991;19(2):226–233.

25. Ekwall B, Gomez-Lechon MJ, Hellberg S, et al. Preliminary results from the Scandinavian multicentre evaluation of *in vitro* cytotoxicity (MEIC). *Toxicol In Vitro* 1990;4(4/5):688–691.

26. Ekwall B, Johannson A. Preliminary studies on the validity of *in vitro* measurement of drug toxicity using HeLa cells: I. Comparative *in vitro* toxicity of 27 drugs. *Toxicol Lett* 1980;5:299–307.

27. Ekwall B, Nordensten C, Albanus L. Toxicity of 29 plasticizers to HeLa cells in the MIT-24 systems. *Toxicology* 1982;24:199–210.

28. Fautrel A, Chesne C, Guillouzo A, et al. A multicentre study of acute *in vitro* cytotoxicity in rat liver cells: validation of *in vitro* cytotoxicity tests. *Toxicol In Vitro* 1991;5:543–547.

29. Fedoroff S. Proposed usage of animal tissue culture terms. *In Vitro* 1966;2:155–159.

30. Fry JR, Garle MJ, Hammond AH. Choice of acute toxicity measures for comparison of *in vitro/in vivo* toxicity. *ATLA* 1988;16:175–179.

31. Fry JR, Garle MJ, Hammond AH, Hatfield A. Correlation of acute lethal potency with *in vitro* cytotoxicity. *Toxicol In Vitro* 1990;4(3):175–178.

32. Gad SC, Chengelis CP. *Acute Toxicology*, 2nd ed. San Diego, CA: Academic Press, 1997.

33. Gillette JR. Factors affecting drug metabolism. *Ann NY Acad Sci* 1971;179:43–66.

34. Gillette JR. Problems in correlating *in vitro* and *in vivo* studies of drug metabolism. In: Benet LZ, Levy G, Farraiolo BL, eds. *Pharmacokinetics: A Modern View*. New York: Plenum, 1984:235–252.

35. Gillette JR. Solvable and unsolvable problems in extrapolating toxicological data between animal species and strains. In: Mitchell JR, Horning G, eds. *Drug Metabolism and Drug Toxicity*. New York: Raven, 1984:237–260.

36. Gillette JR. Pharmacokinetics of biological activation and inactivation of foreign compounds. In: Anders MW, ed. *Bioactivation of Foreign Compounds*. New York: Academic Press, 1985.

37. Gillette JR. On the role of pharmacokinetics in integrating results from *in vivo* and *in*

vitro studies. *Food Chem Toxicol* 1986;24(6/7):711–720.

38. Goto Y, Dujovne CA, Shoeman DW, Arakawa K. Liver cell culture toxicity of general anaesthetics. *Toxicol Appl Pharmacol* 1976;36:121–130.

39. Gratzner HG. Monoclonal antibody to 5-bromo- and 5-iododeoxyuridine: a new reagent for detection of DNA replication. *Science* 1982;218:474–478.

40. Gulden M, Finger J. Effects of the first twenty MEIC reference chemicals on viability, glucose consumption and spontaneous contractility of primary cultured rat skeletal muscle cells. *ATLA* 1992;20:222–225.

41. Hansch C, Maloney PP, Fujita T, Muir RM. *Nature* 1962;194:178–180.

42. Hayflick L, Moorhead PS. The serial cultivation of human diploid cell strains. *Exp Cell Res* 1961;25:585–589.

43. Hellberg S, Eriksson L, Jonsson J, et al. Analogy models for prediction of human toxicity. *ATLA* 1990;19:103–116.

44. Holm G, Perlmann P. Quantitative studies on phytohaemagglutinin-induced cytotoxicity by human lymphocytes against homologous cells in tissue culture. *Immunology* 1967;12:525–536.

45. Homer SA, Fry JR, Clothier RH, Balls M. A comparison of two cytotoxicity assays for the detection of metabolism-mediated toxicity *in vitro*: a study with cyclophosphamide. *Xenobiotica* 1985;15:681–686.

46. Ivett JL, Tice RR. Average generation time: a new method of analysis and quantitation of cellular proliferation kinetics. *Environ Mutagen* 1982;4:358–370.

47. Jackson TR, Hallam J, Downes CP, Hanley MR. Receptor coupled events in bradykinin action rapid production of inositol phosphates and regulation of cytosolic Ca^{2+} in a neural cell line. *EMBO J* 1987;6:49–54.

48. Jover R, Ponsoda X, Castell JV, Gomez-Lechon MJ. Evaluation of the cytotoxicity of ten chemicals on human cultured hepatocytes: predictability of human toxicity and comparison with rodent cell culture systems. *Toxicol In Vitro* 1992;6(1):47–52.

49. Kemp RB, Cross DM, Meredith RWJ. Adenosine triphosphate as an indicator of cellular toxicity *in vitro*. *Food Chem Toxicol* 1986;24(6/7):465–466.

50. Kemp RB, Cross DM, Meredith RWJ. Comparison of cell death and adenosine triphosphate content as indicators of acute toxicity *in vitro*. *Xenobiotica* 1988;18(6):633–639.

51. Khochbin S, Chabanas A, Albert P, Albert J, Lawrence J-J. Application of bromodeoxyuridine incorporation measurements to the determination of cell distribution within the S phase of the cell cycle. *Cytometry* 1988;9:499–503.

52. Knox P, Uphill PF, Fry JR, Benford J, Balls M. The FRAME multicentre project on *in vitro* cytotoxicology. *Food Chem Toxicol* 1986;24(6/7):457–463.

53. Laughton G. Quantification of attached cells in micrometer plates based on Coomassie brilliant blue G-250 staining of total cellular protein. *Anal Biochem* 1983;140:417–423.

54. Litterst CL, Lichtenstein EP. Effects and interactions of environmental chemicals on human cells in tissue culture. *Arch Environ Health* 1971;22:454–461.

55. Lowry OH, Rosebrough NJ, Farr AL, Randall RJ. Protein measurements with the Folin

phenol reagent. *J Biol Chem* 1951;193:265–271.

56. Maduk EU, Baskin ST, Salem H. Attaining a rational refinement and reduction in the use of live animals: studies on cyanide toxicity. In: Salem H, Kat EA, eds. *Advances in Animal Alternatives for Safety and Efficacy Testing*. Philadelphia: Taylor & Francis, 1999:433–436.

57. Maile W, Lindl T, Weiss DG. New methods for cytotoxicity testing: quantitative video microscopy of intracellular motion and mitochondria-specific fluorescence. *Mol Toxicol* 1987;1:427–437.

58. Mosmann T. Rapid colorimetric assay for cellular growth and survival: application of proliferation and cytotoxicity assays. *J Immunol Methods* 1983;65:55–63.

59. Nagele RG, Lee H. A new method for the preparation of "double-fixed" quick-frozen, freeze-substituted cells for whole-cell transmission electron microscopy. *J Microsc* 1987;148:89–95.

60. Olson R, Santone K, Medler D, Oakes SG, Abraham R, Powis G. An increase in intra-cellular free Ca2+ associated with serum free growth stimulation of Swiss 3T3 fibroblasts by epidermal growth factor in the presence of bradykinin. *Biol Chem* 1988;263:18,030–18,035.

61. Otoguro K, Komiyama K, Omura S, Tyson CA. An *in vitro* cytotoxicity assay using rat hepatocytes and MTT and Coomassie blue dye as indicators. *ATLA* 1991;19:352–360.

62. Parke DV, Ioannides C, Lewis DFV. Computer modeling and *in vitro* tests in the safety evaluation of chemicals-strategic applications. *Toxicol In Vitro* 1990;4(4/5):680–685.

63. Parke DV, Lewis DFV, Ioannides C. Current procedures for the evaluation of chemical safety. In: Richardson ML, ed. *Risk Assessment of Chemicals in the Environment*. London: Royal Society of Chemistry, 1988:45–72.

64. Peloux A-F, Federici C, Bichet N, Gouy D, Cano J-P. Hepatocytes in primary culture: an alternative to DL50 testing? Validation of a predictive model by multivariate analysis. *ATLA* 1992;20:8–26.

65. Phillips JC, Gibson WB, Yam J, Alden C, Hard GC. Survey of the QSAR and *in vitro* approaches for developing non-animal methods to supersede the *in vivo* LD_{50} test. *Food Chem Toxicol* 1990;28(5):375–398.

66. Pomerat CM, Leake CD. Short term cultures for drug assays: general considerations. *Ann NY Acad Sci* 1954;58:1,110–1,124.

67. Purchase IFH, Farrar DG, Whicaker IA. Toxicology profiles on substances used in the FRAME cytotoxicology research project. *ATLA* 1987;14:184–242.

68. Putney JW. Calcium-mobilizing receptors. *Trends Pharmacol Sci* 1987;8:481–486.

69. Rasmussen H. The cycling of calcium as an intracellular messenger. *Sci Am* 1989;261(4):66–73.

70. Reinhardt Ca, Pelli DA, Zbinden G. Interpretation of cell toxicity data for estimation of potential irritation. *Food Chem Toxicol* 1985;231–247.

71. Richards GP, Bemis JA, Sample JD. A simple planimetric method to quantify cytotoxicity in cell culture monolayers. *J Virol Methods* 1988;20:33–38.

72. Riddell RJ, Clothier RH, Balls M. An evaluation of three *in vitro* cytotoxicity assays.

Food Chem Toxicol 1986;24(6/7):469–471.

73. Shirhatti V, Krishna G. A simple and sensitive method for monitoring drug-induced cell injury in cultured cells. *Anal Biochem* 1987;147;410–418.

74. Shrivastava R, Delomenie C, Chevalier A, John G, Ekwall B, Walum E, Massingham R. Comparison of *in vivo* acute lethal potency and *in vitro* cytotoxicity of 48 chemicals. *Cell Biol Toxicol* 1992;8(2):157–170.

75. Shrivastava R, John GW, Rispat G, Chevalier A, Massingham R. Can the *in vivo* maximum tolerated dose be predicted using *in vitro* techniques? A working hypothesis. *ATLA* 1991;19:393–402.

76. Sorenson EMB. Validation of a morphometric analysis procedure using indomethacin-induced alterations in cultured hepatocytes. *Toxicol Tell* 1989;45:101–110.

77. Stark DM, Shopsis C, Borenfreund E, Babich H. Progress and problems in evaluating and validating alternative assays in toxicology. *Food Chem Toxicol* 1986;24(6/7):449–455.

78. Stoker M, O'Neill C, Berryman S, Waxman V. Anchorage and growth regulation in normal and virus-transformed cells. *Int J Cancer* 1968;3:683–689.

79. Takasugi M. An improved fluorochromatic cytotoxic test. *Transplant* 1971;12:148–151.

80. Toplin I. A tissue culture cytotoxicity test for large-scale cancer chemotherapy screening. *Cancer Res* 1959;19:959–965.

81. Walker PMB. The mitotic index and interphase processes. *J Exp Biol* 1954;31:8–15.

82. Walum E, Balls M, Bianchi V, et al. ECITTS: an integrated approach to the application of *in vitro* test systems to the hazard assessment of chemicals. *ATLA* 1992;20:406–428.

83. Westphal C, Horler H, Pentz S, Frosch D. A new method for cell culture on an electron-transparent melamine foil suitable for successive LM, TEM and SEM studies of whole cells. *J Microsc* 1988;150:225–231.

84. Wold S, Albano S, Dunn JR III, et al. Multivariate data analysis in chemistry. In: Kawalski Br, ed. *Chemometrics: Mathematics and Statistics in Chemistry.* Dordrecht: D. Reidel, 1984:17–96.

85. Wold S, Dunn JR III. Multivariate quantitative structure-activity relationships (QSAR): conditions for their applicability. *J Chem Inform Comp Sci* 1983;23:6–13.

86. Wold S, Albano S, Dunn JR III. The colinearity problem in linear regression: the partial least squares (PLS) approach to generalized inverses. *SIAM J Stat Comput* 1984;5:735–743.

87. Yuhas JM, Toya RE, Pazmino NH. Neuraminidase and cell viability: failure to detect cytotoxic effects with dye-exclusion techniques. *JNCI* 1974;53(2):465–468.

88. Zanetti C, Angelis ID, Stammati A-L, Zucco L. Evaluation of toxicity testing of 20 MEIC chemicals on Hep-2 cells using two viability endpoints. *ATLA* 1992;20:120–125.

6

In Vitro Genetic Toxicity Testing

Ann D. Mitchell

Genesys Research, Incorporated, Research Triangle Park, North Carolina

In vitro genetic toxicity tests, concerned with the genetic effects of toxic materials *in vitro*, are the forerunner of all *in vitro* toxicity testing. In contrast to *in vitro* tests measuring only toxic effects, but often in cellular systems selected to mimic target tissue and organ toxicity, genotoxicity tests are selected for their precision in assessing genetic effects in surviving cells. Because genetic effects are most frequently observed immediately below toxic levels, all *in vitro* genotoxicity tests include a measurement of *in vitro* toxicity. For this reason, negative genotoxicity results are not conclusive unless testing has been performed to toxic levels, to a concentration at which the test material is insoluble, or to a concentration above which the test results may lack biologic relevance.

A major impetus for the development and use of genetic toxicology test systems has been a recognition that long-term animal testing resources are insufficient for evaluating the universe of chemicals to which humans and the environment may be exposed. Thus, genetic toxicology tests assessing specific mechanisms observed in the whole animal (including humans) are used as initial tests for regulatory submissions because they are useful in predicting the outcome of long-term animal tests. In addition, they are used to conserve resources, reduce animal usage, provide rapid assessments of potential risk, including the risk of exposing human volunteers to drugs under development, and, generally, to assist in determining whether additional testing would be productive.

A wide range of *in vitro* and *in vivo* genetic toxicology assay systems has been defined, and many of these tests have both scientific and regulatory acceptance. The first genotoxicity tests were *in vivo* tests because methodology was not then available to mimic *in vivo* metabolism in assay systems using bacteria or cell cultures. However, shortly after *in vitro* metabolic activation systems were defined, a plethora of *in vitro* tests was developed, including the tests most frequently used today. Approximately a decade ago, over 100 genetic toxicology tests were available (see Table 6.1), and, as described at that time [1,2], a "cafeteria-style" testing approach was often followed, in which tests developed to evaluate different theoretical aspects of genetic alterations were selected to address various regulatory requirements. Further, because testing requirements had evolved independently within separate governmental agencies and in different geographical regions, numerous tests, and sometimes several modifications of the same tests, were required for companies wishing to market their products internationally.

TABLE 6.1.   Representative genetic toxicology assays[a]

Assay type	Organism or cell type	Endpoints measured
Bacterial DNA repair	*Escherichia coli polA⁺/polA⁻* by	Differential toxicity caused by DNA damage not repaired
	Bacillus subtilis Rec⁺/Rec⁻ by	Differential toxicity caused by DNA damage not repaired
Bacterial mutation	*Salmonella typhimurium*	Reverse mutation (Ames assay)
	Salmonella typhimurium	Forward mutation
	Escherichia coli (WP2)	Forward and reverse mutations
Fungal/yeast mutation	*Saccharomyces cerevisiae*	Reverse mutation
	Schizosaccharomyces pombe	Forward mutation
	Neurospora crassa	Forward and reverse mutation
Yeast chromosome effects	*Saccharomyces cerevisiae*	Mitotic recombination
	Saccharomyces cerevisiae	Mitotic aneuploidy and meiotic nondisjunction
Mammalian cell DNA damage	Various cells (*in vitro* and *in vivo*)	Adduct formation
	Various cells, e.g., lymphocytes	DNA damage: Comet assay
Mammalian cell DNA repair	Human fibroblasts	Unscheduled DNA synthesis (UDS)
	Rat and mouse hepatocytes	UDS
	Chinese hamster cells and human lymphocytes	Sister chromatid exchange (SCE)
Mammalian cell gene mutation	L5178Y mouse lymphoma cells	Forward mutation
	Chinese hamster cells (CHO/V79/AS52)	Forward mutation
	Human lymphoblasts (TK6)	Forward mutation

Assay type	Organism or cell type	Endpoints measured
Mammalian cell cytogenetic damage	Chinese hamster cells (CHO/V79)	Chromosomal aberrations
	Human lymphocytes	Chromosomal aberrations
	L5178Y mouse lymphoma cells	Small colony formation
Mammalian cell transformation	Mouse or hamster cells	Change in growth characteristics *in vitro*
Insect mutation	*Drosophila melanogaster*	Sex-linked recessive lethal mutations
Insect chromosomal effects	*Drosophila melanogaster*	Translocation and sex chromosome loss
Whole-animal DNA repair	Mouse or rat hepatocytes	Unscheduled DNA synthesis
	Mouse bone marrow cells	Sister chromatid exchange
Whole-animal mutation	Transgenic mice and rats	Reverse mutation of (bacterial) transgene
	Mouse/specific locus	Forward mutation and deletions
	Mouse/spot test	Forward mutation and deletions
Whole-animal cytogenetic damage	Mouse or rat	Bone marrow chromosomal aberrations
	Mouse or rat	Lymphocyte chromosomal aberrations
Whole-animal germ cell DNA damage	Mouse or rat	Spermatocyte unscheduled DNA synthesis
Whole-animal germ cell cytogenetic damage	Mouse or rat/dominant lethal	Nonviable fetuses
	Mouse/heritable translocation translocations	Sterility and detection of
	Mouse (with heritable translocation)	Spermatocyte chromosomal rearrangements
Host-mediated	Mouse or rat	Gene mutation in bacterial or mammalian cell test system
Body fluids analysis	Urine samples from mouse, rat, or human	Various genetic endpoints in bacterial cells or mammalian cell culture

ᵃAdapted from ref. 2.

However, the field of genetic toxicology testing has undergone a dramatic change in the last decade. Because of a recognized need to harmonize testing requirements for national and international regulatory submissions, only a few of the developed tests are routinely used today, with their selection based on considerations such as the mechanistic relevance of the effects measured, the extent that the test systems have been defined and evaluated, and the experience, ease, and economy in performance of the tests for laboratories worldwide. Therefore, this chapter will not attempt to provide detailed descriptions of genetic toxicology tests that, although used for prior assessments and important in establishing the relevance of the field, are seldom used for current regulatory submissions.

Instead, to provide a practical guide for professionals responsible for evaluating the safety of current products, this chapter begins with a review of cellular and molecular processes important for understanding currently used genetic toxicology tests and then provides a chronologic overview of the development and application of the field of genetic toxicology, including a summary of the current status of regulatory requirements. This overview is followed by a detailed description of the tests most frequently used today to address these regulatory requirements, together with a summary of the advantages and limitations of these tests and a discussion of their predictivity.

Should additional information be required concerning other genetic toxicology tests, e.g., for interpreting previously obtained test results, or for selecting additional tests to augment those most frequently used, the reader may wish to consult genetic toxicology texts [3,4], the series *Chemical Mutagens* [5], and a number of established journals in the field, including *Environmental and Molecular Mutagenesis* (Wiley–Liss), *Mutagenesis* (IRL Press), and *Mutation Research* (Elsevier). In the latter, the series of U.S. Environmental Protection Agency (EPA) Gene-Tox reviews of specific test systems may be found. Internet resources for additional information include http://www.fda.gov/ for the Food and Drug Administration (FDA) homepage; http://www.pharmweb.net/pwmirror/pw9/ifpma/ich5.html for the International Council on Harmonization (ICH) of Technical Requirements of Pharmaceuticals for Human Use guidelines; http://www.epa.gov/epahome/rules. html for EPA environmental regulations; and http://www.oecd.org/ehs/test/ testlist.htm for the Organization for Economic Cooperation and Development (OECD) guidelines.

6.1 BASIC MECHANISMS ASSESSED IN THE GENETIC TOXICOLOGY TESTS MOST FREQUENTLY USED FOR REGULATORY SUBMISSIONS

Genetic toxicology test systems measure the outcome of damage or alterations in DNA, which is the basic blueprint for transmission of hereditary information to daughter cells. If DNA is damaged, it may (1) be correctly repaired with no genetic consequences; (2) lead to cell death, again with no genetic consequences; or (3) be replicated with the damage incorrectly repaired. Only the third consequence leads to mutations, DNA alterations propagated through subsequent generations of cells or individuals.

Hence, genetic toxicology tests assessing DNA damage and repair, such as SCE tests and tests for UDS (the repair of DNA damage at times other than the scheduled phase of the mitotic cycle, i.e., the S-phase), are less frequently used for regulatory submissions today because they provide only indirect evidence of mutagenesis. Instead, the harmonized testing approaches for regulatory submissions, which will be described, consist of tests for gene and chromosomal mutations.

6.1.1 Gene Mutations

Gene mutations can be assessed in bacteria, mammalian cells in culture, and whole organisms. A gene is the simplest functional unit in a DNA molecule. Gene (or point) mutations are changes in the nucleotide sequence at one or a few coding segments. Molecular genetic techniques have now become sufficiently powerful to allow routine identification of the specific DNA sequence changes responsible for mutant phenotypes. About 5,000 diseases in humans are now known to be caused by defective genes. These inherited disorders cause 20% of all infant mortalities, half of all miscarriages, and 80% of all cases of mental retardation [6]. In addition, considerable research during the last two decades to identify the genes involved in the alteration of normal cellular processes has resulted in the identification of over 100 different cancer genes [7].

Gene mutations in bacteria.   Gene mutations in bacteria can occur in a number of ways. In one of the simplest cases, a *base-pair substitution mutation*, a single nucleotide is changed, which is followed by a subsequent change in the complementary nucleotide on the other strand of the DNA double helix. Such mutations are deleterious when they alter a protein coding sequence to conclude translation prematurely (a *nonsense mutation*) or to incorporate a different amino acid (a *missense mutation*). Similarly, *frameshift mutations* occur after the deletion or insertion of one nucleotide, which then changes the "reading frame" for the remainder of the gene, or even for multiple genes. Both base-pair substitution and frameshift mutations are routinely measured in bacterial cells by the cells' acquisition of the capability of growth in an environment containing a missing amino acid, and for these tests, a large number of bacteria is examined to demonstrate significant increases over spontaneous mutation frequencies. A *forward mutation* occurs when a change takes place in the native DNA; *a reverse mutation* occurs when a mutated cell is returned to its initial phenotype. The currently used bacterial tests are reverse mutation assays.

Gene mutations in mammalian cells.   Gene mutations in mammalian cells are generally forward mutations and include base-pair substitution and frameshift mutations. Measurements of gene mutations in mammalian cells reflect the greater complexity of mammalian cells and chromosomes in comparison to those of prokaryotes, and, thus, they more closely approximate the genetic effects of chemicals in rodent species and humans.

In contrast to bacteria, mammalian cells are essentially diploid ($2n$, with two copies of each chromosome). Mammalian chromosomes are located within the nucleus of the cell and contain nonfunctional and noncoding, as well as functional, coding sequences. In addition, whereas in bacterial cells all genes are usually expressed, in diploid cells, usually two copies of each gene exist (one on each chromosome). One (dominant) form of the gene may be expressed while the other (recessive) gene remains unexpressed, unless both copies are recessive. If both copies of the gene are the same, the cell is homozygous for that trait; a heterozygous condition exists if the copies are different, and if only one chromosome is present to carry the trait, the condition is hemizygous. Hemizygous traits are found on the X chromosome in mammals because males have only one X chromosome, and only one X chromosome is expressed in female cells. The first mammalian cell mutation assays that were developed [8,9] used a gene found on the X chromosome of Chinese hamster cells, i.e., a hemizygous gene; the mouse lymphoma cell mutation assay [10] uses heterozygous cells.

Gene mutations are routinely measured in mammalian cells after the mutant cells' acquisition of the capability of growth in the presence of a selective agent, an otherwise toxic drug that can no longer be used by the mutated cell. As for the bacterial assays, a large number of cells is examined to demonstrate significant increases over spontaneous mutation frequencies. However, in mammalian cell gene mutation assays, the chemical exposure step must be followed by an expression period, during which mutant (and nonmutant) cells increase in number and the nonmutant protein (enzyme) present in the mutated cells and the RNA coding for that protein are depleted. Only then can the selective agent be added to permit only the mutated cells to grow and form colonies.

Gene mutations *in vivo*. On a theoretical basis, some justification exists for determining whether test materials inducing gene mutations *in vitro* are also capable of inducing gene mutations *in vivo*. For this reason, numerous test systems have been developed to assess mutations of bacterial transgenes *in vivo*, in transgenic mice and rats. However, in spite of the early promise of transgenic mutagenesis systems, they have been found to lack sensitivity and to be capable of detecting only a few of the chemicals known to be rodent carcinogens. Therefore, such tests are currently identified as optional, but not required, systems for regulatory submissions, and no universally accepted and routinely used test system exists for assessing gene mutations *in vivo*.

6.1.2 Chromosomal Mutations

Chromosomal mutations are large-scale numerical or structural alterations in eukaryotic chromosomes—including small and large deletions (visualized as breaks), translocations (exchanges), nondisjunction (aneuploidy), and mitotic recombination—that may affect the expression of numerous genes with gross effects, or be lethal to affected cells. Chromosomal abnormalities are associated with neoplasia, spontaneous abortion, congenital malformation, and infertility, which occur in approximately 0.6% of live births in humans. It has been estimated that up to 40% of spontaneous abortuses have chromosomal defects [11], and essentially all tumors harbor chromosomal mutations.

Chromosomal mutations in mammalian cells. Chromosomal mutations can be measured in several mammalian cell mutation assays, but the L5178Y mouse lymphoma assay is routinely used because it is the most extensively characterized of the several assays. This assay was first developed as a test for gene mutations at the *tk* (thymidine kinase) locus; however, it was subsequently noted that a biphasic curve of colony sizes is obtained, and that the biphasic size distribution is resolved into small (□) colonies of slowly growing mutant cells and large (□) colonies of more rapidly growing mutant cells. Banded karyotype and molecular analyses of the □ and □ mouse lymphoma colony mutants have revealed that whereas most □ colony mutants have only gene mutations or small chromosomal deletions, most □ colony mutants represent a full array of possible mutational damage, including gene mutations, small and large deletions, translocations, nondisjunction, and mitotic recombination.

Chromosomal aberrations. *In vitro* and *in vivo*. In contrast to the described assays, which assess gene and chromosomal mutations at only one or a few genes, in millions of cells per treatment, the *in vitro* and *in vivo* assays for chromosomal aberrations assess mutagenic events in multiple genes, but usually for no more than 200 cells per culture, or for up to 100 cells per animal. However, because these cells must be arrested at the appropriate interval after exposure, and because of the need for an experienced cytogeneticist to carefully evaluate each of the cells, the chromosomal aberration assays are often more time consuming and costly. Chromosome breakage, necessary for chromosomal rearrangements, is the classic endpoint in chromosomal aberration assays. To visualize chromosomes and chromosomal aberrations with a light microscope after *in vitro* or *in vivo* treatment with a chemical, cells are arrested in metaphase, treated with a hypotonic solution to swell the chromosomes, fixed, transferred to microscope slides, and stained. The first metaphase (M) after chemical exposure, M1, is the time when the greatest number of chromosomally damaged cells may be observed, and the extent of damage declines rapidly after M1 because of the greatly extended cell-cycle times of some cytogenetically damaged cells (e.g., while the cells attempt to repair the damage), and because the most severely damaged cells are often incapable of progressing through another cell cycle.

When the chromosomes of diploid somatic cells are replicated, each chromosome then consists of two (sister) chromatids separating at mitosis to become the chromosomes of the daughter cells. If chromosomal mutations occur before replication (DNA synthesis), both chromatids will be affected. This damage will be visualized as *chromosomal* breaks (deletions) and exchanges (translocations). However, if these mutations occur during replication (the most sensitive stage), or after replication, the damage is visualized as *chromatid* breaks and exchanges. Hence, by enumerating chromatid and chromosome breaks and exchanges, an index can be obtained of the time that the damage occurred. Very large deletions are tolerated only if they do not incapacitate essential genes. In diploid cells, this process is usually accomplished when a normal gene is retained by the homologous chromosome of paired chromosomes, yielding a heterozygous condition for that trait. Exchanges result when the broken ends of the same or different chromatids or chromosomes rejoin in an aberrant manner.

Although not currently used in cytogenetic testing for regulatory submissions, fluorescence *in situ* hybridization (FISH) staining techniques have been recently developed for human and mouse chromosomes, in which each chromosome can be differentially stained, revealing chromosomal rearrangements not apparent with conventional staining techniques [9]. When FISH staining is translated from a research approach to a testing protocol, it may be possible to reduce the number of chromosomes to be analyzed and, hence, the time for chromosomal aberration tests.

Micronuclei. Micronuclei result when nuclear membranes form around broken pieces of chromosomes or around chromosomes failing to separate at cell division. Therefore, micronucleus tests measure chromosome breakage, the classic endpoint for chromosomal aberration assays, and aneuploidy, the loss or gain of a chromosome or a chromosome segment. *In vitro* micronucleus tests are currently under development in a number of laboratories as a less subjective and more economical alternative to *in vitro* chromosomal aberration tests. For these approaches, cytochalasin B is used to arrest cell division (cytokinesis) but not nuclear division, and up to 1,000 binucleate cells are examined for the presence or absence of micronuclei. However, because the *in vitro* micronucleus tests have yet to be validated and shown to be at least as effective as tests for chromosomal aberrations *in vitro*, none is currently recommended for regulatory submissions.

In vivo micronucleus tests are justified for regulatory submissions for assessing chromosomal breakage and aneuploidy in an environment including *in vitro* metabolic reactions. Micronuclei are readily observed microscopically in stained preparations of (otherwise anucleate) polychromatic erythrocytes (PCEs) from the bone marrow of rats or mice or from the peripheral blood of mice; the latter because, in mice, the spleen does not remove micronucleated cells from the blood. With appropriate staining techniques, the PCEs can be differentiated from the more mature normochromatic erythrocytes (NCEs) because the PCEs still contain RNA, which has been lost by the NCEs. For example, with Giemsa staining, the PCEs are blue and the NCEs are salmon pink or red.

Peripheral blood erythrocytes can be obtained for micronucleus evaluations without sacrificing the animal, e.g., from animals under treatment for long-term studies; however, a greater number of cells must be evaluated because the newly formed erythrocytes (PCEs, the cells of interest) are diluted in the population of preexisting erythrocytes. Hence, bone marrow cells, which give a more informative index of toxicity, are routinely used for the micronucleus test as well as for *in vivo* chromosomal aberration assays. Because micronuclei can be evaluated more rapidly and economically than can chromosomal aberrations, the micronucleus test in rodents is now used more extensively than is the rodent bone marrow chromosomal aberration test.

6.2 DEVELOPMENT AND APPLICATION OF GENETIC TOXICOLOGY TESTS

The origins of genetic toxicology testing date from 1900 and the rediscovery of Mendel's classic paper on the basis of inheritance [10], which was closely followed, in 1901, by the first use of the term "mutation" to signify changes in hereditary material [11]. As illustrated in Table 6.2, which highlights some of the relevant events in the development and application of genetic toxicology testing, the focus of mutagenesis research during the first six decades of the century was directed toward using recently developed techniques to gain an increased understanding of the nature of genetic material. However, the relationship between cancer and an abnormal chromosomal constitution was proposed as early as 1914 [12], and in 1927, it was suggested that mutations could cause cancer [13].

TABLE 6.2. **Development and application of genetic toxicology testing**

1900	Rediscovery of Mendel's classic 1865 paper on the basis of inheritance [10].
1901	de Vries [11] applies the term "mutation" to changes in hereditary material.
1914	Boveri [12] proposes the somatic mutation theory of cancer: that the primary cause of cancer may be abnormal cell division resulting in an abnormal chromosomal constitution.
1927	Muller publishes the first demonstration of the induction of mutations by X-rays, in *Drosophila* [13], and he shows that an induced response must be evaluated in relation to background mutations. He also suggests the possibility that mutations may cause cancer.
1938	Federal Food, Drug, and Cosmetic Act enacted to provide for the regulation of foods (except meat and poultry products), food and color additives, human and animal drugs, medicated animal feeds, medical devices, and cosmetics.
1938	Sax shows that X-rays can induce chromosomal alterations in plant pollen cells [24].
1941	Auerbach and Robson demonstrate that nitrogen mustards can induce mutations in *Drosophila*, similar to the mutations induced by X-rays [25]. (Because of censorship, publication was delayed until 1947, after World War II, although nitrogen mustards were not used in World War II.)
1943	Earle et al. establishes continuous cultures of rodent cell lines [26].
1947	Federal Insecticide, Fungicide, and Rodenticide Act (FIFRA) enacted.

1951   Russell [27] demonstrates that X-rays can induce genomic mutations in mice.

1952   Gey et al. [28] establishes a continuous cell line (HeLa) from a human tumor.

1953   Watson and Crick elucidate the double-helical structure of DNA [29].

1956   Tjio and Levan [30] use colchicine to arrest human cells in metaphase and establish 46 chromosomes as the normal (diploid) compliment of humans.

1959   Lejeune et al. [31] discover that children with Down syndrome have an extra chromosome (number 21).

1960   Nowell and Hungerford [32] discover a consistent chromosome change (the Philadelphia chromosome) associated with a specific type of human tumor (chronic granulocytic leukemia).

1960   In a paper entitled "Chemical mutagenesis in animals," Auerbach [33] states that as more and more chemicals are used in therapeutics, food processing, and other industries, the testing of the substances for mutagenic ability will become a necessary protective measure. During the remainder of this decade, numerous similar warnings were issued by the scientific community, and by the end of the 1960s, the first *in vivo* tests for mutagenic effects had been established.

1961   Hayflick and Moorehead [34] demonstrate that normal cells have a finite lifespan (~50 generations) before the cells exhibit abnormal growth. It has recently been found that maintenance of normal cell growth is related to telomerase, which facilitates DNA replication at the ends of chromosomes.

1962   Publication of *Silent Spring* by Rachel Carson [35] warns of the hazards of radiation and chemicals, especially pesticides, and calls for more regulation of industry.

1968   Two teams of scientists, Kao and Puck [36], and Chu and Malling [37], independently demonstrate the chemical induction of increased mutation frequencies in mammalian (Chinese hamster) cells.

1970   The EPA is established to regulate matters concerning air and water pollution, the use of pesticides, and other matters affecting the environment.

1970   Occupational Safety and Health Act (OSHA) enacted.

1971   James and Elizabeth Miller demonstrate that many chemicals become carcinogens only after they are metabolized in animals [38].

1971   Malling [16] demonstrates that chemicals can become mutagenic *in vitro* by the addition of mammalian liver enzymes to mimic *in vivo* metabolism. During the remainder of the 1970s, numerous research approaches for examining mutations, DNA damage and repair, and chromosome alterations were used to define a wide range of *in vitro* and *in vivo* genetic toxicology tests.

1971   Knudson [39] explains the origin of retinoblastoma tumors by proposing that two sequential mutations are required. This two-stage genetic model has been found to apply to additional forms of cancer.

1972   Clive et al. [40] define the L5178Y/*tk*$^{+/-}$ mouse lymphoma cell mutation assay.

1972   Consumer Product Safety Act enacted.

1973   Ames et al. [42] introduce a mammalian liver homogenate into a *Salmonella* reverse mutation test system, test 18 carcinogens and publish the results in a paper entitled "Carcinogens are mutagens."

1973   The first widely publicized alarm about mutagenicity, an unpublished report of an association between human exposure to aerosols of spray adhesives and the appearance of chromosomal abnormalities and birth defects, which led the Consumer Product Safety Commission (CPSC) to remove spray adhesives from the market. However, technically poor cytogenetic preparations rather than chromosomal abnormalities were found in an independent examination of the original slides, and the CPSC rescinded the ban after 6 months [21].

1973   The National Cancer Institute (NCI) initiates the first contracts to evaluate the utility of *in vitro* genetic toxicology tests, which eventually included the "Ames" test, UDS, the mouse lymphoma cell mutagenesis assay, and tests for mammalian cell transformation.

1974   Safe Drinking Water Act enacted.

1975   Rodent bone marrow micronucleus testing protocol defined by Schmid [41].

1975   McCann et al. [17] publish results for 300 chemicals tested in *Salmonella* and find that the system correctly identifies ~90% of the animal carcinogens while maintaining an equally high specificity for noncarcinogens. Although the concordance of results has been lower as additional chemicals have been tested, this initial finding became a goal for evaluating additional tests, as justification for genotoxicity tests shifted from predicting effects on the human gene pool to predicting carcinogenicity.

1976   The Second Task Force for Research Planning in Environmental Health Science estimates that 50–90% of the total cancer incidence is dependent on known or unknown environmental factors, and, citing a backlog of up to 30,000 agents for the costly long-term testing for carcinogenesis, identifies a requirement of short term [genetic toxicology] tests to screen chemicals for carcinogenic activity [43].

1977   Toxic Substances Control Act (TSCA) enacted to provide the EPA with the authority "to regulate commerce and protect human health and the environment by requiring testing and necessary use restrictions on certain chemical substances." This was the first Federal law that identified mutagenicity as an endpoint of toxicological concern.

1978   The EPA proposes guidelines for the mutagenicity testing of pesticides [44].

1978   The National Toxicology Program (NTP) is established, and responsibility for the rodent carcinogenesis bioassay and contracts to evaluate *in vitro* genetic toxicology tests is transferred from the NCI to the NTP.

1981   Publication of the first in a series of EPA Gene-Tox Program genetic toxicology test system reviews. Each review consists of an evaluation by an expert working group of a database of published literature for that test system together with recommendations for use of the test system. The Gene-Tox database currently contains mutagenicity information for about 3,000 chemicals from 39 assay systems.

1981   OECD Genetic Toxicology Guidelines for the Testing of Chemicals and the OECD Principles of Good Laboratory Practice (GLP) are published. The approximately 30 industrialized member countries of the OECD agree that data generated in the testing of chemicals in an OECD member country in accordance with OECD test guidelines and OECD GLP principles shall be accepted in other member countries for the purposes of assessment and other uses relating to the protection of humans and the environment.

1984   The National Research Council publishes a report for the NTP entitled *Toxicity Testing. Strategies to Determine Needs and Priorities* [45], which concludes that of the universe of over 5,000,000 chemicals, 65,725 substances are of possible concern because of their potential for human exposure and that, for the majority, a lack of sufficient data exists for conducting human health hazard assessment.

1987   Tennant et al. [19] publish an assessment of NTP test results obtained in four short-term *in vitro* genetic toxicology tests—the *Salmonella* and mouse lymphoma mutagenesis assays and tests for chromosomal aberrations and sister chromatid exchanges—which finds their tests to be poorly predictive of rodent carcinogenicity. This finding resounded throughout the scientific community and led to an immediate decline in usage for some of the tests and a reassessment of testing methods for others.

1989   Ames [20] reports that a wide variety of edible plants have high levels of toxins and carcinogens, that these natural toxins may have evolved to protect the plants from insects and other predators, and that humans are consuming 10,000 times more natural than man-made pesticides.

1993   FISH techniques are developed [9] for detecting specific chromosomes, chromosomal alterations, and specific genes. Currently, each of the human chromosomes and most of the mouse chromosomes can be individually identified.

1997   A multilaboratory comparison of *in vitro* tests for chromosome aberrations [21] demonstrates that several chemicals, including carcinogens, that were previously positive in the mouse lymphoma assay but negative for chromosomal aberrations with the NTP protocols were positive for aberrations when a longer treatment time and different sampling times were used.

1997   Publication of the Gene-Tox review of data for over 600 chemicals tested in the L5178Y mouse lymphoma cell mutation assay, which finds >90% predictivity of rodent carcinogenicity for adequately tested chemicals [22]. In a separate publication [48], the cells used for this assay are shown to harbor mutant *p53*, the tumor suppressor gene found in >50% of human tumors.

1997   Revised TSCA test guidelines published, including four genetic toxicity testing guidelines adopted from the revised OECD guideline series: the bacterial reverse mutation test, the *in vitro* mammalian cell gene mutation test, the mammalian bone marrow chromosomal aberration test, and the mammalian erythrocyte micronucleus test.

1997   ICH of Technical Requirements for Registration of Pharmaceuticals for Human Use guideline, entitled "Genotoxicity: A standard battery for genotoxicity testing of pharmaceuticals" identifies a standard three test battery: I(1) a test for gene mutation in bacteria; (2) an *in vitro* test with cytogenetic evaluation of chromosomal damage with mammalian cells or an *in vitro* mouse lymphoma *tk* assay; and (3) an *in vivo* test for chromosomal damage using rodent hematopoietic cells, e.g., a test for chromosomal aberrations or a micronucleus test. The ICH also states that the *in vitro* tests should be completed before the first human exposure of candidate pharmaceuticals, and the standard three-test battery should be completed before the initiation of phase II clinical trials.

1998   Revised OECD Genetic Toxicology Guidelines for the Testing of Chemicals approved and published.

1998   The EPA's Office of Pollution Prevention and Toxics urges chemical companies to voluntarily provide physical and chemical data and to conduct five types of screening tests (mutagenicity, ecotoxicity, environmental fate, acute toxicity, and subchronic reproductive and developmental toxicology), which comprise the OECD's screening information data sets (SIDS) Program, for ~3,000 high-production volume chemicals classified as "existing" chemicals when TSCA became effective in 1977 and, hence, which did not require testing for a premanufacture notice (PMN). Concerns have been expressed that resources may be insufficient for completing this testing by the target date of 2005.

Because of the concerns that developed especially in the years after World War II that many chemical products of benefit to industrialized society might adversely affect human health and the environment [14,15], numerous regulations have been promulgated, from 1970 until the present, which include a mandate that industry assess the potential adverse genetic effects of their products. In the 1960s, several laboratories developed methods suitable for testing radiation and chemicals for mutagenicity, and by the end of the decade, the first *in vivo* tests——the heritable translocation, specific locus, and dominant lethal tests, the host mediated assay, and *in vivo* cytogenetics—had been defined [16].

Exponential growth of the field of genetic toxicology testing occurred during the 1970s, shortly after Malling [16] demonstrated, in 1971, that a mammalian liver homogenate could be added to *in vitro* systems to mimic *in vivo* metabolism. Then, a wide range of *in vitro* tests, as well as additional *in vivo* methods, to evaluate the genotoxic effects of chemicals was developed and published. The goal of using such tests to predict carcinogenicity gained momentum in the mid-1970s when McCann et al. published test results suggesting that most carcinogens were mutagens in *Salmonella* [17]. Support for the role of mutagenesis in carcinogenesis has also included the clonal origin of tumors, the association of specific chromosomal abnormalities with certain cancers, information about genetically determined cancer-prone conditions, and recent knowledge about the correlation of mutations in oncogenes and tumor suppressor genes, with key steps leading to tumor formation.

Although a number of chemicals are known to be human carcinogens, and although many of the same mutagenic processes are directly involved in inducing both heritable effects and carcinogenesis in animals, it is virtually impossible to obtain direct evidence for specific chemicals inducing inherited defects in humans. Thus, the *Salmonella* results led to a shift from justifying the genotoxicity testing of chemicals based on potential risk to future generations to primarily, if not exclusively, justifying the tests as screens for carcinogenicity. Consequently, to date, positive mutagenicity results in the absence of proven carcinogenicity have been considered insufficient justification for regulating chemicals already on the market [18].

When positive mutagenicity results are obtained for a new product, however, regulatory agencies may request additional information. The product can often be marketed if further testing yields a negative outcome for genomic mutations or carcinogenicity or, even then, if the product is a drug for a life-threatening condition with no alternative therapies. In practice, however, because of the time and expense involved in this additional testing, coupled with the uncertain outcome of the additional test results, the development of new pharmaceuticals and industrial chemicals is often discontinued if the initial mutagenicity test results are positive [18].

The rapid expansion of the field of genetic toxicology came to an abrupt halt after the publication, in 1987, of an assessment [19] of NTP test results for 73 chemicals evaluated in four short-term *in vitro* genetic toxicology tests in which the NTP found the tests to be only about 60% accurate in predicting rodent carcinogenicity. In particular, only about 50% of the carcinogens were found positive in the assays for reverse mutations in *Salmonella* and chromosomal aberrations in mammalian cells, and, although positive predictivity (the percentage of chemicals yielding positive genetic toxicology results that were rodent carcinogens) was 86% in *Salmonella*, positive predictivity was 73% for the aberration assay, 67% for the SCE assay, and only 50% for the mouse lymphoma cell mutation assay. Thus, the *Salmonella* and chromosomal aberration assays appeared to be poorly predictive of rodent carcinogenicity, and the SCE assay and, particularly, the mouse lymphoma assay appeared to yield an unacceptably high number of "false-positive" results.

Then, less than 2 years later, Ames tested a wide variety of edible plants in the *Salmonella* assay and reported that most, if not all, contained mutagens evolved as natural toxins to protect the plants from insects and other predators and, further, that although even higher levels of some mutagens may be found in the workplace, humans consume 10,000 times more natural than man-made pesticides. In addition, Ames noted that the risk of carcinogenesis may be negligible at the levels of chemicals to which humans are usually exposed, which are far below the maximum tolerated doses used for the rodent carcinogenesis bioassay. He then suggested that basic, instead of applied, research should be emphasized to minimize cancer and other degenerative diseases of aging. Thus, the NTP findings, followed shortly by Ames' assertion [20] that concerns with the risk of exposing humans to mutagens and carcinogens were grossly magnified, resounded throughout the scientific community. This response led to a pronounced lack of enthusiasm and subsequent government funding for developing and evaluating additional tests, an immediate decline in the use of the mouse lymphoma assay, in particular, and, essentially, a discontinuation of the SCE assay, the latter because direct mechanistic relevance of the SCE test has yet to be established.

Although only one of the four *in vitro* tests evaluated by the NTP—the measurement of reverse mutations in *Salmonella*—is currently used by the NTP, in the most recent decade, independent reassessments of the chromosomal aberration and mouse lymphoma assays have been published [21–23] clearly illustrating multiple deficiencies in the testing protocols used by the NTP and justifying the current recommendation and use of these tests for regulatory submissions. Further, recent dramatic research advances in defining the molecular basis of genetic alterations have led to an enhanced understanding of the theoretical basis of genotoxicity test systems, which will be discussed in the following section. At the same time, as noted above, recent concerted efforts have tried to identify the most reliable genetic toxicology tests and to harmonize testing methodology and agency requirements.

Thus, although the tests used for regulatory submissions through most of the 1980s were often based as much on theoretical considerations as on the actual performance of the tests, with increasing use and evaluation of the test systems, only a few have gained universal acceptance and are used today for initial assessments for regulatory agencies, with the selection of protocols usually based on the experience of industry and current regulatory guidance. These few tests, which will be described in greater detail with the reasons that they are used, include bacterial reverse mutation, the mouse lymphoma assay, *in vitro* chromosomal aberrations, and the chromosomal aberration and micronucleus tests in rodent bone marrow cells.

Recent molecular biology advances have led to an expanded awareness of the importance of genotype in susceptibility to cancer, and they promise to yield more reasoned and informed assessments of risk to present and future generations. However, public concerns with the potential hazards of chemicals have not abated, and, from a regulatory perspective, perhaps the most significant recent change at the time of this writing is that the pendulum appears to be again swinging toward an increased volume, if not variety, of genetic toxicology testing. The dramatic growth of the biotechnology industry during the past decade is now yielding numerous potentially useful pharmaceutical products that, under ICH guidelines, must be tested for genotoxicity before they are administered to human volunteers. In addition, at least 3,000 high-production volume chemical products were in existence when TSCA became effective in 1977, and hence, they did not require testing for a PMN. The OECD and the EPA are now urging that, consistent with industry's product stewardship programs, chemical companies voluntarily test these chemicals for several endpoints, including mutagenicity, in the OECD's SIDS program. The EPA has also announced plans to issue a "test rule" to be finalized by the end of the century that will require chemical-specific testing of existing chemicals if it has not been done on a voluntary basis and to require testing beyond the SIDS level for almost 500 high-production volume chemicals used in consumer products [46].

6.3 HARMONIZED GUIDANCE FOR REGULATORY SUBMISSIONS

Under the auspices of the OECD, extensive international efforts have been directed toward defining protocols for the genetic toxicology tests that have been used for product registration. Agreement has been reached that the results obtained with an OECD-defined protocol in one country will be accepted internationally. Thus, the expectation exists that testing for the registration of chemical products and pharmaceuticals will be conducted according to the OECD guidelines and that any deviations from these guidelines will be justified.

Until recently, however, the selection of specific protocols for regulatory submissions has been a separate issue, with different groups of test systems used by chemical and pharmaceutical companies, and within different geographical regions. Thus, the harmonization of regulatory requirements has been considered necessary to facilitate more efficient and economic testing strategies for companies that market their products worldwide. To address this need, harmonized testing guidelines have now been published by the OECD for high-production volume chemicals, by the EPA for chemicals tested under TSCA, and, for pharmaceuticals, by the ICH. These guidelines, which are summarized in Table 6.3, are discussed below.

TABLE 6.3. Harmonized testing guidelines for regulatory submissions

Type of Test	OECD/SIDS	TSCA	ICH
Bacterial reverse mutation test	4	4	4
Nonbacterial *in vitro* test			
Mammalian cell mutation	4	4	4[a]
	or		or
Chromosomal aberration	4		4
In vivo test			
Chromosomal aberration	4	4	4
	or	or	or
Micronucleus	4	4	4

[a]Only the L5178Y mouse lymphoma gene and chromosomal mutation assay are recommended in the ICH guidelines.

6.3.1 Organization for Economic Cooperation and Development Screening Information Data Set Guidelines

The OECD SIDS voluntary testing program for international high-production volume chemicals began in 1989 and includes obtaining six types of data: physical/chemical, mutagenicity, ecotoxicity, environmental fate, acute toxicity, and subchronic reproductive and developmental toxicology. The mutagenicity tests identified by OECD SIDS include a test for gene mutations in bacteria, a nonbacterial *in vitro* test (e.g., a mammalian cell gene mutation assay or a chromosomal aberration assay), and an assessment of genetic toxicity *in vivo* (e.g., a micronucleus or chromosomal aberration test). These tests are most frequently identified for testing chemicals in the guidance provided by regulatory bodies in the United States and other nations; however, in some countries, fewer tests may be identified for chemicals with a lower level of concern, which is based on production volume and use. Then, only one or two of the tests, i.e., a test for gene mutations in bacteria and possibly an *in vitro* chromosomal aberration assay, may be required.

The SIDS program is currently focused on developing base-level test information on approximately 600 poorly characterized international high-production volume chemicals, but only about 80 of the chemicals have been assessed to date. In addition, the EPA has estimated that currently ~3,000 high-production volume chemicals were classified as "existing" chemicals when TSCA became effective in

1977, and hence, they did not require testing for a PMN and are currently produced or imported in amounts of over 1 million pounds per year. At the time of this writing, the U.S. government is preparing to encourage OECD countries to speed testing by chemical companies, and the EPA is also planning to issue a "test rule" that will require chemical-specific testing. However, the EPA and the Chemical Manufacturers Association (CMA) are currently meeting to address concerns that resources may be insufficient for completing this testing within the first decade of the 21st century [46].

6.3.2 Toxic Substances Control Act Guidelines

Historically, because of mandates under TSCA and FIFRA applying to domestically produced and imported chemicals, the EPA has often taken the lead in providing guidance for the registration of chemicals for companies planning to market their products worldwide. In a series of TSCA test guidelines established in 1985, two TSCA Section 4 mutagenicity testing schemes were followed, one for gene mutations and one for chromosomal mutations, but by the end of the decade, it had been proposed that they be combined into one revised mutagenicity testing scheme [47]. In 1997, the EPA Office of Pollution Prevention and Toxics (OPPT) published a Final Rule for TSCA guidelines, including mutagenicity, to harmonize the testing guidelines developed by the Office of Pesticide Programs (OPP), the earlier TSCA guidelines, and the OECD guidelines. The revised guidelines, which incorporate the OECD protocols, establish the following four initial genetic toxicology approaches: a bacterial reverse mutation test, an *in vitro* mammalian cell mutation test, and either a mammalian bone marrow chromosomal aberration test or a mammalian erythrocyte micronucleus test.

By examination of Table 6.3, it may be noted that the TSCA guidelines differ from the OECD SIDS guidelines in that an *in vitro* chromosomal aberration test is not identified as an alternative for the *in vitro* mammalian cell mutation test. Although the EPA's harmonized guidelines may be modified at some future date to accept *in vitro* chromosomal aberration test results as an alternative to the *in vitro* mammalian cell mutation test, this decision may be delayed until the EPA's Gene-Tox review of *in vitro* and *in vivo* chromosomal aberration assays (which is currently in progress) has been completed. It should additionally be noted that although the OECD guidelines, and hence the TSCA guidelines, for the *in vitro* mammalian cell gene mutation test indicate that L5178Y mouse lymphoma cells; a Chinese hamster ovary (CHO), AS52, or V79 line of Chinese hamster cells; or TK6 human lymphoblastoid cells may be used, the mouse lymphoma assay is the only one of the identified tests that has been extensively defined and evaluated and shown to be mechanistically relevant, i.e., to detect a full range of chromosomal mutations as well as gene mutations.

6.3.3 International Council on Harmonization Guidelines

Historically, in the United States, the FDA has identified tests that might be appropriate for the registration of pharmaceuticals, food additives, biomedical materials, and some veterinary products, e.g., in Red Books I and II. However, it was the responsibility of the pharmaceutical industry to select the tests used, and then to conduct additional testing if the initial tests were not sufficiently informative. Thus, often based on the prior experience of specific companies, new products could be evaluated in five or more genotoxicity test systems, including tests for bacterial reverse mutation, mammalian cell mutagenesis, *in vitro* or *in vivo* cytogenetics, DNA damage and repair, and mammalian cell transformation (a change from normal to abnormal growth characteristics of cultured cells). However, at the same time, initially, only a bacterial reverse mutation test, and later a bacterial reverse mutation test plus *in vitro* cytogenetics, were required for some European submissions, and only a bacterial reverse mutation test plus *in vitro* cytogenetic testing, but with different protocols, were required in Japan.

This situation presented obvious problems for pharmaceutical companies wishing to market their products internationally. Hence, after a series of formal deliberations by the six-member ICH Expert Working Group—consisting of one representative each from the appropriate regulatory agency and from the pharmaceutical manufacturers from the United States, the European Union (EU), and Japan—a consensus was reached that pharmaceuticals should be evaluated for genotoxicity in three tests: a test for gene mutations in bacteria, an *in vitro* test with cytogenetic evaluation of chromosomal damage in mammalian cells or an *in vitro* mouse lymphoma *tk* assay, and an *in vivo* test for chromosomal damage (either chromosomal aberrations or micronuclei) using rodent hematopoietic cells.

The ICH Working Group considered the *in vitro* cytogenetic evaluation and the mouse lymphoma assay to be interchangeable because of recent evidence that they yield congruent results and because both detect genetic effects that cannot be measured in bacteria. An *in vivo* test was considered necessary to provide a test model that included factors such as absorption, distribution, metabolism, and excretion that could influence genotoxicity. The ICH Expert Working Group stated that for compounds giving negative results, the completion of this three-test battery, performed in accordance with current (OECD and ICH) recommendations, will usually provide a sufficient level of safety to demonstrate the absence of genotoxic activity, but compounds giving positive results in the standard battery may, depending on their therapeutic use, need to be tested more extensively. The latter is interpreted as requiring testing in additional genotoxicity test systems to resolve incongruent initial results or testing for carcinogenicity in rodents.

As may be noted in Table 6.3, the first and third of the three approaches identified for testing pharmaceutical products are the same as those defined by OECD SIDS and TSCA for chemicals; the second approach is the same for the three sets of guidelines only if the L5178Y mouse lymphoma cell mutagenesis assay is used for testing chemical products as well as for pharmaceuticals.

6.4   GENETIC TOXICOLOGY TESTS FOR CURRENT PRODUCT REGISTRATION

Current OECD SIDS and ICH regulatory guidance has identified the following five basic genetic toxicology approaches: (1) a test for bacterial reverse gene mutations, (2) the L5178Y mouse lymphoma cell assay for gene and chromosomal mutations, (3) an *in vitro* chromosomal aberration test, and either (4) an *in vivo* chromosomal aberration test or (5) an *in vivo* micronucleus test, both for chromosomal mutations. TSCA regulatory guidance has identified all but the *in vitro* chromosomal aberration test. The five identified tests will be described, based on the current OECD guidelines and the author's experience, and then the advantages and limitations of these tests will be summarized.

To establish that a chemical is negative, *in vitro* tests must be conducted in the absence and presence of exogenous metabolic activation, positive and negative controls must be within historical ranges for the testing laboratory, and testing must be conducted to a level at which toxicity or precipitation is observed or to a level for which higher concentrations would not yield biologically relevant results. The latter two requirements are also applicable for *in vivo* tests. In addition, the animals must be maintained under conditions minimizing the influence of environmental variables, and bioavailability must be considered when selecting the route of administration. It should also be noted that, although some prior testing guidelines specified that test results should be reproducible in independent experiments, under current guidelines, no requirement exists for repeating an appropriately conducted test yielding clearly positive results. Appropriately conducted *in vivo* tests, e.g., tests in which dosing is by the appropriate route and to a sufficiently high-dose level, yielding negative results are seldom, if ever, repeated. However, for *in vitro* tests, repeat testing with an alternate metabolic activation system or different exposure conditions may be required if negative results are obtained, and ICH guidance for *in vitro* cytogenetics tests specifies that, in addition, a later harvest time should be used. Repeat testing is clearly required if the initial results are equivocal, although, replication for its own sake (i.e., an exact repeat of the first test) is usually not informative; instead, the repeat test should be designed to correct any deficiencies of the first.

Finally, it is often helpful to evaluate the results in the different test systems for redundancy. For example, chemicals yielding positive results in a bacterial reverse mutation test often yield positive results in the mouse lymphoma assay, and chemicals yielding positive results in an *in vivo* test for chromosomal mutations usually yield positive results in an *in vitro* test for chromosomal mutations (either an aberration assay or the mouse lymphoma assay).

6.4.1   Reverse Gene Mutations in Bacteria

The most extensive testing for gene mutations is in bacteria, particularly using reverse mutation in *Salmonella typhimurium* and *Escherichia coli*. Therefore, bacterial reverse mutation assays are considered by many researchers to be the cornerstone of genetic toxicology testing. Bacterial tests present the advantages of relative ease of performance, economy, efficiency, and the ability to identify specific DNA damage that is induced, e.g., frameshift or base-pair substitution mutations. Bacterial tests can also provide information on the mode of action of the test chemical, because the bacterial strains used vary in their responsiveness to different chemical classes. Many of the tester strains have features making them more sensitive for the detection of mutations, including responsive DNA sequences at the reversion sites, increased cell permeability to large molecules, and the elimination of DNA repair systems or the enhancement of error-prone DNA repair processes.

The *S. typhimurium* strains routinely used were designed for sensitivity in detecting gene mutations reverting the bacteria to histidine independence, and the *Salmonella* strains are histidine auxotrophs by virtue of mutations in the histidine operon. The *E. coli* WP2 *uvrA* strains recommended for initial tests are tryptophan auxotrophs by virtue of a base-pair substitution mutation in the tryptophan operon. When these histidine- or tryptophan-dependent cells are grown on minimal medium agar plates containing a trace of histidine or tryptophan, only those cells reverting to histidine or tryptophan independence are able to form colonies. The small amounts of these amino acids allow all plated bacteria to undergo a few divisions; in many cases, this growth is essential for mutagenesis to occur, and the revertants are easily visible as colonies against a slight lawn of background growth.

In addition to the histidine and tryptophan operons, most of the indicator strains carry a deletion covering genes involved in the synthesis of the vitamin biotin (*bio*), and all carry the *rfa* mutation leading to a defective lipopolysaccharide coat and making the strains more permeable to many large molecules. The strains also carry the *uvrB* mutation, which results in impaired repair of ultraviolet (UV)-induced DNA damage and renders the bacteria unable to use accurate excision repair to remove certain chemically or physically damaged DNA, thereby enhancing the strains' sensitivity to some mutagenic agents. In addition, the more recently developed strains contain the resistance transfer factor, plasmid pKM101, which is believed to cause an additional increase in error-prone DNA repair, leading to even greater sensitivity to most mutagens.

In bacterial reverse mutation testing, usually one strain is used for a preliminary concentration range-finding assay, and then mutagenesis assays are conducted with five strains. The strains recommended by the OECD guidelines are:

1. *S. typhimurium* TA1535
2. *S. typhimurium* TA1537 or TA97 or TA97a
3. *S. typhimurium* TA98
4. *S. typhimurium* TA100
5. *E. coli* WP2 *uvr*A or *E. coli* WP2 *uvr*A (pKM101) or *S. typhimurium* TA102

Salmonella strains TA1535 and TA100 are reverted to histidine independence by base-pair mutagens; strains TA1537, TA97, TA97a, and TA98 are reverted by frameshift mutagens; and all six strains have guanine cystine (GC) base-pairs at the primary reversion site. *Salmonella* strain TA102 and the *E. coli* strains have an adenosine tyrosine (AT) base-pair at the primary reversion sites, and they detect a variety of oxidants, cross-linking agents, and hydrazines not detected by the other strains.

Basic microbial mutagenesis protocols use the plate incorporation approach or a preincubation modification of this approach. In the standard plate incorporation protocol, the test material, bacteria, and either a metabolic activation mixture [9,000-g postmitochondrial supernatant (S9)] or a buffer are added to liquid top agar in a disposable glass tube, which is held at 45°C in a heating block while the components are added, then mixed, and the mixture is immediately poured on a plate of bottom agar. After the agar gels, the bacteria are incubated, at 37°C, for 48–72 hours; then, the resulting colonies are counted. In a typical mutagenesis assay with and without metabolic activation, each of the five strains of bacteria is exposed to the solvent control (with six cultures per strain and activation condition), and to five concentrations of the test chemical (with three cultures per concentration, strain, and activation condition), and to the appropriate positive controls for that strain (with three cultures per activation condition). This process yields a total of 240 bacterial plates and additional plates used to check the sterility of the components.

The preincubation modification is used for materials that may be poorly detected in the plate incorporation assay, including short chain aliphatic nitrosamines, divalent metals, aldehydes, azo-dyes and diazo compounds, pyrollizidine alkaloids, alkyl compounds, and nitro compounds. In this protocol, the test material, bacteria, and S9 mixture (when used) are incubated for 20–30 minutes at 37°C before top agar is added, mixed, and the mixture is poured on a plate of bottom agar. Increased activity with preincubation in comparison to plate incorporation is attributed to the fact that the test chemical, bacteria, and S9 are incubated at higher concentrations (without agar present) than used in the standard plate incorporation test. Other modifications of the assays provide for exposing the bacteria to measured concentrations of gases in closed containers or the use of metabolic activation systems from a variety of species.

For an acceptable assay, the test chemical should be tested to a toxic level, as evidenced by a reduction in colonies or a reduced background lawn, to a level at which precipitated test material precludes visualization of the colonies, or to 5 mg or 5 µl/plate, whichever is lower. To evaluate a result as positive requires a concentration-related or a reproducible increase in the number of revertant colonies per plate for at least one strain with or without activation.

6.4.2   The L5178Y/Thymidine Kinase$^{+/-}$ Gene and Chromosomal Mutation Assay

The L5178Y mouse lymphoma assay measures gene and chromosomal forward mutations at the *tk* locus, *tk*$^{+/-}$ —> *tk*$^{-/-}$ and as indicated in the recent EPA Gene-Tox review of published results for over 600 chemicals tested in this assay [22], when used with appropriate protocols and evaluation criteria, the mouse lymphoma assay yields results at least 95% concordant with the outcome of the rodent carcinogenesis bioassay. The Gene-Tox review, which contains a detailed description of the assay, was published in late 1997, after the most recent OECD guidelines were adopted. Because of the order of publication, the review could not be cited in the guidelines; therefore, for additional information about this assay, the OECD guidelines should be supplemented with information from the Gene-Tox review.

More recently, the L5178Y mouse lymphoma cells were found to harbor gene mutations in *p53* [48], which, in the mouse, is found on the same chromosome as *tk*. The *p53* tumor suppressor gene is considered to be the "guardian of the genome" because its function is to delay the cell-cycle progression of cells that have acquired chromosomal mutations until the damage has been repaired. Thus, the presence of mutant *p53* in the mouse lymphoma cells renders the assay more similar to mutation assays in repair-deficient bacteria. In addition, this finding is not only consistent with the sensitivity of this assay for detecting chromosomal mutations, but it enhances the relevance of the assay for predicting carcinogenicity, as mutant *p53* is found in over 50% of human tumors.

L5178Y mouse lymphoma cells, specifically, clone 3.7.2C, grow in suspension culture with a relatively short cell generation time, 9–10 hours. A few days before use in an assay, a culture is "cleansed" of preexisting spontaneous *tk*$^{-/-}$ mutants by growing the cells for about 24 hours in medium containing methotrexate. After the cells have recovered from cleansing, they are exposed to a series of concentrations of the test chemical, usually for 4 hours, in the absence and presence of metabolic activation. Testing under nonphysiologic conditions must be avoided in this assay and in other *in vitro* mammalian cell assays, as acidic pH shifts, to ≤6.5, and high salt concentrations have been shown to produce physiologically irrelevant positive results. Conversely, if the pH of the medium used to culture the cells is ≥7.5, cell growth in suspension culture may be depressed, and small colony mutants, in particular, may not be detected [22].

Because chromosomal mutations are usually associated with slower growth rates and because the induction of both gene and chromosomal mutations are associated with cytotoxicity, a chemical cannot be considered to be nongenotoxic in this assay unless testing is performed to concentrations producing significant cytotoxicity, e.g., 10–20% relative total growth (RTG) = cloning efficiency x relative suspension growth (RSG). On the other hand, responses observed only at extreme cytoxicity (<10% RTG) are considered to be biologically irrelevant. Therefore, exposure concentrations for each assay are selected, based on the results of a preliminary range-finding experiment, to span a range of anticipated survival from nontoxic or weakly toxic to 10–20% RTG, with the concentrations selected to emphasize the lower RTG values. For relatively noncytotoxic chemicals, the maximum concentration should be 5 μl/ml, 5 mg/ml, or 10 mM, or a concentration evidencing insolubility, whichever is lower. However, scientific judgment should be exercised in evaluating results appearing to be positive only in the presence of a precipitate. In this case, the test chemical may be present in suspension culture during the 2-day expression period, thus, lengthening the exposure time and leading to erroneous cell counts if cell density values are obtained with a cell counter instead of with a hemacytometer. In addition, erroneous colony counts can be obtained if an automated colony counter is used to enumerate the colonies.

After chemical exposure, the cells are rinsed and then grown in suspension culture for a 2-day expression period, with the cells in each culture diluted as necessary to maintain a rapid growth rate; then, the cells are cloned to measure mutagenesis and survival. Two methods are currently used for cloning, and for both approaches, the selective agent (TFT) is present in the cloning medium for the mutant colonies. In the original approach (which was used for virtually all test results included in the Gene-Tox review of this assay, and which is used by most, if not all, U.S. laboratories), the cells are cloned in culture dishes in a medium containing sufficient soft agar to immobilize the cells. The second approach, cloning the cells without agar in microwell plates [49], was developed in the United Kingdom by a laboratory that had difficulty in obtaining small colony mutants with the original approach, and it has been used by Japanese laboratories that recently gained experience with the mouse lymphoma assay [50]. A number of laboratories experienced with both approaches obtain equivalent results with either, and they find the soft agar cloning approach to be less labor-intensive and less subjective.

After cloning, an approximately 10–12-day incubation period is used for the microwell method, and a 2-week incubation period is required for the soft-agar approach before the colonies are counted and sized. With shorter times, the σ mutant colonies may not grow to a sufficient size to be visualized; however, even with only a 10-day incubation period, it is sometimes difficult to determine whether only one or more than one large colony mutant is in a microwell. Because most of the colonies enumerated to obtain cloning efficiency are large colonies, this difficulty can lead to lower cloning efficiency values, which, in turn, can lead to an artificially increased spontaneous mutant frequency or an incorrect conclusion of a positive mutagenic response. Colony counting and sizing are accomplished using colony counters that can discriminate among the size ranges of the objects (colonies) that are counted, and both σ and λ colony mutant frequencies, as well as total mutant frequencies, are reported for positive and negative controls and for treated cultures yielding a positive response. However, many of the older colony counters do not have sufficient resolution to detect all of the small mutant colonies that may be present, a problem circumvented by counting the colonies by hand.

For acceptable assays, absolute cloning efficiencies for the solvent controls are expected to be between 80 and 120%; however, lower and higher cloning efficiencies may be acceptable in individual experiments, especially if a test chemical is unambiguously positive. Solvent control mutant frequencies should be consistent with the testing laboratory's historical ranges and, generally, in a range from $\geq 20 \times 10^{-6}$ to $\leq 120 \times 10^{-6}$. As discussed in the Gene-Tox review [22], no statistical method currently exists that is considered to be appropriate for this assay. In fact, in a number of published mouse lymphoma assay results, e.g., for the NTP, chemicals were erroneously called positive because of inappropriate reliance on a statistical analysis system [23]. Therefore, the Gene-Tox evaluation criteria are recommended for evaluating the results, with the caveat that these criteria were developed for the agar cloning approach and may require modification if results are obtained with the microwell approach. (These criteria are summarized in Table 6.4.)

TABLE 6.4. Gene-Tox evaluation criteria for the mouse lymphoma assay[a]

++	Definitive positive response with evidence of a dose response and an induced (treated minus spontaneous) mutant frequency of at least 100×10^{-6} at RTG $\geq 20\%$.
+	Limited positive response with evidence of a dose response and an induced mutant frequency of at least 70×10^{-6} at a RTG $\geq 10\%$.
-	Negative response for which toxicity is evidenced by a RTG of 10–20%, and the positive control mutant frequency demonstrates that no inherent problems exist with the assay.
- -	Negative response with no toxicity, and the positive control mutant frequency demonstrates that no inherent problems exist with the assay.
I	Inconclusive. Insufficient concentrations tested over the critical range of 10–20% RTG to be able to evaluate the result as clearly positive or negative. The portion(s) of the assay yielding an inconclusive result must be repeated.
E	Equivocal. A negative or inconclusive result is obtained in one mutagenesis assay, and a positive result is obtained another, and no apparent reason exists to give greater weight to either result. In this case, the testing must be repeated a third time. However, some chemicals yield equivocal results regardless of the number of times an assay is repeated. (Repeat testing is not required if the initial result is clearly positive.)
NT	Not testable. Applied to chemicals that cannot be tested to sufficiently high concentration to obtain a conclusive result in the mouse lymphoma assay because of limited solubility, acidic pH shifts, increased osmolality, or the test material's physical properties, such as dissolving plastic, and this limitation cannot be overcome by use of an alternate solvent or by adjusting the exposure conditions to ensure that they are physiologically relevant.

[a]From ref. 22.

6.4.3 Chromosomal Aberrations: *In Vitro*

In vitro chromosomal aberration assays for regulatory submissions are routinely conducted in the absence and presence of exogenous metabolic activation and may use a variety of established cell lines, cell strains, or primary cell cultures, selected on the basis of factors, such as growth ability in culture, stability of the karyotype, chromosome number, chromosome diversity, and spontaneous frequency of chromosome aberrations. The most frequently used cells are Chinese hamster fibroblasts (either CHO or Chinese hamster lung (CHL) cells) or human or rat lymphocytes stimulated to divide synchronously *in vitro*. Established cell cultures present the advantages of minimal variability among experiments and, for cells growing as monolayers, the option of using mitotic shakeoff procedures to obtain repeated samples of cells in metaphase. The cells are propagated from stock cultures and seeded in a culture medium at a density such that the cultures will not reach confluency before the time of harvest, which should be ~1.5 hours after the addition of a mitotic spindle inhibitor (Colcemid or colchicine). Although donor variability in mitogenic is often encountered, human lymphocyte cultures are used because of their perceived relevance to the human condition, in addition to presenting the advantage of synchronous cell division. Variability in mitogenic response can be minimized by using cultures of rat lymphocytes. The lymphocytes are usually obtained from whole blood from healthy (human or rodent) subjects, treated with heparin (an anticoagulant), and stimulated to divide with a mitogen (e.g., phytohemagglutinin). After a prolonged G1 stage, the cells enter S-phase, which should be the time of addition of the test chemical to the cells. For a typical protocol, chemical exposure is initiated at 48 hours, a mitotic spindle inhibitor is added at 70.5 hours, and cells in metaphase are harvested at 72 hours.

After the cells are harvested, they are treated with a hypotonic solution to swell the chromosomes, fixed, and dropped onto prelabeled slides. Then, the chromosomes are stained and coverslips are attached to permit microscopic analysis with oil immersion (100X) objectives. At least three concentrations are analyzed that are selected based on a preliminary evaluation of uncoded slides. All slides to be analyzed, including positive and negative controls, should be coded by an individual not involved in the analysis, and if the slides are to be evaluated by more than one cytogeneticist, the selected concentrations and controls should be divided into equivalent sets before coding. At least 200 well-spread metaphases should be scored per concentration and controls; however, this number can be reduced when a high number of aberrations is observed.

Among the criteria to be considered when determining the highest concentration to be tested for chromosomal aberrations are cytotoxicity, solubility, and changes in osmolality or pH, to ensure that exposure conditions will be in a physiologically relevant range (as was described for the mouse lymphoma assay). These criteria are assessed in preliminary concentration, range-finding assays, with mitotic index determinations or assessments of cell growth (e.g., generation times) in culture used to indicate cytotoxicity and cell-cycle delay. As a general rule-of-thumb, to ensure that a sufficient number of mitotic cells for analyses exist, mitotic indices need not be depressed more than 50%. For relatively noncytotoxic chemicals, the maximum concentration should be 5 µl/ml, 5 mg/ml, or 10 mM, whichever is lowest. For relatively insoluble chemicals, OECD guidelines advise testing one or more than one concentration in the insoluble range as long as a precipitate does not interfere with the analysis.

OECD guidelines recommend that the cells should be exposed to the test chemical both with and without metabolic activation for 3–6 hours and sampled at a time equivalent to about 1.5 times the normal cell cycle after the beginning of treatment. If the result is unambiguously positive, no further testing is needed. However, if the chemical yields negative results with and without activation, a second experiment without metabolic activation is recommended with continuous treatment until a sampling time equivalent to about 1.5 times the normal cell cycle after the beginning of treatment. The ICH guidelines go one step further and specify that if the second experiment is also negative, a third experiment without metabolic activation is needed, with a continuous 24-hour treatment time, because only with this approach were some carcinogens that had previously been evaluated as negative for aberrations by the NTP detected as positive in this assay [21].

6.4.4   Rodent Bone Marrow Chromosomal Effects

In vivo tests for chromosomal effects for regulatory submissions consist of the *in vivo* chromosomal aberration test and the micronucleus test. They are routinely conducted using bone marrow cells from rodents, e.g., mice or rats, because the bone marrow is highly vascularized and contains a population of rapidly dividing cells that can be readily isolated and processed. However, the micronucleus test can also be conducted with sampling of cells from the peripheral blood that were exposed to the test chemical while in the bone marrow. Because the target cells are exposed to the products of *in vivo* metabolism under physiologic conditions, these tests are particularly useful if positive results have been obtained with an *in vitro* test for chromosomal effects. However, they are not appropriate tests if evidence exists that the test substance, or a reactive metabolite will not reach the bone marrow. In this case, other tests may be required.

Healthy, young adult animals are acclimated before testing and, when weighed before dosing, the weight variation of the animals should be minimal and not exceed ±20% per sex. The temperature in the room in which the animals are housed should be 22°C (±3°C), and humidity should be at least 30% and preferably not exceed 70%. Artificial lighting is used with 12 hours light and 12 hours dark. If group housing is used (for each sex), the number of animals per cage should be consistent with the appropriate guidelines, and the animals should be identified individually before they are weighed and dosed.

Dosing is routinely accomplished by intraperitoneal injection or gavage, but subcutaneous and inhalation exposures are also used, if justified. The maximum dose level required for relatively nontoxic chemicals is 2,000-mg/kg body weight, and the maximum volume administered by injection or gavage should not exceed 2-ml/100-g body weight. If available pharmacokinetic or toxicity data demonstrate that no substantial differences exist between sexes, testing in a single sex, preferably males, is sufficient. However, if human exposure may be sex-specific, the test should be performed in animals of the appropriate sex. Preliminary dose range-finding tests for toxicity are necessary, sometimes using an evaluation of mitotic indices for the chromosomal aberration test or enumerating PCE ratios for the micronucleus test, to select a high dose in which at least five analyzable animals per sex are together, with a sufficient number of cells for analysis. Experience has shown that this dose is often significantly lower than is the LD$_{50}$ defined in acute toxicology testing, and it is often prudent to use extra animals for toxic chemicals, particularly for the highest dose group.

Positive and negative (solvent/vehicle) controls are used for each sex and, except for treatment, the positive and negative control animals should be handled in the same way as the treated animals. The positive control may be administered by a different route than is the one used for the test chemical, and only one sampling time is required. The negative control is administered by the same route as the test chemical, and negative controls are used for each sampling time, unless it can be demonstrated by the testing laboratory's historical control data that minimal interanimal variability is obtained for negative control animals irrespective of the sampling time.

6.4.5   Chromosomal Aberrations: *In Vivo*

For the chromosomal aberration test, the highest dose is defined as 2,000-mg/kg body weight or the dose-producing signs of toxicity, such that higher dose levels, based on the same dosing regimen, would be expected to produce lethality. The highest dose may also be defined as a dose that produces some indication of toxicity in the bone marrow (e.g., a greater than 50% reduction in the mitotic index). Chemicals are preferably administered as a single treatment, and samples are taken at two separate times after treatment. The first sampling interval is 1.5 times the normal cell-cycle length; therefore, the first samples are obtained 12–18 hours after treatment, and a minimum of three dose levels plus the appropriate controls are required. A second sampling time, 24 hours after the first, is recommended to allow for the time required for uptake and metabolism of the test chemical as well as its effect on cell-cycle kinetics; however, only the highest dose is used for the second time.

The animals are injected intraperitoneally, approximately 1.5–2 hours before the sampling time, with an appropriate dose of Colcemid or colchicine. Then, the animals are sacrificed, cells are removed from the bone marrow, treated with a hypotonic solution, and fixed; then, slides are prepared, stained, given coverslips, and coded before analysis, as described for the *in vitro* chromosomal aberration test. Mitotic indices are obtained based on at least 1,000 cells per animal, and at least 100 cells per animal should be analyzed for chromosomal aberrations, unless a high number of aberrations is observed. Criteria for a positive response include a dose-related increase in the number of cells with chromosomal aberrations or a clear increase in the number of cells with aberrations in a single dose group at a single sampling time.

6.4.6 Micronucleus Tests.

For micronucleus tests, the highest dose is defined as 2,000-mg/kg body weight or the dose-producing signs of toxicity, such that higher dose levels, based on the same dosing regimen, would be expected to produce lethality. The highest dose may also be defined as a dose producing some indication of toxicity in the bone marrow (e.g., a reduction in the percentage of PCEs in the bone marrow). The test may be performed in two ways. In the first way, animals are treated with the test chemical once, and samples of bone marrow cells are obtained at least twice, with the first samples obtained no earlier than 24 hours after treatment and the last samples no later than 48 hours after treatment. Samples of peripheral blood are obtained at least twice, with the first samples obtained no earlier than hours after treatment and the last samples no later than 72 hours after treatment. Three dose levels, plus negative and positive controls are required for the first sampling time, but only the highest dose may be required for the second sampling time. In the second approach, the animals are treated on each of 2 or more consecutive days to achieve steady-state kinetics, and samples are obtained once, between 18 and 24 hours after the final treatment for the bone marrow, or between 36 and 48 hours after treatment for the peripheral blood.

Bone marrow cells are usually obtained from the femurs or tibias immediately after sacrifice, e.g., by CO_2 asphyxia followed by cervical dislocation; peripheral blood is routinely obtained from the midventral tail vein and, unless the animals are to be continued on test, immediately after they are sacrificed, e.g., by cervical dislocation. The proportion of PCEs among total erythrocytes (PCEs plus NCEs), which is a measure of toxicity, is obtained for at least 200 bone marrow erythrocytes, or for at least 2,000 erythrocytes from the peripheral blood. Then at least 2,000 PCEs per animal are evaluated to obtain the percentage with micronuclei. Criteria for a positive response include a dose-related increase in the number of PCEs with micronuclei or a clear increase in the number of micronucleated PCEs at a single sampling time.

6.5 ADVANTAGES, DISADVANTAGES, AND PREDICTIVITY OF THE GENETIC TOXICOLOGY TESTS USED FOR CURRENT PRODUCT REGISTRATION

The objective of this chapter has been to provide practical guidance to professionals responsible for evaluating the safety of current products. Thus, the harmonized testing guidelines have been stressed, and current developments in more theoretically oriented genetic toxicology research have been largely omitted. Table 6.5

summarizes some of the advantages and disadvantages of the genetic toxicology tests identified in the harmonized guidelines and, hence, the tests currently used for product registration. Other factors to be considered may include previously obtained data for the test chemical, the properties of the test chemical, and the experience of the sponsoring organization and the testing laboratory. The tests most frequently used at the present time in the United States are the bacterial reverse mutation assay, the mouse lymphoma assay, and the micronucleus test; these tests comprise the most economical and rapidly performed set of tests.

TABLE 6.5. Advantages and disadvantages of currently used tests

Test System	Advantages	Disadvantages
Bacterial Reverse Mutation	The "cornerstone" of genotoxicity tests. Efficiency (a relatively rapid test). Economy. The strains are sensitive to specific types of mutations. Several modifications of the test have been sufficiently evaluated for use. A large database of results is available.	Does not detect chemicals that break mammalian chromosomes, i.e., that induce chromosomal mutations. Toxic chemicals often yield a background of small colonies that can be misinterpreted as a positive result.
Mouse Lymphoma Assay	Detects the full range of gene and chromosomal mutations. Highly predictive of carcinogenicity in rodents. An extensively characterized assay. Can be conducted more efficiently and economically than is the *in vitro* chromosomal aberration assay.	Requires careful adherence to a specific protocol. More expensive and time consuming than is the bacterial reverse mutation assay. Chromosomal mutations are assessed indirectly (as small colony mutants).
In Vitro Chromosomal Aberrations	Permits direct visualization of chromosomal events and yields information on cycle effects. More consistent with the expertise available in some laboratories than is the mouse lymphoma assay.	Requires specialized expertise. A more subjective test than is the mouse lymphoma assay. A negative result can be obtained if the cells are not sampled at the appropriate time after exposure.
In Vivo Chromosomal Aberrations	Permits direct visualization of chromosomal events and yields information on cycle effects. Historically, one of the first genetic toxicology tests to be developed. Yields information on the effects of *in vivo* metabolism.	Requires specialized expertise. A more subjective test than is the micronucleus assay. More expensive and time consuming than is the micronucleus test. Sampling times are critical for ensuring valid results.
Micronucleus Tests	More rapid, less expensive, and less subjective than is the *in vivo* chromosomal aberration test. Less-specialized expertise required than for the *in vivo* chromosomal aberration test. Yields information on the effects of *in vivo* metabolism.	Does not permit direct visualization of the full range of chromosomal effects as does the *in vitro* chromosomal aberration assay, and, as a result, it is less informative of cell-cycle effects.

Each of the initial genetic toxicology tests described has been shown to be useful to varying degrees in predicting the outcome of the long-term animal tests, and when used together, these tests provide a higher degree of confidence that chemicals yielding consistently negative results will be found to be negative in other tests for carcinogenesis and heritable genetic effects. However, as has been previously noted, it is virtually impossible to obtain direct evidence for specific chemicals' inducing inherited defects in humans. In addition, *in vivo* tests for gene mutations are insensitive to many chemicals (e.g., transgenic mouse mutagenesis assays) or are so time consuming and costly (e.g., the spot test and the specific locus test) that an insufficient number of chemicals have been tested for the results to be useful in assessing the predictivity of *in vitro* gene mutation assays for mutagenesis *in vivo*. When the outcome of *in vitro* and *in vivo* tests for chromosomal effects are compared, it is found that a number of chemicals that are positive for clastogenesis *in vitro* yield negative results *in vivo*; however, it is exceedingly rare for chemicals that are positive *in vivo* to yield negative results *in vitro*. This difference is usually attributed to detoxification mechanisms that may be present only in the whole animal. However, *in vivo* toxic effects unrelated to the target cells may preclude testing chemicals to sufficiently high doses to reproduce the results observed *in vitro*. Therefore, evidence of target cell toxicity or pharmacokinetic data is necessary before concluding that cytogenetic results observed *in vitro* are without relevance *in vivo*.

Because the predictivity of genotoxicity tests for *in vivo* mutagenesis cannot be assessed, the accuracy of genotoxicity assays is usually evaluated based on the outcome of testing the same chemicals in rodent carcinogenesis bioassays. However, for these comparisons to be reliable, it is necessary to use the results for relatively large groups of chemicals from diverse chemical classes, to exclude chemicals that have not been tested in a valid manner or that have yielded equivocal results in one or both of the assay systems, and for the comparisons to be based on results, from each test system, that were evaluated with an appreciation of biologic relevance. These issues are discussed in greater detail in ref. 23. Because these factors have not always been carefully controlled, the accuracy of microbial mutagenesis tests for identifying rodent carcinogens dropped from 90% in 1975 [17] to less than 50% in 1987 [19], and it is probable that the actual value is between these extremes. Accuracy of the microbial tests should be better than a flip of a coin (50%) because of the association of mutagenesis with carcinogenesis; however, 90% accuracy is no longer expected because mechanisms of carcinogenesis involving mammalian chromosomes cannot be assessed in microbial systems. In spite of this limitation, microbial mutagenesis tests are used today because of their efficiency and economy and because of the large database of microbial test results available.

In contrast to the relatively limited predictivity of microbial mutagenesis assays, tests for chromosomal effects *in vitro* have a high predictivity for the outcome of rodent carcinogenesis bioassays. The L5178Y mouse lymphoma assay, which measures chromosomal as well as gene mutations, accurately predicts at least 95% of chemicals that are rodent carcinogens [22,23]. In addition, evidence is now accumulating that essentially all chemicals found positive in the mouse lymphoma assay also induce chromosomal aberrations *in vitro* when the cytogenetics testing is conducted with appropriate protocols. Because, as noted above, a number of chemicals found to be clastogenic *in vitro* yield negative results in *in vivo* assays for clastogenesis, the *in vivo* tests appear to be less accurate than are *in vitro* chromosomal tests for predicting rodent carcinogenicity. Therefore, by conducting initial testing in the mouse lymphoma assay or an *in vitro* test for chromosomal aberrations, it may be possible to conserve resources and to reduce animal usage if it is determined that additional testing would not be productive.

However, each genotoxicity testing system has inherent limitations. Thus, it cannot be concluded that if a chemical yields negative results in an appropriately conducted *in vitro* mouse lymphoma mutagenesis or chromosomal aberration assay, no further genotoxicity testing will be required for product registration. Confidence in negative results is gained by finding the same outcome in multiple test systems, and the results from *in vivo* genotoxicity tests, coupled with pharmacokinetic data, are useful for risk estimation. In addition, some chemicals, including some aerosols and gases, that are poorly soluble in cell culture medium and, hence, inappropriate for the assays that measure chromosomal effects *in vitro* can be adequately tested in bacteria and in *in vivo* assays for clastogenesis. For these reasons, representatives from industry and governmental regulatory agencies have recently agreed that, for the purposes of product registration, a defined battery of three genotoxicity tests is needed before concluding that chemicals yielding negative results may be considered to be of lesser concern.

References

1. Gad SC, ed. *Product Safety Evaluation Handbook*. New York, Marcel Dekker, 1988.

2. Barfknecht TR, Naismith RW. Practical mutagenicity testing. In: Gad SC, ed. *Product Safety Evaluation Handbook*. New York, Marcel Dekker, 1988:143–217.

3. Brusick D. *Principles of Genetic Toxicology*. New York: Plenum Press, 1987.

4. Li AP, Heflich RH. *Genetic Toxicology*. Boca Raton, FL: CRC Press, 1991.

5. Hollaender A, de Serres FJ. *Chemical Mutagens. Principles and Methods for their Detection*, Vols. 1–8. New York: Plenum Press, 1971–1983.

6. Anon. The promise of gene therapy. *Newslett Childr Hosp Los Angeles Res Inst* 1998:1.

7. Fox TR, Gonzales AJ. *Cell Cycle Controls as Potential Targets for the Development of Chemically Induced Mouse Liver Cancer*, Vol. 16. Research Triangle Park, NC: CIIT Activities, 1996:1–5.

8. Evans HJ. Molecular mechanisms in the induction of chromosome aberrations. In: Scott D, Bridges BA, Sobels FH, eds. *Progress in Genetic Toxicology*. New York: Elsevier/North-Holland, 1977.

9. Trask BJ, Allen S, Massa H, Fetitta A, Sachs R, van den Engh G, Wu M. Studies of metaphase and interphase chromosomes using fluorescence *in situ* hybridization. *Cold Spring Harb Symp Quant Biol* 1993;58:767–775.

10. Mendel G. Versuche uber Pflanzenhybriden. Vern des Naturf Vereines in Brunn, 4, 1866. English translation in *J R Horticulture Soc* 1901;26:1–32.

11. de Vries H. *The Mutation Theory*. Leipzig: Verlag von Veit, 1901.

12. Boveri T. *Zur Frage der Entwicklung maligner Tumoren*. Jena: Gustav Fischer, 1914.

13. Muller HJ. Artificial transmutation of the gene. *Science* 1927;64:84–87.

14. Efron E. *The Apocalyptics. How Environmental Politics Controls What We Know About Cancer*. New York: Simon & Schuster, 1984.

15. Wassom JS. Origins of genetic toxicology and the Environmental Mutagen Society. *Environ Mol Mutagen* 1989;14(S16):1–6.

16. Malling HV. Dimethylnitrosamine: formation of mutagenic compounds by interaction with mouse liver microsomes. *Mutat Res* 1971;13:425–429.

17. McCann J, Choi E, Yamasaki E, Ames BN. Detection of carcinogens as mutagens in the *Salmonella*/microsome test: Assay of 300 chemicals. *Proc Natl Acad Sci U S A* 1975;72:5,135–5,139.

18. Prival MJ, Dellarco VL. Evolution of social concerns and environmental policies for chemical mutagens. *Environ Mol Mutagen* 1989;14(S16):46–50.

19. Tennant RW, Margolin BH, Shelby MD, Zeiger E, Haseman JK, Spalding J, Caspary W, Resnick M, Stasiewicz S, Anderson B, Minor R. Prediction of chemical carcinogenicity in rodents from *in vitro* genetic toxicity assays. *Science* 1987;236:933–941.

20. Ames BN. What are the major carcinogens in the etiology of human cancer? Environmental pollution, natural carcinogens, and the causes of human cancer: Six errors. In: De Vita VT Jr, Hellman S, Rosenberg SA, eds. *Important Advances in Oncology*. Philadelphia: J. B. Lippencott Company, 1989:237–247.

21. Galloway SM, Sofuni T, Shelby M, Thilagar A, Kumaroo V, Kaur P, Gulati D, Putman DL, Murli H, Marshall R, Tanaka N, Anderson B, Zeiger E, Ishidate M, Jr. Multilaboratory comparison of *in vitro* tests for chromosome aberrations in CHO and CHL cells tested under the same protocols. *Environ Mol Mutagen* 1997;29:189–207.

22. Mitchell AD, Auletta AE, Clive D, Kirby PE, Moore MM, Myhr BC. The L5178Y /tk$^{+/-}$ mouse lymphoma specific gene and chromosomal mutation assay. A phase III report of the U.S. Environmental Protection Agency Gene-Tox Program. *Mutat Res* 1997;394:177–303.

23. Mitchell AD, Auletta AE, Clive D, Kirby PE, Moore MM, Myhr BC. A comparison of the Gene-Tox and NTP evaluations of the utility of the L5178Y mouse lymphoma cell mutation assay for predicting rodent carcinogenicity. *Mutat Res* In press.

24. Sax K. Induction by X-rays of chromosome aberrations in *Tradescantia* microspores. *Genetics* 1938;23:494–516.

25. Auerbach C, Robson JM. The production of mutations by chemical substances. *Proc R Soc Edinb B* 1947;62:271–283.

26. Earle WR, Schilling EL, Stark TH, Straus NP, Brown MF, Shelton E. Production of malignancy *in vitro*. IV. The mouse fibroblast cultures and changes seen in the living cells. *J Nat Cancer Inst* 1943;4:165–212.

27. Russell WL. X-ray-induced mutations in mice. *Cold Spring Harb Symp Quant Biol* V1951;16:327–336.

28. Gey GO, Coffman WD, Kubicek MT. Tissue culture studies of the proliferative capacity of cervical carcinoma and normal epithelium. *Cancer Res* 1952;12:364–365.

29. Watson JD, Crick FHC. Molecular structure of nucleic acids. *Nature* 1953;171:737–738.

30. Tjio JH, Levan A. The chromosome number of man. *Hereditas* 1956;42:1–6.

31. Lejeune J, Gautier M, Turpin R. Etude des chromosomes somatiques de neuf enfants mongoliens. *Compt Rend Acad Sci* 1959;248:721–722.

32. Nowell PC, Hungerford DA. A minute chromosome in human chronic granulocytic leukemia. *Science* 1960;132:1,497.

33. Auerbach C. Chemical mutagenesis in animals. *Abh Deutsch Akad Wiss Berlin Klin Med* 1960;1:1–13.

34. Hayflick L, Moorehead PS. The serial cultivation of human diploid cell strains. *Exp Cell Res* 1961;25:585–621.

35. Carson R. *Silent Spring*. Greenwich, CT: Fawcett, 1962.

36. Kao F-T, Puck TT. Genetics of somatic mammalian cells. VII. Induction and isolation of nutritional mutants in Chinese hamster cells. *Proc Natl Acad Sci U S A* 1968;60:1,275–1,281.

37. Chu EHY, Malling HV. Mammalian cell genetics. II. Chemical induction of specific locus mutations in Chinese hamster cells *in vitro*. *Proc Natl Acad. Sci U S A* 1968;61:77–87.

38. Miller JA, Miller EC. The mutagenicity of chemical carciogens: correlations, problems, and interpretations. In: Hollaender AE, ed. *Chemical Mutagens: Principles and Methods for Their Detection*, Vol. 1. New York: Plenum Press, 1971:83–120.

39. Knudson AG, Jr. Mutation and cancer: Statistical study of retinoblastoma. *Proc Natl Acad Sci U S A* 1971;68:820–823.

40. Clive D, Flamm WG, Machesko MR, Bernheim NJ. A mutational assay system using the thymidine kinase locus in mouse lymphoma cells. *Mutat Res* 1972;16:77–87.

41. Schmid W. The micronucleus test. *Mutat Res* 1975;31:9–15.

42. Ames BN, Durston WE, Yamasaki E, et al. Carcinogens are mutagens: a simple test system combining liver homogenates for activatin and bacteria for detection. *Proc Natl Acad Sci U S A* 1973;70:2,281–2,285.

43. Second Task Force for Research Planning in Environmental Health Science. *Human Health and the Environment—Some Research Needs*. Washington, DC, 1976. DHEW Publication NIH 77-1277.

44. Dearfield KL. Use of mutagenicity data by the U.S. Environmental Protection Agency's Office of Pesticide Programs. In: Li AP, Heflich RH, eds. *Genetic Toxicology*. Boca Raton,

FL: CRC Press, 1991:473–484.

45. National Research Council. *Toxicity Testing. Strategies to Determine Needs and Priorities.* Washington, DC: National Academy Press, 1984.

46. Johnson J. Administration backs chemical testing. *Chem Eng News* 1988;76:7–8.

47. Auletta A. Regulatory perspectives: TSCA. In: Li AP, Heflich RH, eds. *Genetic Toxicology.* Boca Raton, FL: CRC Press, 1991:435–454.

48. Storer RD, Kraynak AR, McKelvey TW, Elia MC, Goodrow TL, DeLuca JG. The mouse lymphoma L5178Y TK+/TK⁻ cell line is heterozygous for a codon 170 mutation in the p53 tumor suppressor gene. *Environ Mol Mutagen* 1996:27(S27):66.

49. Cole J, Arlett CF, Green MHL, Lowe J, Muriel W. A comparison of the agar cloning and microtitration techniques for assaying cell survival and mutatin frequency in L5178Y mouse lymphoma cells. *Mutat Res* 1983;111:371–386.

50. Sofuni T, Honma M, Hayashi M, Shimada H, Tanaka N, Wakuri S, Awogi T, Yamamoto KI, Nishi Y, Nakadate M. Detection of *in vitro* clastogens and spindle poisons by the mouse lymphoma assay using the microwell method: interim report of an international collaborative study. *Mutagenesis* 1966;11:349–355.

7

Pyrogenicity and Muscle Irritation

Shayne Cox Gad

Gad Consulting Services, Raleigh, North Carolina

Pyrogenicity and muscle irritation are product hazard endpoints of concern in those parts of the health care industry in which the body surface is penetrated and the product is introduced into the region of the blood vessels ("fluid path") or into the muscles. This process generally encompasses many medical devices (syringes, catheters, and so on) and biologicals and pharmaceuticals that are administered parenterally.

Pyrogenicity is the induction of a febrile (fever) response by the parenteral [usually, intravenous (IV) or intramuscular (IM)] administration of exogenous material, usually bacterial endotoxins. Pyrogenicity is usually associated with microbiologic contamination of a final formulation or product, but it is now increasingly of concern because of the increase in interest in biosynthetically produced materials. Generally, ensuring the sterility of product and process will guard against pyrogenicity. Pyrogenicity testing is performed extensively in the medical device industry. If a device is to be introduced directly or indirectly into the fluid path, it is required that it be evaluated for pyrogenic potential.

Muscle irritation is the local inflammation, pain, and damage resulting from the parenteral injection of pharmaceuticals into a muscle mass. It is caused by a range of physicochemical factors as well as chemical/biological interactions, and it is particularly of concern with antibiotics.

7.1 PYROGENICITY

Both *in vivo* and *in vitro* tests are currently in use. The *in vitro* test is preferred, except for those cases in which it may not be employed.

7.1.1 *In Vivo*

The *United States Pharmacopoeia* [1] specifies a pyrogen test using rabbits as a model. This test is the standard for limiting risks of a febrile reaction to an acceptable level, and it involves measuring the rise in body temperature in a group of three rabbits for 3 hours after injection of 10 ml of test solution.

Apparatus and diluents. Render the syringes, needles, and glassware free from pyrogens by heating at 250°F for not less than 30 minutes by any other suitable method. Treat all diluents and solutions for washing and rinsing of devices or parenteral injection assemblies in a manner ensuring that they are sterile and pyrogen-free. Periodically perform control pyrogen tests on representative portions of the diluents and solutions for washing or rinsing of the apparatus.

Temperature recording. Use an accurate temperature-sensing device, such as a clinical thermometer, thermistor probes, or similar probes, that has been calibrated to ensure an accuracy of ± 0.1°C and has been tested to determine that a maximum reading is reached in less than 5 minutes. Insert the temperature-sensing probe into the rectum of the test rabbit to a depth of not less than 7.5 cm and after a period of time (not less than that previously determined as sufficient), record the rabbit's temperature.

Test animals. Use healthy, mature rabbits. House the rabbits individually in an area of uniform temperature between 20°C and 23°C and free from disturbances likely to excite them. The temperature should vary no more than 3°C from the selected temperature. Before using a rabbit for the first time in a pyrogen test, condition it not more than 7 days before use by a sham test including all of the steps described under the "Procedure" subsection, except injection. Do not use a rabbit for pyrogen testing more frequently than once every 48 hours, or before 2 weeks after a maximum rise in its temperature of 0.6°C or more while being subjected to the pyrogen test, or after, it has been given a test specimen that was adjusted to be pyrogenic.

Procedure. Perform the test in a separate area designated solely for pyrogen testing and under environmental conditions similar to those under which the animals are housed and free from the disturbance likely to excite them. Withhold all food from the rabbits used during the period of the test. Access to water is allowed at all times, but it may be restricted during the test. If rectal-temperature–measuring probes remain inserted throughout the testing period, restrain the rabbits with light-fitting Elizabethan collars that allow the rabbits to assume a natural testing posture. Not more than 30 minutes before the injection of the test dose, determine the "control temperature" of each of the test dose animals, allowing for later determination of any temperature increase resulting from the injection of a test solution. In any one group of test rabbits, use only those rabbits whose control temperatures do not vary by more than 1°C from each other, and do not use any rabbit with a temperature exceeding 39.8°C.

Unless otherwise specified in the individual protocol, inject into an ear vein of each of 3 rabbits 10 ml of the test solution per kilogram of body weight, completing each injection within 10 minutes after the start of administration. The test solution is either the product, constituted if necessary as directed on the label, or the material under test. For pyrogen testing of devices or injection assemblies, use washings or rinsings of the surfaces coming into contact with the parenterally administered material or with the injection site or internal tissues of the patient. Ensure that all test solutions are protected from contamination. Perform the injection after warming the test solution to a temperature of 37 ± 2°C. Record the temperature at 1, 2, and 3 hours after the injection.

Test interpretation and continuation. Consider any temperature decrease as zero rises. If no rabbit shows an individual rise in temperature of 0.6°C or more above its respective control temperature, and if the sum of the 3 individual maximum temperature rises does not exceed 1.4°C, the product meets the requirements for the absence of pyrogens. If any rabbit shows an individual temperature rise of 0.6°C or more, or if the sum of the 3 individual maximum temperature rises exceeds 1.4°C, continue the test using 5 other rabbits. If not more than 3 of the 8 rabbits show individual rises in temperature of 0.6°C or more, and if the sum of 8 individual maximum temperature rises does not exceed 3.7°C, the material under examination meets the requirements for the absence of pyrogens.

7.1.2   In Vitro

Pyrogenicity (or bacterial endotoxin testing) has been one of the great success stories for *in vitro* testing. In the mid-1970s, the limulus amebocyte lysate (LAL) test was developed, validated, and accepted as an *in vitro* alternative [2,3] to the rabbit test. An *in vitro* test for estimating the concentration of bacterial endotoxins that may be present in or on the sample of the article(s) to which the test is applied uses LAL that has been obtained from aqueous extracts of the circulating amebocytes of the horseshoe crab, *Limulus polyphemus*, and that has been prepared and characterized for use as an LAL reagent for gel-clot formation. The test's limitation is that it detects only the endotoxins of Gram-negative bacteria and β-glycans from fungi. This limitation is generally not significant because most environmental contaminants gaining entrance into sterile products are Gram-negative [4].

When the test is conducted as a limit test, the specimen is determined to be positive or negative to the test judged against the endotoxin concentration specified in the individual monograph. When the test is conducted as an assay of the concentration of endotoxin, with calculation of confidence limits of the result obtained, the specimen is judged to comply with the requirements if the result does not exceed (1) the concentration limit specified in the individual monograph, and (2) the specified confidence limits for the assay. In either case, the determination of the reaction endpoint is made with parallel dilutions of redefined endotoxin units.

Because LAL reagents have also been formulated to be used for turbidimetric (including kinetic) assays or colorimetric readings, such tests may be used if shown to comply with the requirements for alternative methods. These tests require the establishment of a standard regression curve, and the endotoxin content of the test material is determined by interpolation from the curve. The procedures include incubation for a preselected time of reacting endotoxin and control solutions with LAL reagent and reading the spectrophotometric light absorbance at suitable wavelengths. In the case of the turbidimetric procedure, the reading is made immediately at the end of the incubation period. In the kinetic assays, the absorbance is measured throughout the reaction period and rate values are determined from those readings. In the colorimetric procedure, the reaction is arrested at the end of the preselected time by the addition of an appropriate amount of acetic acid solution before the readings. A possible advantage in the mathematical treatment of results, if the test is otherwise validated and the assay suitably designed, could be the confidence interval and limits of potency from the internal evidence of each assay itself.

Reference standard and control standard endotoxins. The reference standard endotoxin (RSE) is the USP Endotoxin Reference Standard, which has a defined potency of 10,000 USP endotoxin units (EU) per vial. Constitute the entire contents of one vial of the RSE with 5 ml of LAL reagent water, vortex for not less than 20 minutes, and use this concentrate for making appropriate serial dilutions. Preserve the concentrate in a refrigerator, for making subsequent dilutions, for not more than 14 days. Allow it to reach room temperature, if applicable, and vortex it vigorously for not less than 5 minutes before use. Vortex each dilution for not less than 1 minute before proceeding to make the next dilution. Do not use stored dilutions. A control standard endotoxin (CSE) is an endotoxin preparation other than the RSE that has been standardized against the RSE. If a CSE is a preparation not already adequately characterized, its evaluation should include characterizing parameters both for endotoxin quality and performance (such as reaction in the rabbit) and for suitability of the material to serve as a reference (such as uniformity and stability). Detailed procedures for its weighing or constitution and use to ensure consistency in performance should also be included. Standardization of CSE against the RSE using an LAL reagent for the gel-clot procedure may be effected by assaying a minimum of four vials of the CSE or four corresponding aliquots, when applicable, of the bulk CSE and one vial of the RSE, as directed under the "Test procedure" subsection, but using four replicate reaction tubes at each level of the dilution series for the RSE and four replicate reaction tubes similarly for each vial or aliquot of the CSE. If the dilutions for the four vials or aliquots of the CSE cannot all be accommodated with the dilutions for the one vial of the RSE on the same rack for incubation, additional racks may be used for accommodating some of the replicate dilutions for the CSE, but all of the racks containing the dilutions of the SE and CSE are incubated as a block. However, in such cases, the replicate dilution series from the one vial of the RSE are accommodated together on a single rack and the replicate dilution series from any one of the four vials or aliquots of the CSE are not divided between racks. The antilog of the difference between the main log 10 endpoint of the RSE and the mean log 10 endpoint of the CSE is the standardized potency of the CSE, which is then converted to and expressed in units/nanograms under stated drying conditions for the CSE, or units per container, whichever is appropriate. Standardize each new lot of CSE before use in the test. Calibration of a CSE by the RSE must be with the specific lot of LAL reagent and the test procedure with which it is to be used. Subsequent lots of LAL reagent from the same source and with similar characteristics need only check the potency ratio. The inclusion of one or more dilution series made from the RSE when the CSE is used for testing will enable observation of whether the relative potency shown by the latter remains within the determined confidence limits. A large lot of a CSE may, however, be characterized by a collaborative assay of a suitable design to provide a representative relative potency and the within-laboratory and between-laboratory variance.

A suitable CSE has a potency of not less than 2 EU/ng and not more than 50 EU/ng, when in bulk form, under adopted uniform drying conditions, e.g., to a particular low moisture content and other specified conditions of use, and a potency within a corresponding range when filled in vials of a homogeneous lot.

Preparatory testing. Use an LAL reagent of a confirmed label or a determined sensitivity. In addition, when a change in lot of CSE, LAL reagent, or another reagent is to occur, conduct tests of a prior satisfactory lot of CSE, LAL, or other reagent in parallel upon changeover. Treat any containers or utensils employed to destroy extraneous endotoxins that may be present, such as by heating in an oven at 250°F or above for sufficient time.

The validity of test results for bacterial endotoxins requires an adequate demonstration that specimens of the article, or of solutions, washings, or extracts thereof to which the test is to be applied, do not inhibit or enhance the reaction or otherwise interfere with the test. Validation is accomplished by testing untreated specimens or appropriate dilutions thereof, concomitantly with and without known and demonstrable added amounts of RSE or a CSE, and comparing the results obtained. Appropriate negative controls are included. Validation must be repeated if the LAL reagent source or the method of manufacture or formulations of the article is changed.

Test for confirmation of labeled LAL reagent sensitivity. Confirm the labeled sensitivity of the particular LAL reagent with the RSE (or CSE) using not less than four replicate vials, under conditions shown to achieve an acceptable variability of the test, viz., the antilog of the geometric mean log 10 lysate gel-clot sensitivity is within 0.5 to 2.0, where the labeled sensitivity is in endotoxin units/milliliter. The RSE (or CSE) concentrations selected to confirm the LAL reagent label potency should bracket the stated sensitivity of the LAL reagent. Confirm the labeled sensitivity of each new lot of LAL reagent before use in the test.

Inhibition or enhancement test. Conduct assays, with standard endotoxin, of untreated specimens in which no endogenous endotoxin is detectable, and of the same specimens to which endotoxin has been added, as directed under the "Test procedure" subsection, but use not less than four replicate reaction tubes at each level of the dilution series for each untreated specimen and for each specimen to which endotoxin has been added. Record the endpoints (E, in units/milliliters) observed in the replicates. Take the logarithms (e) of the endpoints, and compute the geometric means of the log endpoints for the RSE (or CSE) for the untreated specimens and for specimens containing endotoxin by the formula antilog, e/f, where e is the sum of the log endpoints of the dilution series used and f is the number of replicate endspoints in each case. Compute the amount of endotoxin in the specimen to which endotoxin has been added. The test is valid for the article if this result is within twofold of the known added amount of endotoxin. Alternatively, if the test has been appropriately set up, the test is valid for the article if the geometric mean endpoint dilution for the specimen to which endotoxin has been added is within one twofold dilution of the corresponding geometric mean endpoint dilution of the standard endotoxin.

Repeat the test for inhibition or enhancement using specimens diluted by a factor not exceeding that given by the formula, x/y (see the "Maximum valid dilution" subsection). Use the least dilution sufficient to overcome the inhibition or enhancement of the known added endotoxin for subsequent assays of endotoxin in test specimens.

If endogenous endotoxin is detectable in the untreated specimens under the conditions of the test, the article is unsuitable for the inhibition or enhancement test, or it may be rendered suitable by removing the endotoxin present by ultrafiltration or by appropriate dilution. Dilute the untreated specimen (as constituted, when applicable, for administration or use) to a level not exceeding the maximum valid dilution (MVD), at which no endotoxin is detectable. Repeat the test for inhibition or enhancement using the specimens at those dilutions.

Test procedure. In preparing for and applying the test, observe precautions in handling the specimens to avoid gross microbial contamination. Washings or rinsings of devices must be with LAL reagent water in volumes appropriate to their use and, when applicable, of the surface area coming into contact with body tissues or fluids. Use such washings or rinsings if the extracting fluid has been in contact with the relevant pathway or surface for not less than 1 hour at a controlled room temperature (15–30°C). Such extracts may be combined, when appropriate.

For validating the test for an article, for endotoxin limit tests or assays, or for special purposes, testing of specimens is conducted quantitatively to determine response endpoints for gel-clot readings. Usually, graded strengths of the specimen and standard endotoxin are made by multifold dilutions. Select dilutions so that they correspond to a geometric series in which each step is greater than the next lower step by a constant ratio. Do not store diluted endotoxin, because of loss of activity by absorption. In the absence of supporting data to the contrary, negative and positive controls are incorporated into the test.

Use not less than two replicate reaction tubes at each level of the dilution series for each specimen under test. Whether the test is employed as a limit test or as a quantitative assay, a standard endotoxin dilution series involving not less than two replicate reaction tubes is conducted in parallel. A set of standard endotoxin dilution series is included for each block of tubes, which may consist of a number of racks for incubation together, provided the environmental conditions within blocks are uniform.

Preparation. Because the form and amount per container of standard endotoxin and of LAL regent may vary, constitution or dilution of contents should be as directed in the labeling. The pH of the test mixture of the specimen and the LAL reagent is in the range of 6–7.5 unless specifically directed otherwise in the individual monograph. The pH may be adjusted by the addition of sterile, endotoxin-free sodium hydroxide or hydrochloric acid or suitable buffers to the specimen before testing.

Maximum valid dilution. The MVD is appropriate to injections or to solutions for parenteral administration in the form constituted or diluted for administration, or when applicable, to the amount of drug by weight if the volume of the dosage form for administration could be varied. When the endotoxin limit concentration is specified in the individual monograph in terms of volume (in endotoxin units/milliliter), divide the limit by, which is the labeled sensitivity (in endotoxin units/milliliter) of the lysate employed in the assay, to obtain the MVD factor. When the endotoxin limit concentration is specified in the individual monograph by weight or by units of active drug (in endotoxin units/milligram or in endotoxin units/unit), multiply the limit by the concentration (in milligrams/milliliter or in units/milliliter) of the drug in the solution tested or of the drug constituted according to the label instructions, whichever is applicable, and divide the product of the multiplication by to obtain the MVD factor. The MVD factor so obtained is the limit dilution factor for the preparation for the test to be valid.

Procedure. To 10 x 75-mm test tubes add aliquots of the appropriately constituted LAL reagent and the specified volumes of specimens, endotoxin standard, negative controls, and a positive product control consisting of the article, or of solutions, washings, or extracts thereof, to which the RSE (or a standardized CSE) has been added at a concentration of endotoxin of two for LAL reagent (see the "Test" subection for confirmation of labeled LAL reagent sensitivity). Swirl each test tube gently to mix, and place in an incubating device, such as a water bath or a heating block, accurately recording the time at which the tubes are so placed. Incubate each tube, undisturbed, for 60 ± 2 minutes at $37 \pm 1°C$, and carefully remove it for observation. A positive reaction is characterized by the formation of a firm gel remaining when inverted through 180°. Record such a result as a positive (+). A negative result is characterized by the absence of such a gel or by the formation of a viscous gel that does not maintain its integrity. Record such a result as a negative (–). Handle the tubes with care, and avoid subjecting them to unwanted vibrations, or false-negative observations may result. The test is invalid if the positive product control or the endotoxin standard does not show the endpoint concentration to be within twofold dilutions from the label claim sensitivity of the LAL reagent or if any negative control shows a gel-clot endpoint.

Calculation and interpretation. Calculate the concentration of endotoxin (in units/milliliter or in units/gram or milligram) in or on the article under test by the formula: pS/U, where S is the antilog of the geometric mean log of the endpoints, expressed in endotoxin units/milliliter for the standard endotoxin; U is the antilog of ef, where e is the log 10 of the endpoint dilution factors, expressed in decimal fractions, f is the number of replicate reaction tubes read at the endpoint level for the specimen under test, and p is the correction factor for those cases in which a specimen of the article cannot be taken directly into test but is processed as an extract solution or washing.

When the test is conducted as an assay with sufficient replication to provide a suitable number of independent results, calculate for each replicate assay the concentration of endotoxin in or on the article under test from the antilog of the geometric mean log endpoint ratios. Calculate the mean and the confidence limits from the replicate logarithmic values of all obtained assay results by a suitable statistical method.

Interpretation. The article meets the requirements of the test if the concentration of endotoxin does not exceed that specified in the individual monograph, and the confidence limits of the assay do not exceed those specified.

Human leukocyte assay. An additional *in vitro* assay for the induction of fever in humans has recently been proposed [5] using samples of whole human blood. When exposed to various concentrations of bacterial lipopolysaccharide (LPS), blood incubations release several pyrogenic factors within 24 hours, including interleukin 1β (IL-1β). The lower limit for quantitation of LPS is 10 pg/ml, with IL-1β concentration as the measured endpoint. In healthy donors, the interindividual variance of LPS-stimulated IL-1β release was found to be ± 23% [standard error of the mean (SEM), $n = 18$]. Not only endotoxin, but also further bacterial components, such as muramyl dipeptide, various lipoteichoic acids, and the superantigen staphylococcus enterotoxin B, induce a qualitatively similar reaction. With blood from volunteers given the nonsteriodal antiflammatory drug aspirin, the *ex vivo* LPS-stimulated PGE2 release, but not the formation of IL-1β in blood from these donors, was inhibited for several hours. It is proposed that this system can also serve as an *in vitro* method alternative to the rabbit pyrogen test.

7.2 IRRITATION OF PARENTERALLY ADMINISTERED AGENTS

Intramuscular and intravenous injection of parenteral formulations of pharmaceuticals can produce a range of discomfort resulting in pain, irritation, or damage to muscular or vascular tissue [14]. These injections are normally evaluated for prospective formulations before use in humans by histopathologic evaluation of damage in intact animal models, usually the rabbit [6–8]. Attempts have been made to make this *in vivo* methodology both more objective and quantitative based on measuring the creatinine phosphokinase released in the tissue surrounding the injection site [6,9,10]. Currently, a protocol using a cultured rat skeletal muscle cell line (L6) as a model has been evaluated in an interlaboratory validation program among 11 pharmaceutical company laboratories. This methodology [10,18] measures creatine kinase levels in media after exposure of the cells to the formulation of interest, and it predicts *in vivo* IM damage based on this endpoint. It is reported to give excellent rank-correlated results across a range of antibiotics [11–13]. The current multilaboratory evaluation covers a broader structural range of compounds and has shown a good quantitative correlation (with *in vivo* results) for antibiotics and a fair correlation for a broader range of parenteral drug products. Likewise, Kato et al. [15] have proposed a model using cultured rat primary skeletal muscle fibers. Damage is evaluated by the release of creatinine phosphokinase. An evaluation using six parenterally administered antibodies (rashing their EC_{50} values) showed good relative correlation with *in vivo* results.

Another proposed *in vitro* assay for muscle irritancy for injectable formulations is the red blood cell hemolysis assay [16,17]. Water-soluble formulations are gently mixed in a 1:2 ratio with freshly collected human blood for 5 seconds, and then mixed with a 5% weight/volume (w/v) 6 dextrose solution and centrifuged for 5 minutes. The percent red blood cell survival is then determined by measuring differential absorbance at 540 nm, and this result is compared against values for known irritants and nonirritants. Against a small group of compounds (four), this process is reported to be an accurate predictor of muscle irritation.

References

1. United States Pharmacopoeia. Bacterial endotoxins test. *USP XXIII— The United States Pharmacopoeia.* Rockville, MD: USP Convention, 1994:1,718.

2. United States Pharmacopoeia. Bacterial endotoxins text. *USP XXIII—The United States Pharmacopoeia.* Rockville, MD: USP Convention, 1994:1,696–1,697.

3. Cooper JF. Principles and applications of the limulus test for pyrogen in parenteral drugs. *Bull Parenter Drug Assoc* 1975;3:122–130.

4. Weary M, Baker B. Utilization of the limulus amebocyte lysate test for pyrogen testing of large-volume parenterals, administration sets and medical device. *Bull Parenter Drug Assoc* 1977;31:127–133.

5. Devleeschouwer MJ, Cornil MJ, Dony J. Studies on the sensitivity and specificity of the limulus amebocyte lystate test and rabbit *pyrogen* assays. *Appl Environ Microbiol* 1985;50:1,509–1,511.

6. Hartung T, Wendel A. Detection of pyrogens using human whole blood. *In Vitro Toxicol* 1996;9:353–359.

7. Gad SC, Chengelis CP. *Acute Toxicity: Principles and Methods,* 2nd ed., San Diego, CA: Academic Press, 1997.

8. Gray JF. Pathological evaluation of injection injury. In: Robinson J, ed. *Sustained and Controlled Release Drug Delivery Systems.* New York: Marcel Dekker, 1978:351–405.

9. Shintani S, Yarmazaki M, Nakamura M, Nakazama I. A new method to determine the irritation of drugs after intramuscular injection in rabbits. *Toxicol Appl Pharmacol* 1967;11:293–301.

10. Sidell FR, Calver DL, Kaminskis A. Serum creatine phosphokinase activity after intramuscular injection. *JAMA* 1974;228:1,884–1,887.

11. Young MF, Trobetta LD, Sophia JV. Correlative *in vitro* and *in vivo* study of skeletal muscle irritancy. *Toxicologist* 1986;6(1):1,225.

12. Williams PD, Masters BG, Evans LD, Laska DA, Hattendorf GH. An *in vitro* model for assessing muscle irritation due to parenteral antibiotics. *Fundam Appl Toxicol* 1987;9:10–17.

13. Hoover D, Gardner J, Timmerman T, Klepfer A, Laska D, White S, McGrath J, Buening M, Williams P. Comparison of *in vitro* and *in vivo* models to assess venous irritation of parenteral antibiotics. *Fundam Appl Toxicol* 1990;14:578–597.

14. Laska DA, Williams PD, Reboulet JT, Morris RM. The L6 muscle cell line as a tool to evaluate parental products for irritation. *J Parenter Sci Technol* 1991;45(2):77–82.

15. Nelson AA, Price CW, Welch H. Muscle irritation following the injection of various penicillin preparations in rabbits. *J Am Pharmacol Assoc* 1949;38:237–239.

16. Kato I, Harihara A, Mizushima Y. An *in vitro* method for assessing muscle irritation of antibiotics using rat primary cultured skeletal muscle fibers. *Toxicol Appl Pharmacol* 1992;117:194–199.

17.   Brown S, Templeton L, Prater DA, Potter CJ. Use of an *in vitro* hemolysis test to predict tissue irritancy in an intramuscular formulation. *J Parenter Sci Technol* 1989;43:117–120.

18.   Meltzer HY, Morozak S, Bozer M. Effect of intramuscular injections on serum creatinine phosphokinase activity. *Am J Med Sci* 1970;259:42–48.

8

In Vitro Models for Evaluating Developmental Toxicity

Thomas A. Lewandowski, Rafael A. Ponce, Stephen G. Whittaker, and Elaine M. Faustman

Department of Environmental Health, University of Washington, Seattle, Washington

As reported by Warburton and Fraser [1], the frequency of early spontaneous abortion in humans is at least 15% of all recognized pregnancies. Although approximately 3% of liveborns have major congenital malformations and birth defects account for 20% of postnatal deaths (reviewed in ref. 2), most spontaneous abortions are from unknown causes. Thus, crucial questions are raised. First, what fraction of miscarriages and abnormalities reflects intrinsic properties of the human reproductive process? Second, what is the contribution of external factors, such as environmental and industrial exposure, to this high frequency of poor reproductive outcomes?

The sensitivity of the conceptus to the effects of environmental exposures is a well-established principle of developmental toxicology. An example of a classic developmental toxicant is the sedative/hypnotic drug thalidomide, which elicits congenital malformations in humans at dosages considerably lower than those required to induce overt toxicity in the mother. Other drugs prescribed for human use exhibiting patterns of differential maternal and developmental toxicity include certain synthetic retinoids. A more recent and controversial subject of interest is the potential of environmental chemicals to induce subtle effects in hormone response patterns that may impact *in utero* and postnatal functional development. Chemicals such as diethylstilbestrol (DES) and certain phytoestrogens are believed to act by these mechanisms.

Recognizing the need to determine adverse effects of environmental agents on reproduction and development, a series of toxicologic safety evaluation protocols were developed in the mid-1960s. Consisting of three experimental segments, these tests were developed using rats or rabbits to predict potential hazard to human reproduction and development. The segment II protocol was generally recognized as an important experimental component, because the test agent is administered to pregnant females throughout the period of embryonic and fetal development. However, these standard *in vivo* tests are expensive and require extensive facilities and considerable technical expertise [3,4]. Because between 50,000 and 70,000 chemicals are currently in the marketplace, with an additional 700–1,000 agents introduced each year [5], the exhaustive evaluation of these agents is impractical using the standard assays for *in vivo* teratogenicity. Efforts by the Environmental Protection Agency (EPA) to obtain toxicologic data on all high production volume chemicals (HPVC) as well as the initiatives proposed by the EPA's Endocrine Disruptor Screening and Testing Advisory Committee (ED-STAC) [6] reveal the need for various levels of testing of compounds and the need for informative tests at early stages of chemical evaluation.

In vitro models for developmental toxicity have been proposed as a tool for screening and prioritizing chemicals of interest for subsequent *in vivo* evaluation. *In vitro* tests cannot completely replace *in vivo* teratology studies because they do not examine all interactions and conditions present in the whole organism. Rather, data gathered from *in vivo* and *in vitro* methods may be combined to develop the required information on the spectrum of effects in the developing conceptus, while minimizing cost and use of animals. For example, *in vitro* assays can be useful in structure-activity analyses, in which the relative potencies of structurally related agents are assessed and compared with a related compound of known teratogenic potential [7]. Such an approach allows for a determination as to whether a complete segment II study is required for each related agent. Although the screening potential of *in vitro* assays has been noted by a number of authors [8–14], it has proven difficult to develop an accepted, standardized screening battery [15,16]. Reasons for the difficulty include the unrealistic expectation that *in vitro* tests must be as conclusive as *in vivo* methods and an inability to reach a consensus on an appropriate set of validation chemicals. Nonetheless, the value of *in vitro* data and its proper role is gradually being incorporated into regulatory policy and practice. For example, a review of the EPA's published guidelines for evaluating the developmental, reproductive, and neurotoxic risk of chemicals [17–19] indicates an increasing flexibility to permit the use of *in vitro* data to supplement data collected *in vivo*. In addition, new efforts to assess the endocrine-disrupting potential of chemicals will employ a tiered approach with molecular and invertebrate-based assays as an initial step in evaluation [6].

Beyond screening of new compounds for developmental toxicity, *in vitro* models may be used to elucidate the mechanisms of developmental toxicity [20]. Advances in molecular and cellular biology have occurred in the last decade that have facilitated the investigation of gene expression patterns, cell-cycle kinetics, programmed cell death, and a host of other phenomena during development [21]. These methods have also allowed researchers to define cross-species similarities in cell-signaling pathways, in addition to morphologic patterns of cell differentiation [22]. Through these advances, toxicologists have improved our knowledge of the mechanisms of developmental toxicants, the corresponding implications for development across species, and the validity of *in vitro* assays for predicting *in vivo* effects. However, if we hope to develop a strategy for comprehensive toxicant assessment, the significance of changes at basic cellular and subcellular levels for overall development must be further evaluated in systems allowing for the complexities of morphogenesis (i.e., those encompassing cellular migration or organ development), toxicokinetics, and toxicodynamic interactions.

Two major factors contribute to the difficulties associated with designing *in vitro* models for developmental toxicology. The first difficulty is that the mechanisms of teratogenicity are poorly understood. Second, the process of development includes a complex and integrated sequence of cell proliferation, differentiation, and death. No unifying theory exists for the process of teratogenicity corresponding to the "somatic mutation theory" of carcinogenesis. Consequently, *in vitro* systems have been proposed that encompass a variety of developmental processes, including cell death, altered cell–cell interactions, cell migration, reduced biosynthesis, altered cellular, biochemical and morphologic differentiation, mechanical disruption, and growth inhibition. Two major philosophical approaches have been taken to design these *in vitro* models [23]. The first approach involves examining "like" events, in which processes such as regeneration, reaggregation, and cell–cell communication are monitored as a model process that has characteristics similar to those of *in vivo* development. Examples of these *in vitro* systems include the mouse ovarian tumor (MOT) cell assay [24] and the hydra assay [25]. The second approach to *in vitro* testing for teratogenicity is to design systems allowing the investigator to examine limited aspects of embryogenesis (i.e., "windows" of development). Examples of this approach include monitoring chondrogenesis in limb bud cultures and monitoring early organogenesis in rodent postimplantation cultures. In these approaches, either limited time periods during gestation are monitored or limited organ system processes are observed *in vitro*.

The major *in vitro* systems currently being used or developed for evaluating impacts of chemicals on development are summarized in Table 8.1. Because the successful application of *in vitro* techniques is critically dependent on the reliability of the study or test design, the assays should be reproducible between laboratories (high precision) and possess readily quantifiable endpoints accurately reflecting effects *in vivo*. When used for screening purposes, the *in vitro* tests should exhibit high sensitivity (i.e., few false negatives) and high specificity (i.e., few false positives).

This review specifically aims to summarize the approaches in use or under development for evaluating developmental toxicants *in vitro*. Emphasis is placed on the use of these systems for screening potential teratogens, with a particular focus on the monitored endpoints, sensitivity, specificity, and, when available, valida-

TABLE 8.1. Example systems used to assess the
potential impact of toxicants on developmental processes

Intact embryo/fetus	Cell culture
Mammals	*Primary cell cultures*
Preimplantation	Drosophila embryo cells
Postimplantation	Chick embryo neural retina cells
	Micromass (neural and limb bud cells)
Other vertebrates	Aggregate neural cell cultures
Chick	
Fish	*Established cell lines*
Frog	Mouse ovarian tumor cells
	Neuroblastoma cells
Invertebrates	Embryonal stem cells
Insects	Vaccinia-infected cell lines
Hydra	Cell–cell communication
	Teratocarcinoma cell lines
Intact organ culture	Human embryonic palatal mesenchyme cells
	Limb bud
Palatal shelves	

tion data. Examples of how particular models have been used to study mechanisms of developmental toxicity are presented when available.

8.1 PRIMARY CELL CULTURE

Primary cell cultures are obtained by disaggregation of the tissue or organ system from which the cells are derived. Use of these cultures is designed to investigate cell–type-specific cellular or molecular events underlying chemical toxicity. Primary cultures normally are composed of mixed cell types because they are derived directly from the tissue or organ they are intended to model. This process can be advantageous if the investigator desires a model system more accurately reflecting the characteristics and interactions existing *in vivo*, but it may complicate investigations of specific cell types. The dependency of the primary cell culture system on availability of tissues or organs normally means that cells still have to be harvested from animals. However, usually, fewer animals can be used and greater numbers of experimental conditions evaluated than can be used in the standard *in vivo* developmental toxicity assay. Constraints may be imposed on the types of assays that can be conducted if the amount of tissue available for the specific culture is limiting or if culture conditions required to maintain *in vivo*-like responses are difficult to reliably maintain. Despite these potential limitations, the relative convenience of cell culture, and the basic similarities between primary cells in culture and those in the intact animal, provide powerful incentives for using these systems. In the following sections, we will first consider the use of cells derived from nonmammalian species (fly and chick) and then consider those derived from mammals, primarily rodents.

8.1.1 Nonmammalian Species

Drosophila embryo cell culture. Primary cultures of embryonic Drosophila cells have been proposed as an *in vitro* teratogen screening system [26–28]. These cultures differentiate into several cell types: Neuroblasts form neurons and organize into ganglia, and myoblasts form myotubes. Differentiation is quantified using an automated image analyzer; a positive response is indicated by a statistically significant reduction in the number of myotubes or ganglia compared with controls. Over 150 chemicals (including both teratogens and nonteratogens and structurally related compounds) have been tested to date, demonstrating a low false-negative rate (<10%) [29,30]. Microsomal monooxygenase fractions from either rats or Drosophila can also be added to these cultures to investigate the role of metabolic activation in developmental toxicity. Cyclophosphamide is an example of a compound showing such enhanced toxicity *in vitro* after metabolic activation [31].

The utility of this model in genetic toxicology is also well recognized [32,33]. Komori et al. [34] reported using a transgenic Drosophila strain carrying a mammalian cytochrome P-450A1 gene to evaluate the effect of P-450 activation on the mechanism of action of promutagens and procarcinogens. Such a system could also provide a means for evaluating the role of bioactivation in teratogenesis.

Chick embryo neural cell culture. The chick neural cell culture technique [35] has been proposed as a teratogen screen by Daston and Yonker [36]. Neural retinas derived from day-6 chick embryos are dissociated into single cells and incubated in the presence of test chemical for 24 hours. After several days in culture, cells differentiate to form tissue layers comparable with those in the intact retina. The following endpoints may be quantifiably assessed: number and size of aggregated cells, protein content, histology, and gene expression. Daston et al. [37] reported the testing of 17 substances with this system; accuracy was 94% compared with mammalian *in vivo* data. Subsequently, they tested 22 additional chemicals (14 known developmental toxicants and 8 nontoxicants) and reached a 95% concordance with *in vivo* results [38]. The only teratogen that was tested and was not active in the assay, 2-methoxyethanol, was active via a metabolite, 2-methoxyacetic acid.

Despite the interesting characteristics of the system, the chick embryo neural cell culture system appears to have been applied only for a few investigations of teratogenic mechanisms. A review of the literature indicates cultured chick embryo cells have been used to study the mechanisms of developmental toxicity in carbamates [39] and ethanol [40].

8.1.2 Mammalian Species

Limb bud cells. Single-cell suspensions derived from both rodent and avian embryonic limb buds can be cultured via trypsinization and plating at high cell density. The attached cells proliferate and differentiate into chondrocytes over a 5- to 6-day culture period [41,42]. This system is sensitive to teratogens perturbing differentiation-specific processes, in addition to those that inhibit overall cell growth and proliferation.

Numerous differentiation-specific processes can be monitored *in vitro*, including cell aggregation, cell condensation, and cellular differentiation into cartilagenous tissue [23]. General cytotoxicity can be monitored via staining with the vital dye neutral red. Differentiation can be quantified using the cartilage-specific stain alcian blue or by monitoring $^{35}SO_4^{2-}$ incorporation into proteoglycans. Effects of test compounds on limb cell differentiation are quantitatively expressed as inhibition of alcian blue staining or radiolabel incorporation.

The teratogenic potency of a number of chemicals was assessed in a validation study involving 25 test chemicals by determining the concentration of compound reducing alcian blue staining to 50% of concurrent control values (IC_{50}) [41,43–48]. These laboratories reported that the IC_{50} values of teratogens are similar to the doses required to elicit teratogenicity *in vivo*. Kistler [49] determined the *in vitro* activity of 25 retinoids in the limb bud cell culture system and concluded that this system may be useful as a preliminary screen to identify nonteratogenic retinoids.

Guntakatta et al. [50] employed double radiolabeling techniques in which limb bud cells are cultured in the presence of $[^3H]$-thymidine to label DNA and $^{35}SO_4^2$ to label sulfated proteoglycans. This assay was designed to monitor specific disruption of extracellular matrix protein and glycoprotein synthesis leading to reduction in cartilage proteoglycan. In this assay, a $^3H/^{35}S$ ratio of >1 indicates specific inhibition of proteoglycan synthesis, suggesting teratogenic potential. Nineteen of 22 known mouse teratogens inhibited proteoglycan or DNA synthesis; five nonteratogens failed to elicit selective inhibition. The overall sensitivity of the limb bud cell system was approximately 86%, and the false-negative rate was approximately 14%. Because no false positives were observed, the specificity was 100% in this study.

An additional validation study was carried out by Wise et al. [51] to determine the specificity (ability to correctly identify nonteratogens or chemicals not exerting specific developmental toxicity) of the test. The $^3H/^{35}S$ ratio was determined for 23 compounds, most of which were not considered teratogenic. In this instance, the ^{3}H/^{35}S ratio was taken to indicate the potential of a compound to preferentially inhibit developmental processes (chondroitin sulfate synthesis) relative to overall cell function (DNA synthesis). All compounds produced ^{3}H/^{35}S ratios less than 2 in the assay, whereas in vivo 22 of the 23 compounds tested were classified as nonteratogens or nonselective developmental toxins.

The limb bud cell culture system has also been used to evaluate the mechanisms of teratogenicity. For example, Sass et al. [52] used the system to study the effects of retinoid metabolism on retinoic–acid-induced malformations. The role of the retinoic acid receptor in the teratogenesis of limb development was investigated by Jiang et al. [53]. On a macroscopic level, Renault et al. [54] used ultrastructural characterization of chondrogenesis to evaluate possible mechanisms of teratogenicity. In studying the teratogens retinoic acid and 6-aminonicotinamide and the nonteratogen doxylamine succinate, they observed differences in cellular disruption, suggesting mechanistic differences between the compounds. The system has also been used to explore differences in gene expression in response to toxicants. For example, expression patterns of the cell-cycle arrest genes gadd45 and gadd153 were found to change in response to methylmercury exposure in vitro [55].

Neural. Neural crest and midbrain neural cells have been successfully cultured and undergo differentiation in vitro [41,56]. Neural crest cells are derived from explanted segments of neural tube and are dissociated, trypsinized, and plated for in vitro culture. In the presence of fetal bovine serum, these cells differentiate into pigment-containing melanocytes over 6 days in culture [41,57]. If horse serum is present, they differentiate into neuron-like cells. The effects of 14 known teratogens and nonteratogens have been monitored in this in vitro system. Alterations in the growth and differentiation of these cells included detachment of cells, vacuolation, altered melanocyte or neuronal morphology, and inhibition of differentiation. A positive relationship between teratogenicity in vivo and alterations in these cultures was noted [41,57].

The use of micromass cultures derived from embryonic midbrains was developed and extensively characterized in the laboratory of Oliver Flint while at ICI Pharmaceuticals (Wilmington, DE) [46,56,58]. Neuronal cell cultures are established by isolating day-12 embryonic midbrain (mesencephalon) tissues, preparing single-cell suspensions, and plating at high cell density. Biochemical and morphologic differentiation of neuronal cell foci occurs after 5 days of culture. This differentiation can be monitored by staining the cultures with hematoxylin and quantitating differentiation by automated image analysis or by ^{3}H-GABA uptake. Flint has verified that the foci observed in these cultures are primarily neuronal cells by selective silver-staining techniques, reaction of monoclonal antibodies directed against neuronal ganglioside markers, and radiolabeled neurotransmitter uptake [56]. In our laboratory, immunohistochemical analysis confirmed the presence of differentiated midbrain cells possessing neuron-specific markers, such as GQ ganglioside, neural cell adhesion molecule (NCAM), microtubule-associated protein 2 (MAP2), MAP 5, neuron-specific enolase, and acetylated tubulin [59]. Using differentiation-related expression of the protooncogene *src* and neurofilament protein, Sweeney and Faustman [60,61] were able to show concentration-dependent decreases in protein expression after exposure to ethylnitrosourea. Such markers of neuronal differentiation may prove to be useful indicators of chemical alterations.

In an extensive validation effort, Flint and Orton evaluated the responses of midbrain and limb bud cell cultures to 46 compounds (27 teratogens and 19 nonteratogens) in a blind trial [46]. A complete monooxygenase system was successfully included in these cultures to provide metabolizing enzymes. Inhibition of differentiation (i.e., alcian blue staining) was monitored as an indicator of potential teratogenicity; concentrations of test compounds inhibiting differentiation to 50% of control values were determined. Of 27 teratogens, 25 inhibited either neuronal or limb bud differentiation. Only two false negatives were observed: 2,4-dichlorophenoxyacetic acid and thalidomide. If cell density was reduced, thalidomide did cause inhibition of limb bud differentiation. Eleven percent (2/19) of the nonteratogens inhibited neuronal or limb differentiation. Glutethimide (structurally related to thalidomide) and dimenhydrinate (both nonteratogens *in vivo*) inhibited differentiation in both cultures. If inhibition of differentiation in one cell type alone was examined, predictability decreased from 90% (both assays) to 85% for neuronal cells and 82% for limb bud cells. Flint et al. have examined a series of structurally related teratogenic and nonteratogenic compounds. The accuracy of the assay was sufficient to differentiate among the developmental toxicity of three pairs of structurally related compounds. The contribution of the exogenous monooxygenase system and metabolism by the differentiating cultures themselves is under further investigation [62–65]. The micromass system has been used to evaluate the teratogenic potential of numerous compounds, including benzimidazoles [66,67], pesticides [68], the drinking water contaminant MX [69], alkylating agents [70,71], aromatic compounds [72], and methylmercury [73,55]. The potential impact of using different animal strains to obtain cells for the micromass culture was investigated by Ward and Newall [74]. They used three different rat strains to study the response of cultured neural and limb bud cells to the teratogen transretinoic acid. They reported only minimal strain-specific differences for neural cells, but they found marked differences among limb bud cells. The differences in susceptibility observed with limb bud cells *in vitro* were similar to strain-specific differences in susceptibility noted *in vivo*.

The micromass system has been used in numerous mechanistic studies of developmental toxicity [75,76]. Ribeiro and Faustman [77,78] conducted cell-cycle analyses on the micromass cultures and demonstrated the utility of examining alterations in cellular kinetics in identifying chemical-induced perturbations of these cells. This work was extended by Ponce et al. [73] using the thymidine analog 5'-bromodeoxyuridine (BrdU) and flow cytometry to quantify the changes in cell-cycle phase distribution and cycling rates caused by methylmercury in embryonic rat central nervous system (CNS) cells. Histologic evaluations have also been incorporated into the micromass system. For example, Bolon et al. [79] evaluated cytotoxicity to specific cell types (glia or neurons) and differences in staining of neuron-specific marker proteins as a means of identifying specific mechanisms of toxicity. A thorough review of the micromass test, including updates on protocols and the results of validation and mechanistic studies, is provided by Flint [80].

Although most micromass work has focused on midbrain and limb bud cultures, a similar system has been developed using suspensions of cerebellar cells [81]. This system involves examining cells in the postnatal period. Suspensions of cerebella from 6-day-old mice are prepared and evaluated for aggregation and fiber formation in the presence of teratogens. This system was used to test the teratogenicity of valproic acid (VPA) and two stereospecific analogues and found to match well with reported *in vivo* data. Based on the utility of the micromass assay for identifying developmental toxicants, this system was recognized by the European Federation of Pharmaceutical Industries as having promising potential. The micromass culture system is one of three *in vitro* systems currently undergoing validation in studies sponsored by the European Community [82].

Aggregate cell culture systems. Distinct from the micromass approach, in which cells are plated in specific foci at high density, fetal cells may be grown in aggregate three-dimensional culture [83]. Cells are dissociated into a solution of culture medium and then incubated under constant motion. The cells form three-dimensional aggregates from the presence of specific cell–cell adhesion molecules. Because the cells form three-dimensional structures, this approach allows for the study of cell–cell interactions and models the formation of circuitry in the developing brain.

The system has been used to study the effects of several drugs, such as methamphetamine and fenfluramine [83]. In cultures of chick brain aggregates, results for three teratogens (cadmium chloride, ara-C, and phenytoin) correlated well with plasma levels eliciting toxicity *in vivo* [84]. This process was based on staining of specific neuronal markers, such as glialfibrillary acidic protein (GFAP) and neurofilaments. In another variation on the method, Bolon et al. [79] reported a method for culturing embryonic cerebrocortical cells in tissue culture chamber/slides. These cells undergo reaggregation, neuronal migration and process formation, and selective neuronal death. Using this system, Bolon et al. were able to distinguish among general toxicants, selective neurotoxicants (formate, methyl iodide), and those selectively toxic to dividing cells (cytosine arabinoside).

Other systems. Several other culture systems of primary cells have been proposed as methods for studying or identifying developmental toxicants, but they have not been widely studied. For example, Elstein et al. [85] proposed murine erythroleukemic cells (MELC) as a model for developmental toxicity. These cells exhibited similar cell-cycle perturbations to mouse limb bud cells when exposed to the developmental toxicant 5-fluorouracil. Saito [86] studied the effect of cholinesterase inhibitors on neurite outgrowth and growth cone collapse using briefly (<1 hour) cultured tissue slices from chick dorsal root ganglia (peripheral neurons) and retinae (central neurons). Validation of these systems will be required before they can be proposed for a screening role in toxicologic evaluation.

8.2 ESTABLISHED CELL LINES

Established cell lines are derived from existing tumor cells or by transformation of stem cells or differentiated cells so that the cells can be grown over multiple passages and to sufficiently large quantities to produce a stable stock supply for long-term use. As with primary cell cultures, established cell lines are used to investigate cell-type–specific cellular or molecular events underlying chemical toxicity. Established cell cultures may have advantages over primary cell cultures when cost, availability of cells, or use of animals are issues. Because established cell lines are based on modified versions of the normal cells from which they are derived, they may not be appropriate for investigation of specific endpoints, notably, cell-cycle regulation and intercellular communication. Moreover, established cell cultures may not appropriately reflect intercellular interactions that exist *in vivo* (and perhaps in primary cell cultures), which may play a role in chemically induced developmental toxicity because they are usually based on a single cell type. Despite these limitations, the use of these cultures in chemical screening may provide valuable information at low cost. In this section, we will describe the major types of cell lines used in developmental toxicology, including those derived from rodent ovarian tumors, human embryonic mesenchymal cells, embryonic stem cells, neuroblastomas, and teratocarcinomas.

Mouse ovarian tumor cells. This assay relies on the assumption that teratogens interfering with embryonic cell interactions also inhibit lectin-mediated attachment of cells. Mice are inoculated with ascitic MOT cells 1 week before the experiment is performed. Approximately 12 hours before use, the cells are radioactively labeled by intraperitoneal injection with [³H]-thymidine. Cells are then harvested, exposed to the test chemical, and allowed to sediment on concanavalin-A–coated polyethylene disks. The amount of radioactivity associated with the disks after washing indicates the extent of cell attachment. Inhibition of attachment at noncytotoxic concentrations has been used by the authors to identify potential teratogens. In a validation effort involving 102 test agents, 60 of the 74 teratogens (81%) inhibited cell attachment; 21 of the 28 nonteratogens (75%) were noninhibitory [87,88]. Nineteen percent of the teratogens tested (14/74) did not specifically inhibit attachment, possibly reflecting effects on DNA replication or mitosis.

Thalidomide was scored as a positive in this system when a complete system for monooxygenation was included [89]. Braun et al. [89] investigated the relationship between attachment inhibition and *in vivo* teratogenicity. For 54 of the 60 inhibitory teratogens, they found a significant correlation between the inhibitory *in vitro* dose and the lowest reported *in vivo* teratogenic dose.

Human embryonic palatal mesenchymal cells. A line of human embryonic palatal mesenchymal (HEPM) cells derived from a day-55 human abortus has been extensively characterized and proposed as a screening assay for teratogens [90–93]. The HEPM cells consist of undifferentiated fibroblast-like diploid cells with a stable chromosomal complement, high plating efficiency (>90%), and a cell-cycle transit time of approximately 2 hours.

The theory underlying this assay is that teratogens will cause growth inhibition of these rapidly proliferating embryonic cells. Dose-response curves for cell growth inhibition are generated, and an IC_{50} value is established. Growth inhibition is assayed by determining cell number as a percentage of control in the presence of the test compound. Growth inhibition below a test chemical concentration of 1 mM was postulated to reflect potential teratogenicity. Thirty-five teratogenic and 20 nonteratogenic compounds were tested in a series of validation assessments. The test compounds were compiled from agents suggested by the EPA consensus workshop [94] and included most of the false negatives from the MOT assay validation studies [24,87]. The HEPM assay correctly identified 23 of the 35 (66%) teratogens as true positives in this assay and identified 12 of the 20 (60%) nonteratogens as true negatives. Combining the results from the tumor cell assay of Braun et al. [24,87] and the HEPM assay yielded an overall predictability of 90%; the rate of false negatives was 3%. Incorporation of an exogenous metabolic activation system in the HEPM assay yielded concentration-dependent inhibition of growth by cyclophosphamide. The inhibitory concentrations were comparable with those observed in the whole-embryo culture system.

The National Institute of Environmental Health Services (NIEHS) (through the National Toxicology Program) evaluated the MOT assay [24] and the HEPM assay [91–93]. This evaluation study used 44 coded compounds (30 animal teratogens and 14 nonteratogens). These agents were tested in each assay in two different laboratories [95–96]. The optimal combined MOT/HEPM assay was approximately 70% accurate and had greater than 80% sensitivity [97,98]. These assay results agreed with *in vivo* teratogenicity data for 60 to 70% of test chemicals, depending on the assay used. Draus et al. [45] compared the effects of some 20 compounds in the HEPM growth inhibition assay and the mouse limb bud assay. Using rat liver S9 systems when metabolic activation was required, preliminary results suggest that the mouse limb bud assay may be a more sensitive assay.

The combined MOT/HEPM assay was more recently used to examine structure-activity relationships (SAR) in the developmental toxicity of chlorinated phenols [99]. A clear SAR was observed between toxicity in the assay and the degree of chlorine substitution, in agreement with *in vivo* data. However, contrary to the existing *in vivo* data, the test predicted all of the phenols evaluated would be teratogenic, which suggests that the utility of this combined test lies as a screening tool for prioritizing *in vivo* studies and reinforces the notion that *in vitro* methods alone cannot be relied on by to evaluate human health risk.

Neuroblastoma cells. Various cell lines derived from embryonic brain tumors have been used to investigate the possible teratogenic effects of environmental chemicals. These cell lines include the P19, F9, and PC12. Under the appropriate conditions, these cells differentiate into neurites in culture; both growth inhibition and stimulation of differentiation can be monitored in the presence of test chemicals. Mummery et al. [100] proposed using differentiating murine neuroblastoma cells in culture as predictors of teratogenicity. Thirty-nine teratogens and 18 nonteratogens were tested for their ability to induce differentiation and inhibit growth. Eighty-six percent of the test compounds were correctly identified as teratogens. The authors contend that the proportions of both false negatives (10%) and false positives (22%) were comparable with other *in vitro* teratogenicity screening systems.

Various neuroblastoma cell lines have found considerable application in studying the mechanisms of neurotoxicity and developmental toxicity. Weismann and Caldecott-Hazard [101] studied the effects of methamphetamines on cell mortality and cell proliferation using trypan blue exclusion and [^{3}H]-thymidine incorporation, respectively. This work was performed in conjunction with *in vivo* studies that examined changes in neurotransmitter levels. Some concordance existed between *in vivo* and *in vitro* toxicity, although evaluations of different endpoints (e.g., [^{3}H]-thymidine incorporation versus neurotransmitter levels) made comparison difficult. Neuroblastoma cells have been used to study the mechanisms of action of other developmentally relevant compounds, including retinoids [102], metals [103–106], alkylating agents [107–109], and pesticides [110]. It should also be kept in mind that because immortalized cell lines have altered proliferation characteristics, their utility for studying the mechanisms of toxicity may be limited; for example, cell-cycle perturbations in this cell line may have little relevance to similar effects *in vivo* in the normal organism.

Embryonal stem cells. In 1991, Laschinski et al. [111] described an *in vitro* developmental toxicity screen using embryonal stem cells (ESCs) derived from mouse blastocysts. The potential use of this system in developmental toxicology has subsequently been described by a number of authors [112–116]. ESCs can be isolated from the inner cell mass of the murine blastocyst and maintained indefinitely in an undifferentiated state in the presence of inhibitory factors. After removal of the inhibitory factors, the cells undergo differentiation into various cell types; e.g., muscle cells and hematopoietic cells. The differentiating cells can then be exposed to suspected teratogens in a manner analogous to the micromass test system. The concentration of test chemical eliciting toxicity (i.e., cell death or inhibition of differentiation) may be compared with corresponding data from mouse fibroblasts.

Laschinski et al. determined that the ESCs showed a higher sensitivity to known teratogens than did fibroblast cultures [111]. An *A/D* ratio can therefore be calculated, in which *A* corresponds to the fully differentiated fibroblasts and *D* corresponds to the embryonal stem cells. Comparing the *in vitro* ESC results with *in vivo* teratogenicity data from mice revealed a 31% false-negative rate (i.e., 5 of 16 chemicals tested) [111]. The authors noted that inclusion of an exogenous biotransformation system could constitute a valuable refinement of the assay.

Additional validation of the ESC model has been conducted by Newall and Beedles [112, 113]. Using a test battery of 25 teratogens and nonteratogens, they reported a sensitivity of 73% and a specificity of 70% [113]. The test proved more reliable at identifying highly potent teratogens (e.g., cyclophosphamide, retinoic acid) than are those that are less potent (e.g., caffeine). Additional evaluation of this promising model has been reported in several recent conference abstracts [114, 115]. The methods of ESC culture and the parallel nature of stem cell differentiation *in vivo* and *in vitro* have been well described by Heuer et al. [116]. A current European validation study is evaluating these cells along with micromass cultures and whole-embryo cultures for their use in developmental toxicity assessment [82].

Vaccinia-infected cell lines. The ability of primate-derived cell cultures to suppress infection by vaccinia virus has been proposed as a teratogen-screening test [117]. Vaccinia-infected monolayers are exposed to the test compound. After a 24- to 46-hour incubation at 37°C, the number of active viruses released from infected cells is determined by counting plaques. Infected monolayers that are either untreated or treated with vehicle alone serve as controls. The rationale underlying this system is that viruses will only reproduce if they infect actively proliferating cells. Furthermore, virion replication involves a complex sequence of genetic and molecular events postulated to reflect differentiation-like processes. In validation studies, vaccinia-WR-infected BSC 40-cell monolayers were exposed to 42 teratogens and nine nonteratogens. Thirty-three of the 42 (79%) teratogens tested inhibited virus proliferation, and eight of the nine (89%) nonteratogens were noninhibitory. Six of the 42 (14%) teratogens were false negatives, and three teratogens stimulated the virus. One out of nine (11%) nonteratogens was a false positive. The assay was sensitive to two agents known to require metabolic activation. Qualitative comparisons between the dose-producing 50% inhibition/stimulation compared with a control-infected culture (RD_{50}) and the lowest reported teratogenic dose *in vivo* suggested a linear correlation ($r = 0.98$). To determine whether this viral assay detects differentiation-specific effects rather than general growth inhibition, the cytotoxic effects of each teratogen at the RD_{50} were assayed on uninfected cell cultures. Only six of the 42 teratogens were positive for cytotoxicity. The viral endpoint was considerably more reliable than was any of these cytotoxicity endpoints in predicting teratogenicity, suggesting that the assay is monitoring specific rather than general effects.

The teratogens used in this assay were separated into (1) "molecular teratogens," agents that acted on DNA information, storage, and transfer, and (2) "morphogenetic teratogens," agents that interfered with cytoplasmic metabolism, cytoskeletal function, or intercellular signaling. The highest correlation to *in vivo* results was seen with the "molecular teratogens"; however, 82% of the "morphogenetic teratogens" were also correctly assigned.

Cell–Cell communication assay. Inhibition of intercellular communication during critical periods of organogenesis has been proposed as the mechanism of action for certain teratogens. Inhibition of gap-junction–mediated communication can be monitored using the metabolic cooperation assay [118–120]. This system involves assaying the recovery of a few (usually 100) 6-thioguanine-resistant cells (HGPRT$^-$) in the presence of an excess (usually 4 x 10^5 cells) of 6-thioguanine-sensitive cells (HGPRT$^+$). Under control conditions, a phosphorylated 6-thioguanine product from HGPRT$^+$ cells is transferred via gap junctions to HGPRT$^-$ cells. Incorporation of this phosphorylated product into cellular DNA results in the death of both cell populations. However, inhibition of cell–cell communication results in the recovery of HGPRT$^-$ cells. Such assays are interpreted as positive only when chemical concentrations eliciting rescue of HGPRT$^-$ cells are neither cytostatic nor cytotoxic. Originally proposed as an assay for tumor promoters, it is evident that many other critical biologic processes are mediated via cell–cell communication [119,120]. A series of compounds with varying teratogenic potential have been tested [121–126], including a series of ethylene glycol derivatives, diphenylhydantoin, warfarin, phorbol esters, phenoxyacetic acid derivatives, retinoids, and nonteratogens, such as saccharin and ascorbic acid. Known teratogens were shown to inhibit metabolic cooperation at concentrations equivalent to those effective in *in vivo* teratogenicity studies; the nonteratogens saccharin and ascorbic acid failed to inhibit metabolic cooperation over these concentration ranges.

Because a variety of well-established nonteratogenic tumor promoters also inhibit metabolic cooperation *in vitro*, the use of this system as a screen for teratogens is still uncertain [123]. Some teratogens may also elicit developmental toxicity primarily through inhibition of cell–cell communication. Modifications of the metabolic cooperation assay, such as monitoring intercellular communication in human cell lines, including HEPM cells, may improve the information obtainable from this assay for human evaluation [127].

Teratocarcinoma cell lines. Hulme et al. [128] described the use of a differentiating teratocarcinoma cell line as an *in vitro* developmental toxicity screen. Originally derived from a mouse testicular teratocarcinoma, these cells can be stimulated to differentiate by retinoic acid treatment. Differentiation is monitored by determining synthesis of laminin, allowing evaluation of test agents in undifferentiated, differentiating, and differentiated cells. Few chemicals have been tested thus far in this system, so its general utility is unclear. Skladchikova et al.

[129] found that the teratogen VPA, but not its nonteratogenic analog, 2-iso-propylpentanoic acid, blocked the effects of retinoic acid on cell proliferation, viability, and neuronal differentiation in the NTERA-2 cell line. VPA did not dedifferentiate already differentiated NTERA-2 cells, suggesting a narrow window of susceptibility in which retinoic acid exerts its effect. One potential confounding aspect of these teratocarcinoma cell-line systems is that retinoic acid, an *in vivo* teratogen, is required to stimulate differentiation.

8.3 WHOLE-ANIMAL SYSTEMS

Of all *in vitro* testing systems used to identify potential human teratogens, the culturing of whole embryos, or, in certain cases, whole organisms, may provide the most direct and meaningful evidence of risk to humans. These experiments can be conducted to largely maintain the critical spatiotemporal interactions underlying normal development and that would be at least temporarily disrupted in preparing a primary cell culture and would be largely absent when using a cell line. On the other hand, whole-animal culture is perhaps the most complex and demanding of the *in vitro* techniques, requiring considerable skill and resources. Nonetheless, several whole-embryo/whole-animal systems are perhaps the most promising candidates for use as screening tools and provide a valuable contribution to evaluation of teratogenic mechanisms. In this section, we first discuss nonmammalian systems (chick, frog, fish, invertebrates) before turning to the use of preimplantation and postimplantation rodent embryos.

8.3.1 Nonmammalian Systems

Chick. The chick embryo has been extensively used in developmental biology studies and early teratogen evaluation. Chemically induced perturbation of development is readily amenable to analysis after injection of test agent into the subgerminal yolk or chorioallantoic vessels. Embryos or chicks (in egg hatchability studies) are subsequently examined for structural and skeletal abnormalities.

In 1967, the World Health Organization recommended against the use of this *in ovo* system because of its nonmammalian pharmacokinetics and high, nonspecific sensitivity. Subsequently, considerable effort has been expended in characterizing and standardizing the chick embryotoxicity screening test (CHEST) [130–138]. Modifications of the *in ovo* system have facilitated the simultaneous screening of genotoxicity and embryotoxicity [139,140]. Jelinek et al. [131] reported the testing of 130 substances for their ability to cause death, malformations, and growth retardation *in ovo*. The lower portion of the embryotoxicity dose-response curve was determined for 117 out of the 130 test agents. Of the 13 "nonembryotoxic" agents, four had solubility limitations and nine reached biologically implausible concentrations. Numerous agents requiring metabolic activation were detected as positive in this system.

Despite the fact that this system has found extensive use and has been well characterized in the field of developmental biology, the absence of maternal metabolism and drug distribution, in addition to differences in the functional properties of the membranes surrounding the chick embryo, represent severe disadvantages [141]. This disadvantage undoubtedly has limited the use of this model in both screening and mechanistic evaluations when other mammalian models are available. The model has found limited use in studying the developmental toxicity of pesticides, a field in which the avian model is well established [142,143].

Fish. A variety of fish systems have been used in the study of reproductive and developmental toxicity [144–147]. Fish embryos represent a simple vertebrate system in which the embryo can be manipulated and monitored throughout the entire developmental period. Fish species used for such studies include Japanese medaka, zebra fish, rainbow trout, and fathead minnows. Assays can be performed in static systems (in which eggs are placed in deep Petri dishes) or in flow-through systems (in which water is continually circulated) [144]. Alternatively, embryos may be microinjected with test agent [148]. Endpoints such as hatchability, larval mortality, and teratogenicity may be monitored, and a standardized morphologic scoring system has been developed [149]. Exogenous monooxygenase systems may be added or endogenous piscine monooxygenase systems may be induced to examine the effects of chemical metabolism on teratogenic potency [148,150,151].

A variety of compounds have been tested for developmental toxicity in these fish systems, including metals [144,145,152,153], pesticides [144–154], carcinogens [155–160], and complex mixtures [148]. Birge et al. [145] tested 34 inorganic elements and 25 organic compounds in multiple fish and amphibian populations. These investigators reported that the rainbow trout was generally the most sensitive species to teratogenic effects when compared with goldfish, largemouth bass, sunfish, and amphibians. However, channel catfish embryos were more sensitive to mercury than were rainbow trout embryos. Although a variety of malformation types were observed, upward of 80–90% of the gross anomalies identified in fish embryos after exposure to either inorganic or organic toxicants involved skeletal system alterations. VanLeeuwen et al. [161] compared teratogenicity test results derived from trout, zebrafish, and chick for ten compounds used in the tire industry. A good correlation ($r = 0.9$) was observed among the test results in the three species. Further characterization of dose–exposure–time relationships in fish systems would be required before these models could be used for routine screening purposes.

A concerted multiphase effort has been made at assay validation of the frog embryo teratogenesis assay-xenopus (FETAX) assay. Seven different laboratories evaluated three chemicals (hydroxyurea, isoniazid, and 6-aminonicotinamide) in the first phase of validation [167]. Considerable interlaboratory variation was observed in test results; for example, the coefficient of variation (CV) in reported EC_{50} (malformation) values ranged from 32.7 to 70.1, depending on the compound studied. Differences in technician training and problems in microbial contamination of cultures were identified as major contributors to interlaboratory variance. In phase II, additional training reduced the level of interlaboratory variation somewhat, with CVs ranging from 7.3 to 54.7. In phase III, six compounds were tested (monosodium glutamate, ascorbic acid, β-aminopropionitrile, sodium acetate, sodium arsenate, and copper sulfate), and the CV in reported EC_{50} (malformation) values ranged from 53.0 to 134.9%, depending on the compound studied [168]. The high degree of interlaboratory variability in this phase may have resulted from allowing technicians to choose their own set of test concentrations, with some selecting broader ranges than others [168]. This result suggests that fairly rigid guidelines may be required if the FETAX assay is to achieve reproducible results across different laboratories. The FETAX system is currently being reviewed for possible regulatory applications by the NIEHS [171].

The FETAX system has been used to study the mechanistic aspects of several developmental toxicants, including nicotine [172], urethane [173], coumarin [174], acetominophen [175], benzo(a)pyrene [176], ethanol [177], DMSO [177], formamide [177], trichloroethylene [178], and theophylline [179]. Many of these studies focused on the role of biotransformation or bioactivation in developmental toxicity. For example, Fort et al. [179] used the FETAX system with exogenous metabolic activation to evaluate the teratogenic potential of theophylline and its major metabolites demethyluric acid and methylxanthine. FETAX has also been used for studies with mixtures of toxicants [180].

The American Society for Testing and Materials (ASTM) has developed a *Standard Guide for Conducting Frog Embryo Teratogenesis Assay-Xenopus* [181], and an *Atlas of Abnormalities* has been published [182] to assist in embryo staging and identification of malformations. These guidelines should increase interlaboratory consistency and acceptance of this test as a standard method.

Sabourin et al. [166] compared the efficacy of the frog system with the hydra and planaria systems for four known mammalian teratogens. The frog embryo system offered several advantages over these other systems because of its numerous endpoints, and it was distinctly more sensitive than was the planarian assay. However, because the FETAX assay is run for only 96 hours, late developmental effects, such as limb dismorphogenesis, are not monitored.

Insects. The Drosophila melanogaster screen involves deposition of eggs onto a nutrient medium-containing test chemical [183]. The developing fly feeds on this medium; exposure is throughout the metamorphosis period from egg to pupa formation. Numerous morphologic abnormalities and distinct abnormality patterns can be monitored in adult flies. Advantages to this system include the capacity to absorb, circulate, metabolize, and excrete chemicals [184]. However, it is impossible to precisely define the chemical concentrations to which the eggs are exposed. Several chemicals and physical treatments (heat [185]) have been used to induce abnormalities in Drosophila. However, most compounds tested have been water soluble [183]. This system has not been extensively used for mechanistic toxicity studies, although early work in Drosophila genetics defined early research in mutagens and developmental biology [186–188]. Eisses [189–191] used the system to examine the effects of the differing alcohol dehydrogenase genotypes in the toxicity of several teratogenic chemicals.

Cricket embryos have been examined as an invertebrate teratology model [192]. Endpoints, such as number of compound eyes, are monitored after exposure to a variety of individual compounds, as well as to complex organic mixtures.

Hydra. The hydra reaggregation and differentiation assay evaluates the effects of test chemicals on adult hydra, as well as on an "artificial hydra embryo" [25,193,194]. Adult hydra are exposed to a test substance, and the minimal toxic concentration is determined. Subsequently, dissociated hydra cells are allowed to reform adult, freestanding hydra. The "embryo" is exposed to the test chemical during this process of whole-body regeneration. The dose of the test compound causing toxicity in the adult hydra (A) is then compared with the dose causing developmental toxicity in the "embryo" (D). The developmentally toxic dose is typically defined as the dose preventing successful reaggregation of the dissociated hydra cells into an adult. An A/D ratio greater than 1 has been used by the authors to identify compounds selectively more toxic to the developing organism, and it has been proposed as a method for identifying teratogenic hazard. Thalidomide and vinblastine yielded A/D ratios of 60 and 30, respectively [193]. However, most substances yield ratios near unity, suggesting a lack of selective developmental toxicity. A modification of the hydra assay compares the effects of test substances on intact adult polyps with effects on the regenerating isolated adult digestive system [195]. Thus, effects on redifferentiation and reorganization can be investigated.

Of 30 agents tested, close agreement of the A/D ratio in the *in vitro* assay to *in vivo* A/D ratios in mammals was observed [166,193,196,197]. The SARs of 38 chemicals, primarily chlorophenol analogs, were studied in both the hydra assay and whole-embryo culture [198]. In both systems, all of the chemicals studied were found to be coeffective developmental hazards (chemicals perturbing fetal development only at maternally toxic doses) and not teratogens. Xylenes and related compounds have also been evaluated using this model [199]. SARs for a series of glycols and glycol ethers were examined in the hydra assay, and a series of predictions of their potential *in vivo* teratogenicity was suggested [200]. However, Welsch [122] cautions that *in vivo* studies with glycol ethers show dramatically different A/D ratios for such compounds compared with the A/D ratios observed in the hydra assay. Other investigators have also questioned the utility of A/D ratio for hazard prediction [201,202]. Clearly, an extensive validation effort is required to critically determine the utility of this screening assay.

Other invertebrates. Several other invertebrates have been proposed for teratogenicity testing, including Planaria [166,203–205], Daphnia [206], sea urchin embryos [207–210], and brine shrimp larvae [211,212]. Both Daphnia and mysid shrimp have been proposed as possible systems for evaluating possible endocrine-disrupting chemicals [6]. However, none of these systems has been well characterized or validated for evaluating developmental toxicity.

8.3.2 Mammalian Whole Embryos

Preimplantation embryos. The preimplantation stage represents a critical period in mammalian embryonic development because the organism consists of relatively few pluripotent cells. Consequently, the preimplantation embryo is a potentially sensitive target for developmental toxicants.

Fertilized eggs of mice, rats, and rabbits are readily cultured from the one-cell stage through blastocyst stages [23,213–219]. A variety of experimental manipulations are possible, including (1) treatment *in vivo* followed by culturing *in vitro*, (2) culturing in serum from treated rodents, or (3) treatment *in vitro* followed by reimplantation into foster dams [213,216–222]. Endpoints monitored in preimplantation embryo culture can include lethality, morphologic alterations, growth retardation, micronucleus formation, sister chromatid exchange, and biochemical changes, such as oxygen utilization, nucleic acid synthesis, specific protein synthesis, and isozyme pattern changes [213,216,218,222,223]. Because cytochrome P450 monooxygenases have been detected in preimplantation mouse embryos [224], it may be possible to detect both direct- and indirect-acting agents in these cultures. Although the effects of chemical agents on preimplantation embryos depend on the number of affected cells [225], exposure to mutagens early in development has been shown to yield abnormalities [23,221,226–228]. However, the primary effect of exposure during preimplantation is lethality and growth retardation [23,217,218]. Agents tested in the preimplantation embryo culture system include cancer chemotherapeutic compounds [229,230], hormones [231], polycyclic aromatic hydrocarbons [232], metals [233,233,234], ethanol and acetaldehyde [235], miscellaneous therapeutic agents [236,237], ethylene oxide [227], and radiation [223]. Recent modifications of the culture technique have improved the performance of the methodology [238].

Postimplantation embryos. The whole-embryo culture technique has been used extensively as a testing system for developmental toxicants [239–250]. Rat or mouse embryos are explanted from the uterus at early somite stages, placed into culture medium, and maintained in culture from 24 hours to 4 days. During this period of early organogenesis, embryonic growth and development can be supported over most of the postimplantation embryonic phase up to 72 hours [242,249,251]. Modifications of the culture technique have allowed extension of the culture period [249–257], but the success rate is relatively low and impractical for routine teratogenicity testing.

The endpoints routinely monitored in whole-embryo culture include viability, gross and histologic abnormalities, growth, and macromolecular content. Although several morphologic scoring systems have been developed in attempts to standardize, quantify, and objectively evaluate the extensive embryonic development occuring in postimplantation culture [258–260], further efforts are required before general embryotoxicity and specific teratogenic effects can be distinguished [4].

Chemicals requiring metabolic activation can be assessed in this system by inclusion of exogenous biotransformation systems (postmitochondrial supernatant or microsomes) [239,261–264], cocultivation with hepatocytes [265,266], or culture in the presence of serum from treated humans and monkeys for embryo culture medium [267,268]. However, Juchau et al. [269,270] reported that cultured rat conceptuses contain cytochrome P450 in sufficient quantities to elicit biotransformation. Both Ferrari et al. [271] and Flynn et al. [272] reported that sera from women with a history of chronic spontaneous abortion affects *in vitro* development of postimplantation embryos when used as an embryo culture medium.

Agents that have been investigated in the postimplantation culture system include vitamins and antimetabolites [273,274], metals [275–279], ethanol and its metabolites [280,281], caffeine [282,283], salicylates [284–286], cyclophosphamide and its metabolites [261,288–291], hydroxyurea [292], chlorambucil [293], DES and E2 [294,295], retinoic acid and its derivatives [296–299], solvents [300], saccharin [301], immunosuppressive agents [302], miscellaneous cancer chemotherapeutic agents [303,304], antibiotics [305,306], amphetamine [307], diphenylhydantoin [308], thalidomide [309], urethane [310], VPA [308,311,312], hyperthermia [313–318], fungal toxins [318–323], ethylating and methylating agents [324,325], aromatic amines and their metabolites [263,264,269,326–329], glycol ethers [330], nitroheterocyclic compounds [331], halogenated aliphatic and aromatic hydrocarbons [332,333], haloacetic acids [334], substituted phenols [335], and other chemicals [336]. Bechter et al. [296] showed excellent qualitative and quantitative correlations between the whole-embryo culture and *in vivo* teratogenicity for five structurally related retinoids.

The majority of the studies referred to above have focused primarily on mechanistic teratology and assessments of SARs. For example, Brown-Woodman et al. [332] examined the individual and combined effects of chloroform, dichloromethane, and dibromoethane on embryonic development. They found a consistent deficit of yolk sac blood vessels compared with controls and cell death in the neuroepithelium of the developing neural tube. The chemicals appeared to act in an additive manner, suggesting a common mechanism of action.

A number of studies have been conducted that examined the effects of particular chemicals in whole-embryo culture as well as in the micromass system [46,338–341]. This approach allows for the examination of cellular and organism level effects. For example, the effects of styrene and styrene oxide were examined in this combined system [341]. Data on cytotoxicity, cellular differentation and growth, whole-embryo growth and dysmorphogenesis were collected. Differentiation of limb bud cells was found to be a particularly susceptible endpoint, whereas differentiation of neural cells occurred at similar levels that elicited cytotoxicity. The concentration-producing neural malformations in the whole embryo culture was about twice that found to affect differentation in the CNS micromass cultures, though serum concentration difference between the two culture systems may have played a role in this effect. Carefully controlled validation studies using the whole-embryo culture system are rare and limit its utility for compound screening. However, the current European validation studies evaluating this system with the micromass and embryonic stem cells culture systems should help to address this deficiency [82].

The embryo culture system has proven useful in screening multiple, structurally related chemical products to prioritize those for further product development and for providing mechanistic clues for *in vivo* follow-up. Like the preimplantation system, the postimplantation system still relies on animal use, but it can minimize the number of animals used in testing. It does, however, require considerable skill in terms of embryo culture and evaluation. Schmid [240] tested 38 chemicals in this system, including 19 known *in vivo* teratogens and 19 compounds reported to have no adverse effects on *in utero* development, and reported good agreement of *in vivo* and *in vitro* results. This article was followed by a study by Cicurel and Schmid [342] that investigated 17 known *in vivo* teratogens and ten nonteratogens. All 17 teratogens tested positive in the test, whereas eight of ten nonteratogens tested negative. Similarly, Sadler et al. [343] compared the *in vivo* and *in vitro* results for four compounds and reported concordance with *in vivo* responses for these agents. However, *in vivo* pharmacokinetic considerations are essential when using this system as a screen for developmental toxicants [343]. Thus, this system has been useful for the screening of close structural analogues when data for *in vivo* developmental toxicity of related compounds exist [344].

8.4 ORGAN CULTURE

Limb buds. Originally applied in the laboratories of Shepard and Kochhar, the limb bud culture system is one of the most well-characterized organ culture systems. Rodent limbs are excised from embryos at approximately 10–11 days (33–45 somite mouse). Morphologic differentiation *in vitro* is highly consistent, reproducible, and complex [23,250]. At the 10- to 11-day stage, the limb buds are largely undifferentiated, containing only blastema and surrounding epithelium but with minimal collagen or glycosaminoglycans. After 7 days in culture, these explants develop cartilaginous bone structures representing scapula, humerus, ulna, radius, and hand skeleton [345]. The following developmental endpoints can be monitored: cell proliferation, differential growth, nucleic acid and protein content, morphogenetic cell death, size and shape of limb parts, chondrogenesis, and collagen or proteoglycan biosynthesis [346].

Concentration-response relationships for abnormal development have been established using quantitative assessments of limb development. Neubert et al. [347] described a scoring system for morphogenetic differentiation that assigns a score to each bone rudiment and digit. Thus, a statistical analysis of treatment effects in addition to the detection of gross abnormalities may be performed [250].

Numerous *in vivo* teratogens induce abnormal development in limb buds *in vitro*. These agents include antimetabolites [347], alkylating agents [346,348,349], thalidomide derivatives [350], pesticides [351], analgesics and antirheumatics [353], caffeine [353], sulfonamides [354], and vitamins and derivatives [346]. For example, Shury et al. [355] studied the interaction between cadmium and retinoic acid in mouse limb bud development. Both compounds resulted in distal aphalangia (absence of bones in the digits), whereas retinoic acid alone also caused preaxial forelimb extrodactyly (extra digits) and fusion of carpal bones. They concluded that the effects of combined cadmium and retinoic acid treatment were morphologically attributable to retinoic acid.

The main advantage of organ culture is that complex interactions of different mechanisms can be studied in the developing organ. However, this technique is laborious and lacks complete metabolic activation, although exogenous biotransformation systems have been incorporated into the system [348,356].

Palatal shelves. Both Shiota et al. [357] and Buckalew and Abbott [358] described a technique for the cultivation of explanted palates from mouse fetuses by a suspension culture technique. The palates of day-12 fetuses close within 72 hours of culture, and those of day-13 fetuses within 48 hours. Both groups report that the *in vitro* fusion of palatal shelves accurately mimics *in vivo* development. Abbott [359–363] and Couture et al. [364] have investigated and compared gene expression and other responses of human and mouse embryonic palatal shelf cultures to retinoic acid, dexamethasone, and 2,3,7,8-tetrachlorofibenzo-p-dioxin (TCDD). Additional work has been conducted with methanol [365] and 5'-fluorouracil [366].

8.5 CELL-SIGNALING ASSAYS

Increased attention has recently been given to the use of cell-free assays and transfected cells to examine the effects of environmental chemicals on specific receptor-mediated cell-signaling pathways involved in development. Assays, such as the estrogen and androgen nuclear receptor binding assays, can be used to obtain information on the potential of a chemical to perturb estrogen- or androgen-stimulated signaling pathways *in vivo* [367]. These assays have been used to obtain extensive SAR data for endocrine-altering compounds, particularly in the pharmaceutical industry. These assays have recently been proposed by the EPA as part of an initial tier of an endocrine disruptor screening process [6]. The EPA has also discussed the use of transfected cells to examine cell-signaling implications in developmental toxicity. Such systems can examine not only receptor binding, but also downstream effects on gene transcription and translation. Because it examines an actual effect of receptor binding, this type of assay can distinguish between the effects of receptor agonists and antagonists and might be useful for evaluating potential synergistic effects among compounds.

Support for the application of these approaches in early screening of potential developmental toxicants is provided by Gerhart and Kirschner [22], who have noted the remarkable similarity in basic cell-signaling pathways found across cell types and species. Advances in molecular biology should allow us to capitalize on such cross-species homology in signaling pathways to design more effective early screening systems for developmental toxicology. Although such systems are not yet clearly defined, the implications of using these approaches for human health risk assessment has been discussed by Lee and Taylor [367].

8.6 RELATING *IN VITRO* RESULTS TO RISK FOR THE INTACT ORGANISM

The ultimate goal of *in vitro* toxicology is to provide a simplified testing scheme that can contribute useful information about the effects of exposure to a particular chemical to the whole organism and to provide a useful format for mechanistic evaluations. In order to limit time and cost, the *in vitro* approach simplifies various aspects of the whole organism. In order to interpret *in vitro*-derived results for the *in vivo* condition, these simplifications must be taken into account.

8.6.1 *In Vitro–In Vivo* Dose Conversion

A primary problem in extrapolating the results of *in vitro* testing to the whole animal involves the conversion of the *in vitro* dose of the chemical (typically given in culture medium) to the corresponding concentration present in the intact organism. The simplest approach is to assume that the concentration in the culture media can be related to the concentration in plasma. However, even this approach requires an understanding of the toxicokinetics of the chemical involved, that is, the distribution and metabolism of the chemical among the various body tissues [368]. For example, if a chemical is cleared from blood quickly, the *in vitro* dose may overestimate the toxic effect. The toxicokinetic behavior of a chemical may be estimated from the chemical's properties (e.g., lipid solubility, presence of functional groups, similarity to known compounds), but it cannot be certain that the chemical will behave as predicted. A more rigorous approach would be to collect the cultured tissue and assay it for the chemical of interest, obtaining a concentration value (e.g., microgram/gram), equivalent to target organ concentrations, which can be compared with similar results *in vivo*. However, because a fairly significant amount of sample is usually required (e.g., 1 g) for routine chemical analysis, this approach has rarely been done.

A number of other *in vitro—in vivo* differences should be considered when attempting to interpret *in vitro* findings. These differences include the presence of conjugating/inactivating ligands that may be present in reserve in the intact organism (e.g., glutathione), the effects of *in vivo* transport-facilitating carriers (e.g., transferritin), differences in serum protein concentrations between culture systems and between culture media and plasma, differences in metabolizing enzyme status between cells/tissue in culture and those in the intact organism (e.g., inducible CYP-450s), barriers to chemical transport, such as the blood–brain barrier, or sources of chemical metabolism or sequestration (e.g., cadmium and the placenta).

One quantitative solution to these issues is to develop a physiologically based toxicokinetic model of chemical distribution in the maternal and fetal tissues [369]. Models have been developed for a number of compounds of concern for developmental toxicology [370–374]. However, the problem of obtaining the necessary data for model development and validation remains. The requirements for this amount of data would tend to negate the advantages of *in vitro* testing for screening purposes, unless the test was used widely enough to make the effort to gather kinetic information worthwhile. In general, before attempting to include *in vitro* data in assessments of human health, a qualitative assessment of the chemicals' metabolism and distribution should be obtained.

8.6.2 Limitations of *In Vitro* Data for Assessing Health Risks

We have broadly reviewed the *in vitro* systems used for evaluating potential chemical teratogens, including the specificity and the sensitivity of the methods when available. Some of the methods are valuable in qualitatively assessing the teratogenic risks of a particular chemical, especially when several tests are used in series or where structurally related compounds are evaluated and qualitative comparisons are made. However, from regulators' standpoint, the current status of *in vitro* systems limit their utility for answering questions for risk evaluation. A number of instances occur, in which *in vivo* versus *in vitro* differences in susceptibility are likely. In general, cells or tissues in culture experience a different milieu than do cells or tissues *in vivo*. For example, the higher oxygen tension *in vitro* may predispose cells to oxidative stress or may conversely upregulate enzymes (i.e., those involved in glutathione synthesis) protecting the cells from oxidative stress. In addition, in the intact organism, a reserve capacity of protective enzymes or ligands may exist, drawn from both the fetus and the mother, which may not be present *in vitro*. Whole-embryo culture systems would be expected to most closely resemble the *in vivo* condition, but even this system is not a true representation of the *in vivo* condition. Again, *in vitro* analyses for screening applications are best used for preliminary investigation or prioritization of substances before *in vivo* evaluation.

8.7 CONCLUSIONS

The need to prioritize substances according to their developmental toxicity hazard potential has been the driving force behind the development of *in vitro* models for developmental toxicity. Clearly, the standard multisegment *in vivo* teratology tests cannot be applied to the many thousands of untested substances to which humans are exposed.

The *in vitro* systems of most use are those allowing determinations of specific effects on development. Assays in which generalized cytotoxicity can be distinguished from differentiation-specific effects are also of considerable use. Examples of this latter class of assay are whole-embryo culture, limb bud organ culture, and primary cell cultures of limb buds and neural tissue.

As stated in the beginning of this chapter, such assays are useful only when test procedures are proven to be reliable in the hands of investigators other than those of the originator. Despite the large number of reviews [5,23,216,239,344,374–379] and conferences [380–384] held on the topic of *in vitro* developmental toxicity test validation, the utility of many of the *in vitro* assays described in this review is unknown, because the validation efforts have not been performed. The problems associated with validation of short-term assays in general are delineated earlier in this book, and many of those principles apply here. Issues specific for *in vitro* developmental toxicity assays are as follows: (1) selection of chemicals is problematic because the assignment of potential developmental toxicity is usually not a simple yes or no, but is dose dependent; (2) a lack of identified human teratogens and high-quality *in vivo* studies makes a rigorous evaluation of different *in vitro* methods difficult; and (3) a lack of consensus exists as to which *in vitro* endpoints/alterations should be included in the screening battery. Major efforts have been made to explicitly define the validation criteria required to produce broad acceptance of these types of tests, including efforts in the United States [171] and within the European Community [385,386].

In vitro systems do appear to be useful for prioritizing chemicals or determining gradients of chemical potencies and for examining mechanisms of toxicity and teratogenicity. A powerful use is in the testing of structurally related compounds with different teratogenic potencies, and reliable and validated assays may prove highly valuable as prescreens for prioritizing chemicals for subsequent testing in pregnant animals. This approach would apply to both traditional teratogens and compounds operating by nontraditional mechanisms, such as endocrine-active compounds. However, at the current time, it is generally recognized that the use of *in vitro* systems alone as predictors of animal or human teratogenicity is not practicable until further validation studies are performed.

ACKNOWLEDGMENTS

Appreciation is expressed for the valuable editorial and secretarial assistance of Zamyat Kirby, Terra Krebsbach, and Azure Morgan Skye; for support by the Institute for Risk Analysis and Risk Communication; and for the following grants: NIEHS P30-ES-07033 and P01-ES09601, USDOE DE-FC01-95EW55084, EPA R826886, and NIH ES-03157 and ES-07032.

References

1. Warburton D, Fraser FC. Spontaneous abortion risks in man: data from reproductive histories collected in a medical genetics unit. *Hum Genet* 1964;16:1–12.

2. Manson JM, Wise LD. Teratogens. In: Amdur MO, Doull J, Klaassen CD, eds. *Casarett and Doull's Toxicology*. New York: Pergamon Press, 1991:226–254.

3. Palmer AK. The design of subprimate animal studies. In: Wilson JG, Fraser FC, eds. *Handbook of Teratology*, Vol 4. New York: Plenum Press, 1978:215–253.

4. Brown NA, Freeman SJ. Alternative tests for teratogenicity. *ATLA* 1984;12:7–23.

5. United States Environmental Protection Agency (USEPA). TSCA at twenty, the toxic substances control act at twenty. In: *Chemicals In The Environment: Public Access Information*. Washington, DC: Office of Pollution Prevention and Toxics, 1996.

6. United States Environmental Protection Agency (USEPA). Endocrine Disruptor Screening Program: Statement of Policy; Notice. *Fed Register* 1998;63(248):71,541–71,568.

7. Moore JA, Daston GP, Faustman EM, Golub MS, Hart WL, Hughes C, Kimmel CA, Lamb JC, Schwetz BA, Scialli AR. An evaluative process for assessing human reproductive and developmental toxicity of agents. *Reprod Toxicol* 1995;9(1):61–95.

8. Kavlock RJ. Structure-activity approaches in the screening of environmental agents for developmental toxicity. *Reprod Toxicol* 1993;7:113–116.

9. Christian MS. Is there any place for nonmammalian *in vitro* tests. *Reprod Toxicol* 1993;7:99–102.

10. Goldberg AM, Frazier JM, Brusick D, Dickens MS, Flint O, Gettings SD, Hill RN, Lipnick RL, Renskers KJ, Bradlaw JA, et al. Framework for validation and implementation of *in vitro* toxicity tests *in vitro*. *Cell Dev Biol Anim* 1993;29A(9):688–692.

11. Schmid BP, Honnegar P, Kucera P. Embryonic and fetal development: fundamental research. *Reprod Toxicol* 1993;7(suppl):155–164.

12. Prati M, Giavini E, Menegola E. Alternatives to *in vivo* tests for teratologic screening. *Ann Ist Super Sanita* 1993;29(1):41–46.

13. Daston GP. The theoretical and empirical case for *in vitro* developmental toxicity screens, and potential applications. *Teratology* 1996;53:339–344.

14. Kimmel GL. Invited perspective: *in vitro* testing in developmental toxicity risk assessment. *Teratology* 1998;58(2):25–26.

15. Schwetz BA. *In vitro* approaches in developmental toxicology. *Reprod Toxicol* 1993;7 (suppl):125–127.

16. Mirkes PE. Prospects for the development of validated screening tests that measure developmental toxicity potential: view of one skeptic. *Teratology* 1996;53(6):334–338.

17. United States Environmental Protection Agency (USEPA). Guidelines for developmental toxicity risk assessment. Office of Research and Development. *Fed Register* 1991;56:63,798–63,826.

18. United States Environmental Protection Agency (USEPA). *Guidelines for Reproductive Toxicity Risk Assessment.* Washington, DC: Office of Research and Development, 1996. EPA Publication EPA/630/R-96/009.

19. United States Environmental Protection Agency (USEPA). Guidelines for neurotoxicity risk assessment. *Fed Register* 1998;63:26,925–26,954.

20. Welsch F. *In vitro* approaches to the elucidation of mechanisms of chemical teratogenesis. *Teratology* 1992;46(1):3–14.

21. Copp AJ. Birth defects: from molecules to mechanisms. *J R Coll Phys Lond* 1994;28(4)294–300.

22. Gerhart J, Kirschner M. *Cells, Embryos and Evolution.* Malden, MA: Blackwell Science, 1997.

23. Neubert D. The use of culture techniques in studies on prenatal toxicity. *Pharmacol Ther* 1982;18:397–434.

24. Braun AC, Emerson DJ, Nichinson BB. Teratogenic drugs inhibit tumor cell attachment to lectin-coated surfaces. *Nature* 1979;282:507–509.

25. Johnson EM, Goman RM, Gabel GEG, George ME. The Hydra attenuata system for detection of teratogenic hazards. *Teratog Carcinog Mutagen* 1982;2:263–276.

26. Bournias-Vardiabasis N, Teplitz RL. Use of drosophila embryo cell cultures as an *in vitro* teratogen assay. *Teratog Carcinog Mutagen* 1982;2:333–341.

27. Bournias-Vardiabasis N, Teplitz RL, Chemoff GF, Seecof RL. Detection of teratogens in the drosophila embryonic cell culture test: assay of 100 chemicals. *Teratology* 1983;28:109–122.

28. Buzin CH, Bornias-Vardiabasis N. Teratogens induce a subset of small heat shock proteins in drosophila primary embryonic cell cultures. *Proc Natl Acad Sci U S A* 1984;81:4,075–4,079.

29. Bournias-Vardiabasis N. Letter to the editor—alternative tests for teratogens. *Reprod Toxicol: Med Lett* 1985;8:410–451.

30. Bournias-Vardiabasis N, Buzin CH, Reilly JG. The effect of 5-azacytidine and cytidine analogs on Drosophila melanogaster cells in culture. *Roux's Arch Dev Biol* 1983;192:299–302.

31. Bournias-Vardiabasis N, Flores J. Drug metabolizing enzymes in Drosophila melanogaster: teratogenicity of cyclophosphamide *in vitro*. *Teratog Carcinog Mutagen* 1983;3:255–262.

32. Foureman P, Mason JM, Valencia R, Zimmering S. Chemical mutagenesis testing in Drosophila. X. Results of 70 coded chemicals tested for the National Toxicology Program. *Environ Mol Mutagen* 1994;23(3):208–227.

33. Foureman P, Mason JM, Valencia R, Zimmering S. Chemical mutagenesis testing in Drosophila. IX. Results of 50 coded compounds tested for the National Toxicology Program. *Environ Mol Mutagen* 1994;23(1):51–63.

34. Komori M, Kitamura R, Fukuta H, Inoue H, Baba H, Yoshikawa K, Kamataki T. Transgenic Drosophila carrying mammalian cytochrome P-4501A1: an application to toxicology testing. *Carcinogenesis* 1993;14(8):1,683–1,688.

35. Moscona A. Rotation-mediated histogenetic aggregation of associated cells. *Exp Cell Res* 1961;22:455–475.

36. Daston GP, Yonker JE. Chick embryo retina cell culture as an *in vitro* teratogen screen. *Toxicologist* 1987;7:141.

37. Daston GP, Yonker JE, Baines D, Poynter JI. Chick embryo cell culture; teratogen screen and mechanistic probe. *Toxicologist* 1988;8:114.

38. Daston GP, Baines D, Yonker JE. Chick embryo neural retina cell culture as a screen for developmental toxicity. *Toxicol Appl Pharmacol* 1991;109(2):352–366.

39. Farage-Elawar M, Rowles TK. Toxicology of carbaryl and aldicarb on brain and limb cultures of chick embryos. *J Appl Toxicol* 1992;12(4):239–244.

40. Heaton MB, Carlin M, Paiva M, Walker DW. Perturbation of target-directed neurite outgrowth in embryonic NS co-cultures grown in the presence of ethanol. *Brain Res Dev Brain Res* 1995;89(2):270–280.

41. Wilk AL, Greenberg JH, Horigan EA, Pratt RM, Martin GR. Detection of teratogenic compounds using differentiating embryonic cells in culture. *In Vitro* 1980;16:269–276.

42. Solursh M, Jansen KL, Singley CT, Linsenmayer TF, Reiter RS. Two distinct regulatory steps in cartilage differentiation. *Dev Biol* 1982;94:311–325.

43. Hassell JR, Horigan EA. Chondrogenesis; a model developmental system for measuring teratogenic potential of compounds. *Teratog Carcinog Mutagen* 1982;2:325–331.

44. Kistler A. Inhibition of chondrogenesis by retinoids: limb bud cell cultures as a test system to measure the teratogenic potential of compounds? *Concepts Toxicol* 1985;3:86–100.

45. Draus MA, Kennedy EM, Tait JD, Farrow MG. A comparison of two *in vitro* assays for the evaluation of teratogenic potential. *Environ Mutagen* 1984;6:460–467.

46. Flint OP, Orton TC. An *in vitro* assay for teratogens with cultures of rat embryo midbrain and limb bud cells. *Toxicol Appl Pharmacol* 1984;76:383–395.

47. Uphill PF, Wilkins SR, Allen JA. *In vitro* micromass teratogen test: results from a blind trial of 25 compounds. *Toxicol In Vitro* 1990;4:623–626.

48. Parsons JF, Rockely J, Richold M. *In vitro* micromass teratogen test—interpretation of results from a blind trial of 25 compounds using 3 separate criteria. *Toxicol In Vitro*

4(4–5):609–611.

49. Kistler A. Limb bud cell cultures for estimating the teratogenic potential of compounds. Validation of the test system with retinoids. *Arch Toxicol* 1987;60:403–414.

50. Guntakatta M, Matthews EJ, Rundell JO. Development of a mouse embryo limb bud cell culture system for the estimation of chemical teratogenic potential. *Teratog Carcinog Mutagen* 1984;4:349–364.

51. Wise LD, Clark RL, Rundell JO, Robertson RT. Examination of a rodent limb bud micromass assay as a prescreen for developmental toxicity. *Teratology* 1990;41(3):41–51.

52. Sass JO, Zimmermann B, Ruhl R, Nau H. Effects of all-trans-retinoyl-beta-D-glucuronide and all-trans-retinoic acid on chondrogenesis and retinoid metabolism in mouse limb bud mesenchymal cells *in vitro*. *Arch Toxicol* 1997;71(3):142–150.

53. Jiang H, Soprano DR, Li SW, Soprano KJ, Penner JD, Gyda M, Kochhar DM. Modulation of limb bud chondrogenesis by retinoic acid and retinoic acid receptors. *Int J Dev Biol* 1995;39(4):617–627.

54. Renault JY, Caillaud JM, Chevalier J. Ultrastructural characterization of normal and abnormal chondrogenesis in micromass rat embryo limb bud cell cultures. *Toxicol Appl Pharmacol* 1995;130(2):177–187.

55. Ou YC, Thompson SA, Kirchner SC, Kavanagh TJ, Faustman EM. Induction of growth arrest and DNA damage-inducible genes Gadd45 and Gadd153 in primary rodent embryonic cells following exposure to methylmercury. *Toxicol Appl Pharmacol* 1997;147(1):31–38.

56. Flint OP. A micromass culture method for rat embryonic neural cells. *J Cell Sci* 1983;61:247–262.

57. Greenberg JH. Detection of teratogens by differentiating embryonic neural crest cells in culture: evaluation as a screening system. *Teratog Carcinog Mutagen* 1982;2:319–323.

58. Flint OP, Boyle FT. An *in vitro* test for its application in the selection of nonteratogenic triazole antifungals. *Concepts Toxicol* 1985;3:29–35.

59. Wroble JT, Whittaker SG, Faustman EM. Characterization of differentiation-specific and cytoskeletal markers in micromass cultures. *Toxicologist* 1992;12.

60. Sweeney C, Faustman EM. Expression of differentiated characteristics in rodent embryo central nervous system cell cultures: effects of teratogen exposure. *Teratology* 1990;39:484–500.

61. Sweeney C, Faustman EM. Differentiation-related expression of protooncogene *src* in CNS micromass cultures: effects of chemical exposure. *Toxicologist* 1990;10:28.

62. Brown LP, Flint OP, Orton TC, Gibson GG. *In vitro* metabolism of teratogens by differentiating rat embryo cells. *Food Chem Toxicol* 1986;24:737–742.

63. Brown LP, Flint OP, Orton TC, Gibson GG. Metabolism of teratogens by differentiating rat embryo cells. *Teratology* 1986;33:52A.

64. Brown LP, Flint OP, Orton TC, Gibson GG. Chemical teratogenesis: testing methods and the role of metabolism. *Drug Metab Rev* 1986;17:221–260.

65. Brown LP, Foster JR, Orton TC, Flint OP, Gibson GG. Inducibility and functionality of rat embryonic/foetal cytochrome P-450: a study of differentiating limb-bud and mid-

brain cells *in vitro*. *Toxicol In Vitro* 1989;3:253–260.

66. Whittaker SG, Faustman EM. Effects of benzimidazole analogs on cultures of differentiating rodent embryonic cells. *Toxicol Appl Pharmacol* 1992;113(1):144–151.

67. Whittaker SG, Faustman EM. Effects of albendazole and albendazole sulfoxide on cultures of differentiating rodent embryonic cells. *Toxicol Appl Pharmacol* 1991;109(1):73–84.

68. Cosenza ME, Bidanset J. Effects of chlorpyrifos on neuronal development in rat embryo midbrain micromass cultures. *Vet Human Tox* 1995;37(2):118–121.

69. Teramoto S, Takahashi K, Kikuta M, Kobayashi H. Potential teratogenicity of 3-chloro-4-(dichloromethyl)-5-hydroxy-2(5H)-furanone (MX) in micromass *in vitro* test. *J Toxicol Environ Health* 1998;53(8):607–614.

70. Seeley MR, Faustman EM. Toxicity of four alkylating agents on *in vitro* rat embryo differentiation and development. *Fundam Appl Toxicol* 1995;26(1):136–142.

71. Kidney JK, Faustman EM. Modulation of nitrosourea toxicity in rodent embryonic cells by O6-benzylguanine, a depletor of O6-methylguanine-DNA methyltransferase. *Toxicol Appl Pharmacol* 1995;133(1):1–11.

72. Tsuchiya T, Oguri I, Yamakushi Y, Miyata N. Novel harmful effects of [60]fullerene on mouse embryos *in vitro* and *in vivo*. *FEBS Lett* 1996;393:139–145.

73. Ponce RA, Kavanagh TJ, Mottet NK, Whittaker SG, Faustman EM. Effects of methyl mercury on the cell cycle of primary rat CNS cells *in vitro*. *Toxicol Appl Pharmacol* 1994;127(1):83–90.

74. Ward SJ, Newall DR. The micromass test—is it subject to strain variation? *Tox In Vitro* 1990;4(4–5):620–622.

75. Tsuiki H, Fukiishi Y, Kishi K. Relation of TGF-beta 2 to inhibition of limb bud chondrogenesis by retinoid in rats. *Teratology* 1996;54(4):191–197.

76. Pignatello MA, Kauffman FC, Levin AA. Multiple factors contribute to the toxicity of the aromatic retinoid, TTNPB (Ro 13-7410): binding affinities and disposition. *Toxicol Appl Pharmacol* 1997;142(2):319–327.

77. Ribeiro PL, Faustman EM. Chemically induced growth inhibition and cell cycle perturbations in cultures of differentiating rodent embryonic cells. *Toxicol Appl Pharmacol* 1990;104:200–211.

78. Ribeiro PL, Faustman EM. Embryonic micromass limb bud and midbrain cultures: different cell cycle kinetics during differentiation *in vitro*. *Toxicol In Vitro* 1990;4:602–608.

79. Bolon B, Dorman DC, Bonnefoi MS, Randall HW, Morgan KT. Histopathologic approaches to chemical toxicity using primary cultures of dissociated neural cells grown in chamber slides. *Toxicol Pathol* 1993;21(5):465–479.

80. Flint OP. *In vitro* tests for teratogens: desirable endpoints, test batteries and current status of the micromass teratogen test. *Reprod Toxicol* 1993;7(suppl 1):103–111.

81. Maar TE, Ellerbeck U, Bock E, Nau H, Schousboe A, Berezin V. Prediction of teratogenic potency of valproate analogues using cerebellar aggregation cultures. *Toxicology* 1997;116(1–3):159–168.

82. Scholz G, Pohl I, Seiler A, Bremer S, Brown NA, Piersma AH, Holzhutter HG,

Spielmann H. Results of the first phase of the ECVAM project "Prevalidation and validation of three in vitro embryotoxicity tests" (in German). *Altex-Alternativen zu Tierexperimenten* 1998;15:3–8.

83. Heller A, Won L, Heller B, Hoffman PC. Examination of developmental neurotoxicity by the use of tissue culture model systems. *Clin Exp Pharmacol Physiol* 1995;22:375–378.

84. Reinhardt CA. Neurodevelopmental toxicity *in vitro*: primary cell culture models for screening and risk assessment. *Reprod Toxicol* 1993;7:165–170.

85. Elstein KH, Zucker RM, Shuey DL, Lau C, Chernoff N, Rogers JM. Utility of the murine erythroleukemic cell (MELC) in assessing mechanisms of action of DNA-active developmental toxicants: application to 5–fluorouracil. *Teratology* 1993;48:75–87.

86. Saito S. Cholinesterase inhibitors induce growth cone collapse and inhibit neurite extension in primary cultured chick neurons. *Neurotox Teratol* 1998;20(4):411–419.

87. Braun AG, Buckner CA, Emerson DF, Nichinson BB. Quantitative correspondence between the *in vivo* and *in vitro* activity of teratogenic agents. *Proc Natl Acad Sci U S A* 1982;79:2,056–2,060.

88. Braun AG, Nichinson BB, Horowicz PS. Inhibition of tumor cell attachment to concanavalin A–coated surfaces as an assay for teratogenic agents: approaches to validation. *Teratog Carcinog Mutagen* 1982;2:342354.

89. Braun AG, Harding FA, Weinreb SL. Teratogenic drugs inhibit tumor cell attachment to lectin coated surfaces P450. *Toxicol Appl Pharmacol* 1986;82:175–179.

90. Yoneda T, Pratt RM. Mesenchymal cells from the human embryonic palate are highly responsive to EGF. *Science* 1981;213:565–568.

91. Pratt RM, Grove RI, Willis WD. Prescreening for environmental teratogens using cultured mesenchymal cells from the human embryonic palate. *Teratog Carcinog Mutagen* 1982;2:313–318.

92. Pratt RM, Willis WD. *In vitro* screening assay for teratogens using growth inhibition of human embryonic cells. *Proc Natl Acad Sci U S A* 1985;82:5,791–5,794.

93. Welsch F, Stedman DB, Willis WD, Pratt RM. Karyotype, growth, and cell cycle analysis of human embryonic palatal mesenchymal cells: relevance to the use of these cells in an *in vitro* teratogenicity screening assay. *Teratog Carcinog Mutagen* 1986;383–392.

94. Smith MK, Kimmel GL, Kochhar DM, Shepard TN, Spielberg SP, Wilson JG. A selection of candidate compounds for *in vitro* teratogenesis test validation. *Teratog Carcinog Mutagen* 1983;3:461–480.

95. Yang LL, Steele VE, Lamb JC, et al. Evaluation and validation of two *in vitro* teratology systems: results from the first twelve coded compounds. *Environ Mutagen* 1986;8:94–107.

96. Steele VE, Elmore EL, Lamb JC, et al. Validation of two short-term *in vitro* assays to identify potential teratogens. *Teratology* 1986;33:610–616.

97. National Toxicology Program, NIEHS. *Evaluation of Two In Vitro Teratology Testing Systems.* December 2, 1986. NTP Publication 86-372.

98. Steele VE, Morrissev RE, Elmore EL, et al. Evaluation of two *in vitro* assays to screen for potential developmental toxicants. *Fund Appl Toxicol* 1988;11:673–684.

99. Zhao F, Mayura K, Hutchinson RW, Lewis RP, Burghardt RC, Phillips TC. Developmental toxicity and structure-activity relationships of chlorophenols using human embryonic palatal mesenchymal cells. *Toxicol Lett* 1995;78:35–42.

100. Mummery CL, Van Den Brink CE, Van Der Saag PT, De Laat SW. A short–term screening test for teratogens using differentiating neuroblastoma cells *in vitro. Teratology* 1984;29:271–279.

101. Weisman AD, Caldecott-Hazard S. Developmental neurotoxicity to methamphetamines. *Clin Exp Pharmacol Physiol* 1995;22:372–374.

102. Kissaalita WS, Bowen JM. Effect of culture age on the susceptibility of differentiating neuroblastoma cells to retinoid cytotoxicity. *Biotech Bioeng* 1996;50(5):580–586.

103. Repetto G, Sanz P, Repetto M. Comparative *in vitro* effects of sodium arsenite and sodium arsenate on neuroblastoma cells. *Toxicology* 1994;92(1–3):143–153.

104. Hare MF, Atchison WD. Methylmercury mobilizes Ca^{++} from intracellular stores zssensitive to inositol 1,4,5-trisphosphate in NG108-15 cells. *J Pharm Exp Ther* 1995;272(3):1,016–1,023.

105. Di Lorenzo D, Ferrari F, Agrati P, de Vos H, Apostoli P, Alessio L, Albertini A, Maggi A. Manganese effects on the human neuroblastoma cell line SK-ER3. *Toxicol Appl Pharmacol* 1996;140(1):51–57.

106. Davey FD, Breen KC.Stimulation of sialyltransferase by subchronic low-level lead exposure in th developing nervous system. A potential mechanism of teratogen action. *Toxicol Appl Pharmacol* 1998;151(1):16–21.

107. Ray R, Legere RH, Majerus BJ, Petrali JP. Sulfur mustard-induced increase in intracellular free calcium level and arachidonic acid release from cell membrane. *Toxicol Appl Pharmacol* 1995;131(1):44–52.

108. Seeley MR, Faustman EM. Effects of O6-benzylguanine on growth and differentiation of P19 embryonic carcinoma cells treated with alkylating agents. *Teratog Carcinog Mutagen* 1998;18(3):111–122.

109. Seeley MR, Faustman EM. Evaluation of P19 cells for studying mechanisms of developmental toxicity: application to four direct-acting alkylating agents. *Toxicology* 1998;127(1–3):49–58.

110. Marinovich M, Ghilardi F, Galli CL. Effect of pesticide mixtures on *in vitro* nervous cells: comparison with single pesticides. *Toxicology* 1996;108(3):201–206.

111. Laschinski G, Vogel R, Spielmann H. Cytotoxicity test using blastocyst-derived euploid embryonal stem cells: a new approach to *in vitro* teratogenesis testing. *Reprod Toxicol* 1991;5:57–64.

112. Newall DR, Beedles KE. The stem-cell test—a novel *in vitro* assay for teratogenic potential. *Toxicol In Vitro* 1994;8:697–705.

113. Newall DR, Beedles KE. The stem-cell test: an *in vitro* assay for teratogenic potential. Results of a blind trial with 25 compounds. *Toxicol In Vitro* 1996;10(2):229–241.

114. Scholz G, Doring B, Moldenhauer F, Pohl I, Seiler A, Speilmann H. *In vitro* embryotoxicity assay using two permanent cell lines: mouse embryonic stem cells and 3t3 fibrob-

lasts (conference abstract). *Eur J Cell Biol* 1997;74 (suppl 47):141.

115. Scholz G, Pohl I, Seiler A, Bremer S, Moldenhauer F, Holzhutter HG, Speilmann H. *In vitro* testing of embryotoxicity using the embryonic stem cell test: an inter-laboratory comparison (conference abstract). *Naunyn Schmiedebergs Arch Pharmacol* 1998:347(suppl 4):606.

116. Heuer J, Bremer S, Pohl I, Speilmann H. Development of an *in vitro* embryotoxicity test using murine embryonic stem cell cultures. *Toxicol In Vitro* 1993;7:551-556.

117. Keller SJ, Smith MK. Animal virus screens for potential teratogens. 1. Poxvirus morphogenesis. *Teratog Carcinog Mutagen* 1982;2:361-374.

118. Trosko JE, Chang C-C, Netzloff M. The role of inhibited cell-cell communication in teratogenesis. *Teratog Carcinog Mutagen* 1982;2:31-45.

119. Trosko JE, Chang CC. Role of intracellular communication in tumor promotion. In: Slaga TJ, ed. *Mechanisms of Tumor Promotion*, Vol. IV. Boca Raton, FL: CRC Press, 1984:119-145.

120. Loch-Caruso R, Trosko JE. Inhibited intercellular communication as a mechanistic link between teratogenesis and carcinogenesis. *CRC Crit Rev Toxicol* 1985;16:157-183.

121. Loch-Caruso R, Trosko JE, Corcos IA. Interruption of cell-cell communication in Chinese hamster V79 cells by venous alkyl glycol ethers: implications for teratogenicity. *Environ Health Perspect* 1984;57:119-123.

122. Welsch F. The applicability of *in vitro* methods to teratogenicity testing and to studies on the mechanism of action of chemical teratogens. *CIIT Activities* 1986;6:1,3-7.

123. Welsch F, Stedman DB. Inhibition of metabolic cooperation between Chinese hamster V79 cells by structurally diverse teratogens. *Teratog Carcinog Mutagen* 1984;4:285-301.

124. Rubinstein C, Jone C, Trosko JE, Chang CC. Inhibition of intercellular communication in cultures of Chinese hamster V79 cells by 2,4-dichlorophenoxyacetic acid and 2,4,5-trichlorophenoxyacetic acid. *Fund Appl Toxicol* 1984;4:731-739.

125. Davidson JS, Baumgarten IM, Harley EH. Effects of 12-O-tetradecanoylphorbol—13-acetate and retinoids on intercellular junctional communication measured with a citrulline incorporation assay. *Carcinogenesis* 1985;6:645-650.

126. Chen TH, Kavanagh TJ, Chang CC, Trosko JE. Inhibition of metabolic cooperation in Chinese hamster V79 cells by various organic solvents and simple compounds. *Cell Biol Toxicol* 1984;1:155-171.

127. Welsch F, Stedman DB, Carson JL. Effects of a teratogen on [^{3}H]-uridine nucleotide transfer between human embryonal cells and on gap junctions. *Exp Cell Res* 1985;159:91-102.

128. Hulme LM, Atkinson KA, Clothier RH, Balls M. The potential usefulness of a differentiating carcinoma cell line in *in vitro* toxicity testing. *Toxicol In Vitro* 1990;4:589-592.

129. Skladchikova G, Berezin V, Bock E. Valproic acid, but not its non-teratogenic analogue 2-isopropylpentanoic acid, affects proliferation, viability and neuronal differentiation of the human teratocarcinoma cell line Ntera-2. *Neurotox* 1998;19(3):357-370.

130. Jelinek R. Use of chick embryo in screening for embryotoxicity. *Teratog Carcinog Mutagen* 1982;2:255-261.

131. Jelinek R, Peterka M, Rychter A. Chick embryotoxicity screening test—130 substances

tested. *Ind J Exp Biol* 1985;23:588–595.

132. Schowing J. Chick embryos as experimental material for teratogenic investigations. *Concepts Toxicol* 1985;3:58–73.

133. Summerbell D, Hornbruch A. The chick embryo: a standard against which to judge *in vitro* systems. In: Neubert D, Merker HJ, eds. *Culture Techniques*. New York: Walter de Gruyter, 1981:529–538.

134. Gebhardt DOE. The use of the chick embryo in applied teratology. In: Woollam DHM, ed. *Advances in Teratology*. London: Academic Press, 1972:97–111.

135. Vesely D, Vesela D, Jelinek R. Nineteen mycotoxins tested on chick embryos. *Toxicol Lett* 1982;13:239–245.

136. Fisher M, Schoenwolf GC. The use of early chick embryos in experimental embryology and teratology: improvements in standard procedures. *Teratology* 1983;27:65–72.

137. Korhonen A, Hemminki K, Vainio H. Application of the chicken embryo in testing for embryotoxicity: thiurams. *Scand J Work Environ Health* 1982;8(1):63–69.

138. Vesely D, Vesela D. Use of chick embryos for prediction of embryotoxic effects of mycotoxins in mammals. *Vet Med* 1991;36(3):175–181.

139. Muscarella DE, Keown JF, Bloom SE. Evaluation of the genotoxic and embryotoxic potential of chlorpyrifos and its metabolites *in vivo* and *in vitro*. *Environ Mutagen* 1984;6:13–23.

140. Bloom SE. Sister chromatic exchange studies in the chick embryo and neonate: actions of mutagens in a developing system. *Basic Life Sci* 1984;29:509–533.

141. Peters PWJ, Piersma AH. *In vitro* embryotoxicity and teratogenicity studies. *Toxicol In Vitro* 1990;4:570–576.

142. Farage-Elawar M. Effects of in ovo injection of carbamates on chick embryo hatchability, esterase enzyme activity and locomotion of chicks. *J Appl Toxicol* 1990;10(3):197–201.

143. Farage-Elawar M. Development of esterase activities in the chicken before and after hatching. *Neurotoxicol Teratol* 1991;13(2):147–152.

144. Birge WJ, Black JA, Ramey BA. The reproductive toxicology of aquatic contaminants. In: Saxena J, Fisher F, eds. *Hazard Assessment of Chemicals: Current Developments*, Vol. I. New York: Academic Press, 1981:59–115.

145. Birge WJ, Black JA, Westerman AG, Ramey BA. Fish and amphibian embryos—a model system for evaluating teratogenicity. *Fundam Appl Toxicol* 1983;3:237–242.

146. Cameron IL, Lawrence WC, Lum JB. Medaka eggs as a model system for screening potential teratogens. *Prog Clin Biol Res* 1985;163C:239–243.

147. Collodi P, Kamei Y, Ernst T, Miranda C, Buhler DR, Barnes DW. Cultures of cells from zebrafish (Brachydanio rerio) embyro and adult tissues. *Cell Biol Toxicol* 1992;8(1):43–61.

148. Medcalfe CD, Sonstegard RA. Oil refinery effluents: evidence of cocarcinogenic activity in the trout embryo microinjection assay. *J Natl Cancer Inst* 1985;75:1,091–1,097.

149. Shi M, Faustman EM. Development and characterization of a morphological scoring system for medaka (*Oryzias latipes*) embryo development. *Aquatic Toxicol* 1989;15:127–140.

150. Binder RL, Stegeman JJ. Microsomal electron transport and xenobiotic monooxygenase activities during the embryonic period of development in the killifish, *Fundulus heteroclitus*. *Toxicol Appl Pharmacol* 1984;73:432–443.

151. Binder RL, Stegeman JJ, Lech JJ. Induction of cytochrome P-450-dependent monooxygenase systems in embryos and eleutheroembryos of the killfish, *Fundulus heteroclitus*. *Chem Biol Interact* 1985;55:185–202.

152. Hiraoka Y, Ishiasawa S, Kamada T, Okuda H. Acute toxicity of 14 different kinds of metals affecting medaka fry. *Hiroshima J Med Sci* 1985;34:327–330.

153. Eckhert CD. Boron stimulates embryonic trout growth. *J Nutr* 1998;128(12):2,488–2,493.

154. Schreiweis DO, Murray GJ. Cardiovascular malformations in *Oryzias latipes* embryos treated with 2,4,5-trichlorophenoxyacetic acid (2,4,5-T). *Teratology* 1976;14:287–290.

155. Hendricks JD, Wales JH, Sinnhuber RO, Nixon JE, Loveland PM, Scanlan RA. Rainbow trout (*Salmo gairdneri*) embryos: a sensitive animal model for experimental carcinogenesis. *Fed Proc* 1980;39:3,222–3,229.

156. Hendricks JD, Scanlan RA,. Williams JL, Sinnhuber RO, Grieco MP. Carcinogenicity of *N*-methyl,*N*-nitro-,*N*-nitroso-Guanidine to the livers and kidneys of rainbow trout (*Salmo gairdneri*) exposed as embryos. *J Natl Cancer Inst* 1980;64:1,511–1,519.

157. Ishikawa T, Masahito P, Takayama S. Usefulness of the medaka, Oryzias latipes, as a test animal: DNA repair processes in medaka exposed to carcinogens. *Natl Cancer Inst Monogr* 1984;65:35–43.

158. Solomon FP, Faustman EM. Developmental toxicity of four model alkylating agents on Japanese medaka fish (*Oryzias latipes*) embryos. *Environ Toxicol Chem* 1987;6:747–753.

159. Faustman-Watts E, Solomon F. Examination of a candidate test compound and structural homolog in two *in vitro* teratogenicity test systems. *Teratology* 1985;31:33A.

160. Llewellyn GC, Stephenson GA, Hofman JW. Aflatoxin B1 induced toxicity and teratogenicity in Japanese Medaka eggs (Oryzias latipes). *Toxicon* 1977;15(6):582–587.

161. VanLeeuwen CJ, Grootelaar EMM, Niebeek G. Fish embryos as teratogenicity screens: a comparison of embryotoxicity between fish and birds. *Ecotox Environ Safety* 1990;20:42–52.

162. Dumont JN, Epler RG. Validation studies on the FETAX teratogenesis assay (frog embryos). *Teratology* 1984;29:27A.

163. Dumont JN, Schultz TW, Buchanan MV, Kao GL. Frog embryo teratogenesis assay: Xenopus (FETAX)—a short-term assay applicable to complex environmental mixtures. In: Waters MD, Sandhu SS, Lewtas J, Claxton L, Chemoff N, Nesnow S, eds. *Short-Term Bioassays in the Analysis of Complex Environmental Mixtures III*. New York: Plenum Press, 1983:393–405.

164. Dumont JN, Schultz TW, Epler RG. The response of the FETAX model to mammalian teratogens. *Teratology* 1983;27:39A.

165. Dumpert K, Zietz E. Platanna (*Xenopus laevis*) as a test organism for determining the embryotoxic effects of environmental chemicals. *Ecotoxicol Environ Safety* 1984;8:55–74.

166. Sabourin TD, Faulk RT, Gross LB. The efficacy of three non-mammalian test systems in the identification of chemical teratogens. *J Appl Toxicol* 1985;5:227–233.

167. Bantle JA, Burton DT, Dawson DA, Dumont JN, Finch RA, Fort DJ, Linder G, Rayburn JR, Buchwalter D. Initial interlaboratory validation study of FETAX: phase I testing. *J Appl Toxicol* 1994;14(3):213–223.

168. Bantle JA, Finch RA, Burton DT, Fort DJ, Dawson DA, Linder G, Rayburn JR, et al. FETAX interlaboratory validation study: phase III—Part 1 testing. *J Appl Toxicol* 1996;16(6):517–528.

169. Bantle JA, Fort DJ, Rayburn JR, DeYoung DJ, Bush SJ. Further validation of FETAX: evaluation of the developmental toxicity of five known mammalian teratogens and non-teratogens. *Drug Chem Toxicol* 1990;13:267–282.

170. Dawson DA, Bantle JA. Development of a reconstituted water medium and preliminary validation of the frog embryo teratogenesis assay-Xenopus (FETAX). *J Appl Toxicol* 1987;7:237–244.

171. National Institute of Environmental Health Sciences (NIEHS). *Validation and Regulatory Acceptance of Toxicological Test Methods: A Report of the Ad Hoc Interagency Coordinating Committee on the Validation of Alternative Methods.* 1997. NIH Publication 97-3981.

172. Dawson DA, Fort DJ, Smith GJ, Newell DL, Bantle JA. Evaluation of the developmental toxicity of nicotine and cotinine with frog embryo teratogenesis assay: Xenopus. *Teratog Carcinog Mutagen* 1988;8(6):329–338.

173. Dawson DA, Fort DJ, Newell DL, Bantle JA. Developmental toxicity testing with FETAX: evaluation of five compounds. *Drug Chem Toxicol* 1989;12(1):67–75.

174. Fort DJ, Stover EL, Propst T, Hull MA, Bantle JA. Evaluation of the developmental toxicities of coumarin, 4-hydroxycoumarin, and 7-hydroxycoumarin using FETAX. *Drug Chem Toxicol* 1998;21(1):15–26.

175. Fort DJ, Rayburn JR, Bantle JA. Evaluation of acetaminophen-induced developmental toxicity using FETAX. *Drug Chem Toxicol* 1992;15(4):329–350.

176. Propst TL, Fort DJ, Stover EL, Schrock B, Bantle JA. Evaluation of the developmental toxicity of benzo[a]pyrene and 2-acetylaminofluorene using Xenopus: modes of bio-transformation. *Drug Chem Toxicol* 1997;20(1–2):45–61.

177. Dresser TH, Rivera ER, Hoffmann FJ, Finch RA. Teratogenic assessment of four solvents using the Frog Embryo Teratogenesis Assay—Xenopus (FETAX). *J Appl Toxicol* 1992;12(1):49–56.

178. Fort DJ, Stover EL, Rayburn JR, Hull M, Bantle JA. Evaluation of the developmental toxicity of trichloroethylene and detoxification metabolites using Xenopus. *Teratog Carcinog Mutagen* 1992;13(1):35–45.

179. Fort DJ, Stover EL, Propst T, Hull MA, Bantle JA. Evaluation of the developmental toxicity of theophylline, dimethyluric acid, and methylxanthine metabolites using Xenopus. *Drug Chem Toxicol* 1996;19(4):267–278.

180. Dawson DA, Wilke TS. Evaluation of the frog embryo teratogenesis assay: Xenopus (FETAX) as a model system for mixture toxicity hazard assessment. *Environ Toxicol Chem* 1991;10:941–948.

181. American Society for Testing and Materials (ASTM). *Standard Guide for Conducting the Frog Embryo Teratogenesis Assay-Xenopus (FETAX).* Philadelphia: ASTM, 1991. ASTM

Publication E1439-91.

182. Bantle JA, Dumont JN, Finch R, Linder G. *ATLAs of Abnormalities: A Guide to the Performance of FETAX*. Oklahoma City: Oklahoma State Publications, 1991.

183. Schuler RL, Hardin BD, Niemeier RW. Drosophila as a tool for the rapid assessment of chemicals for teratogenicity. *Teratog Carcinog Mutagen* 1982;2:293–301.

184. Wilson JG. Survey of *in vitro* systems: their potential use in teratogenicity screening. In: Wilson JG, Fraser FC, eds. *Handbook of Teratology*, Vol. 4. New York: Plenum Press, 1978;135–153.

185. Eberlein S. Stage specific embryonic defects following heat shock in Drosophila. *Dev Genet* 1986;6:179–197.

186. Sobels FH. The advantages of drosophila for mutation studies. *Mutat Res* 1974;26(4):277–284.

187. Kilbey BJ, MacDonald DJ, Auerbach C, Sobels FH, Vogel EW. The use of Drosophila melanogaster in tests for environmental mutagens. *Mutat Res* 85(3):141–146.

188. Lee WR, Abrahamson S, Valencia R, von Halle ES, Wurgler FE, Zimmering S. The sex-linked recessive lethal test for mutagenesis in Drosophila melanogaster. A report of the U.W. Environmental Protection Agency Gene-Tox Program. *Mutat Res* 123(2):183–279.

189. Eisses KT. Teratogenicity and toxicity of ethylene glycol monomethyl ether (2-methoxyethanol) in Drosophila melanogaster: involvement of alcohol dehydrogenase activity. *Teratog Carcinog Mutagen* 1989;9(5):315–325.

190. Eisses KT. Differences in teratogenic and toxic properties of alcohol dehydrogenase inhibitors pyrazole and 4-methylpyrazole in Drosophila melanogaster: II. ADH allozymes in an isogenic background. *Teratog Carcinog Mutagen* 1994;14(6):291–302.

191. Eisses KT. Differences in teratogenic and toxic properties of alcohol dehydrogenase inhibitors pyrazole and 4-methylpyrazole in Drosophila melanogaster: I. ADH allozymes in variable genetic backgrounds. *Teratog Carcinog Mutagen* 1995;15(1):1–10.

192. Walton BT. Use of the cricket embryo (Acheta domesticus) as an invertebrate teratology model. *Fundam Appl Toxicol* 1983;3:233–236.

193. Johnson EM, Gabel BEG. An artificial embryo for detection of abnormal developmental biology. *Fundam Appl Toxicol* 1983;3:243–249.

194. Johnson EM, Christian MS. The hydra assay for detecting and ranking developmental hazards. *Concepts Toxicol* 1985;3:107–113.

195. Wilby OK, Newall DR, Tesh JM. A hydra assay as a pre-screen for teratogenic potential. *Am Coll Toxicol Proc* 1985;Poster 6.

196. Johnson EM. A subvertebrate system for rapid determination of potential teratogenic hazards. *J Environ Pathol Toxicol* 1980;4:153–156.

197. Wiger R, Stottum A. *In vitro* testing for developmental toxicity using the *Hydra attenuata* assay. *NIPH ANN* 1982;8:43–47.

198. Mayura K, Smith EE, Clement BA, Phillips TD. Evaluation of the developmental toxicity of chlorinated phenols utilizing Hydra attenuata and postimplantation rat embryos in culture. *Toxicol Appl Pharmacol* 1991;108:253–266.

199. Johnson EM, Gabel BE, Christian MS, Sica E. The developmental toxicity of xylene and xylene isomers in the Hydra assay. *Toxicol Appl Pharmacol* 1986;82(2):323–328.

200. Johnson EM, Gabel BEG, Larson J. Developmental toxicity and structure/activity correlates of glycols and glycol ethers. *Environ Health Perspect* 1984;57:135–139.

201. Rogers JM. Comparison of maternal and fetal toxic dose responses in mammals. *Teratog Carcinog Mutagen* 1987;7:297–306.

202. Daston GP, D'Amato RA. *In vitro* techniques in teratology. *Toxicol Ind Health* 1989;5:555–585.

203. Best JB, Morita M, Ragin J, Best J, Jr. Acute toxic responses of the freshwater planarian, *Dugesia dorotocephela*, to methylmercury. *Bull Environ Contam Toxicol* 1981;27:49–54.

204. Best JB, Morita M. Planarians as a model system for *in vitro* teratogenesis studies. *Teratog Carcinog Mutagen* 1982;2:277–291.

205. Schaeffer DJ. Planarians as a model system for *in vivo* tumorigenesis studies. *Ecotoxicol Environ Safety* 1993;25(1):1–18.

206. Ohta T, Tokishita S, Shiga Y, Hanazato T, Yamagata H. An assay system for detecting environmental toxicants with cultured cladoceran eggs *in vitro*: malformations induced by ethylenethiourea. *Environ Res* 1998;77(1):43–48.

207. Nacci D, Jackim E, Walsh R. Comparative evaluation of three rapid marine toxicity tests: sea urchin early embryo growth test, sea urchin sperm cell toxicity test and microtox. *Environ Toxicol Chem* 1986;5:521–525.

208. Jackim E, Nacci D. Improved sea urchin DNA-based embryo growth toxicity test. *Environ Toxicol Chem* 1986;5:561–565.

209. Hose JE. Potential uses of sea urchin embryos for identifying toxic chemicals: description of a bioassay incorporating cytologic, cytogenetic and embryological endpoints. *J Appl Toxicol* 1985;5:245–254.

210. Hinkley RE, Jr, Wright BD. Comparative effects of halothane, enflurane, and methoxyflurane on the incidence of abnormal development using sea urchin gametes as an *in vitro* model system. *Anesth Analg* 1985;64:1,005–1,009.

211. Kerster HW, Schaeffer DJ. Brine shrimp (Artemia salina) nauplii as a teratogen test system. *Ecotoxicol Environ Safety* 1983;7:342–349.

212. Sleet RB, Brendel K. Homogeneous populations of Artemia nauplii and their potential use for *in vitro* testing in developmental toxicology. *Teratog Carcinog Mutagen* 1985;5:41–54.

213. Eibs HG, Spielmann H. Preimplantation embryos, part II: culture and transplantation. In: Neubert D, Merker HI, Kwasigroch TE, eds. *Methods in Prenatal Toxicology: Evaluation of Embryonic Effects in Experimental Animals*. Stuttgart: Georg Thieme, 1977:221–230.

214. Spielmann H, Eibs HG. Recent progress in teratology: a survey of methods for the study of drug action during the preimplantation period. *Arzneimittelforschung* 1978;28:1,733–1,742.

215. Spielmann H, Druger C, Vogel R. Embryotoxicity testing during the preimplantation

period. *Concepts Toxicol* 1985;3:22-28.

216. Spielmann N, Kruger C, Tenschen B, Vogel R. Studies in the embryotoxic risk of drug treatment during the preimplantation period in the mouse. *Drug Res* 1986;36:219-223.

217. Fabro S, Mclachlan JA, Dies NM. Chemical exposure of embryos during the preimplantation stages of pregnancy: mortality rate and intrauterine development. *Am J Obstet Gynecol* 1984:929-938.

218. Biggers JD, Borland RM. Physiological aspects of growth and development of the preimplantation mammalian embryo. *Annu Rev Physiol* 1976;38:95-119.

219. Iannaccone PM. Long-term effects of exposure to methyinitrosourea on blastocysts following transfer to surrogate female mice. *Cancer Res* 1984;44:2,785-2,789.

220. Bossert NL, Iannaccone PM. Midgestational abnormalities associated with *in vitro* preimplantation N-methyl-N-nitrosourea exposure with subsequent transfer to surrogate mothers. *Proc Natl Acad Sci U S A* 1985;82:8,757-8,761.

221. Vogel R, Kruger C, Granata I, Spielmann N. Development and sister chromatic exchange of mouse morulae and blastocysts cultured in rat serum containing active metabolites of cyclophosphamide. *Toxicol Lett* 1985;28:23-28.

222. Muller WU, Streffer C. Risk to preimplantation mouse embryos of combinations of heavy metals and radiation. *Int J Radiat Biol* 1987;51:997-1,006.

223. Filler R, Lew K. Developmental onset of mixed-function oxidase activity in preimplantation mouse embryos. *Proc Natl Acad Sci U S A* 1981;78:6,991-6,995.

224. Austin CR. Embryo transfer and sensitivity to teratogenesis. *Nature* 1973;244:333-334.

225. Iannaccone PM, Tsao TY, Stols L. Effects on mouse blastocysts of *in vitro* exposure to methylnitrosourea and 3-methylcholanthrene. *Cancer Res* 1982;42:864-868.

226. Rutledge JC, Generoso WM. Fetal pathology produced by ethylene oxide treatment of the murine zygote. *Teratology* 1989;39:563-572.

227. Generoso WM, Rutledge JC, Aronson J. Developmental anomalies: mutational consequence of mouse zygote exposure. *Banbury Rep* 1990;34:311-319.

228. Spielmann H, Jacob-Muller U, Eibs HG, Beckord W. Investigations on cyclophosphamide treatment during the preimplantation period. I. Differential sensitivity of preimplantation mouse embryos to maternal cyclophosphamide treatment. *Teratology* 1981;23:1-5.

229. Vogel R. Spielmann N. Increased sister-chromatic exchange frequency in preimplantation mouse embryos after maternal cyclophosphamide treatment before implantation. *Toxicol Lett* 1986;32:81-88.

230. Eibs HG, Spielmann H, Hagele M. Effects of sex steroid treatment during the preimplantation period on the development of mouse embryos *in vivo* and *in vitro*. In: Neubert D, Merker HJ, Nau H, Langman J. eds. *Role of Pharmacokinetics in Prenatal and Perinatal Toxicology*. Stuttgart: Georg Thieme, 1978:435-438.

231. Pedersen RA. Benzo(a)pyrene metabolism in early mouse embryos. In: Neubert D, Merker HJ, eds. *Culture Techniques: Applicability for Studies on Prenatal Differentiation and*

Toxicity. Berlin: Walter de Gruyter, 1981:447–454.

232. Lutwak-Mann C, Hay MF, New DAT. Action of venous agents on rabbit blastocysts *in vivo* and *in vitro*. *J Reprod Fertil* 1969:18:735–257.

233. Yu HS, Tam PP, Chan ST. Effects of cadmium on preimplantation mouse embryos *in vitro* with special reference to their implantation capacity and subsequent development. *Teratology* 1985;32(3):347–353.

234. Devine R, Silbergeld E. Effects of methyl mercury on preimplantation mouse embryos (poster). Presented at Society of Toxicology Annual Meeting, Seattle, WA, 1998.

235. Lau CF, Vogel R, Obe G, Spielman H. Embryologic and cytogenetic effects of ethanol on pre-implantation mouse embryos *in vitro*. *Reprod Toxicol* 1991;5(5):405–410.

236. Spielmann H, Eibs HG, Jacob-Muller U. *In vivo* and *in vitro* studies on the effects of cyclophosphamide treatment before implantation in the mouse and rat. In: Neubert D, Merker HJ, Nau H, Langman J, eds. *Role of Pharmacokinetics in Prenatal and Perinatal Toxicology*. Stuttgart: Georg Thieme, 1978:422–433.

237. Shepard TH, Fantel AG, Mirkes PE, et al. Teratology testing: I. Development and status of short-term prescreens. II. Biotransformation of teratogens as studied in whole embryo culture. *Prog Clin Biol Res* 1983;135:147–164.

238. Erbach GT, Lawitts JA, Papaioannou VE, Biggers JD. Differential growth of the mouse preimplantation embryo in chemically defined media. *Toxicol Reprod* 1994;50:1,027–1,033.

239. Schmid BP. Teratogenicity testing of new drugs with the postimplantation embryo culture system. *Concepts Toxicol* 1985;3:46–57.

240. Schmid BP. Action sites of known *in vivo* teratogens in extracorporeally exposed rat embryos. *Concepts Toxicol* 1985;3:74–85.

241. Sadler TW. The role of mammalian embryo culture in developmental biology and teratology In: Kalter H, ed. *Issues and Reviews in Teratology*, Vol. 3. New York: Plenum Press, 1985:273–294.

242. Sadler TW, Horton WE, Warner CW. Whole embryo culture: a screening technique for teratogens? *Teratog Carcinog Mutagen* 1982;2:243–253.

243. Sadler TW, Horton WE, Jr, Hunter ES. Mammalian embryos in culture: a new approach to investigating normal and abnormal developmental mechanisms. *Prog Clin Biol Res* 1985;171:227–240.

244. Fantel AC. Culture of whole rodent embryos in teratogen screening. *Teratog Carcinog Mutagen* 1982;2:231–242.

245. Kochhar DM. The use of *in vitro* procedures in teratology. *Teratology* 1975;11:273–288.

246. Kochhar DM. *In vitro* testing of teratogenic agents using mammalian embryos. *Teratog Carcinog Mutagen* 1980;1:63–74.

247. Kitchin KT, Schmid BP, Sanyal MK. Rodent whole-embryo culture as a teratogen screening method. *Methods Find Exp Clin Pharmacol* 1986;8:291–301.

248. Brown NA, Fabro SE. The *in vitro* approach to teratogenicity testing. In: Snell K, ed. *Developmental Toxicology*. New York: Praeger, 1982:31–57.

249. New DAT. Whole-embryo culture and the study of mammalian embryos during organo-

genesis. *Biol Rev* 1978;53:81–122.

250. Webster WS, Brown-Woodman PD, Ritchie HE. A review of the contribution of whole embryo culture to the determination of hazard and risk in teratogenicity testing. *Int J Dev Biol* 1997;41(2):329–335.

251. Chen LT, Hsu YC. Development of mouse embryos *in vitro*: preimplantation to the limb bud stage. *Science* 1982;218:66–68.

252. Eto K, Takakubo F. Improved development of rat embryos in culture during the period of craniofacial morphogenesis. *J Craniofacial Genet Dev Biol* 1985;5:351–355.

253. Priscott PK, Yeoh GCT, Oliver IT. The culture of 12- and 13-day rat embryos using continuous and noncontinuous gassing of rotating bottles. *J Exp Zool* 1984;230:247–253.

254. Wee EL, Wolfson LG, Zimmermann EF. Palate shelf movement in mouse embryo culture: evidence for skeletal and smooth muscle contractility. *Dev Biol* 1976;48:91–103.

255. Tarlatzis BC, Sanyal MK, Biggers WJ, Naftolin F. Continuous culture of the postimplantation rat conceptus. *Biol Reprod* 1984;31:415–426.

256. Sanyal MK, Naftolin F. *In vitro* development of the mammalian embryo. *J Exp Zool* 1983;228:235–251.

257. Sadler TW, Warner CW. Use of whole embryo culture for evaluating toxicity and teratogenicity. *Pharmacol Rev* 1984;36:1,455–1,505.

258. Klug S, Lewandowski C, Neuben D. Modification and standardization of the culture of early post implantation embryos for toxicological studies. *Arch Toxicol* 1985;58:84–88.

259. Brown NA, Fabro S. Quantitation of rat embryonic development *in vitro*: a morphological scoring system. *Teratology* 1981;24:65–78.

260. Fantel AG, Greenaway JC, Juchau MR, Shepard TH. Teratogenic bioactivation of cyclophosphamide *in vitro*. *Life Sci* 1979;25:67–72.

261. Kitchin KT, Sanyal MK, Schmid BP. A coupled microsomal-activating/embryo culture system: toxicity of reduced-nicotinamide adenine dinucleotide phosphate (NADPH). *Biochem Pharmacol* 1981;30:985–992.

262. Faustman-Watts E, Greenaway JC, Namkung MJ, Fantel AG, Juchau MR. Teratogenicity *in vitro* of 2-acetylaminofluorene: role of biotransformation in the rat. *Teratology* 1983;27:19–28.

263. Faustman-Watts EM, Namkung MJ, Greenaway JC, Juchau MR. Analysis of metabolites of 2-acetylamino-fluorene generated in an embryo culture system: relationship of biotransformation to teratogenicity *in vitro*. *Biochem Pharmacol* 1985;34:2,953–2,959.

264. Brown NA, Kram D. Intact human and rodent hepatic cells used for bioactivation in an embryo culture system. *Teratology* 1982;25:30A.

265. Oglesby LA, Ebron MT, Beyer PE, Carver BD, Kavlock RJ. Co-culture of rat embryos and hepatocytes: *in vitro* detection of a proteratogen. *Teratog Carcinog Mutagen* 1986;6:129–138.

266. Chatot CL, Klein NW, Pierro LJ. Successful culture of rat embryos on human serum: use in the detection of teratogens. *Science* 1980;207:471–473.

267. Brinster RL. Teratogen testing using preimplantation mammalian embryos. In: Shepard

TH, Miller JR, Marois M, eds. *Methods for Detection of Environmental Agents that Produce Congenital Defects*. New York: Elsevier, 1975:113–124.

268. Klein NW, Chatot CL, Plenefisch JD, Carey SW. Human serum teratogenicity studies using *in vitro* cultures of rat embryos. In: Waters MD, Sandhu SS, Lewtas J, Claxton L, Chernoff N, Nesnow S, eds. *Short-Term Bioassays in the Analysis of Complex Environmental Mixtures III*. New York: Plenum Press, 1983:407–415.

269. Juchau MR, Giachelli CM, Fantel AG, Greenaway JC, Shepard TN, Faustman-Watts EM. Effects of 3-methylcholanthrene and phenobarbital on the capacity of embryos to bioactivate teratogens during organogenesis. *Toxicol Appl Pharmacol* 1985;80:137–146.

270. Juchau MR, Harris C, Stark KL, et al. Cytochrome P450-dependent bioactivation of pro-dysmorphogens in cultured conceptuses. *Reprod Toxicol* 1991;5:259–263.

271. Ferrari DA, Gilles PA, Klein NW. Sera teratogenicity to cultured rat embryos in women with histories of spontaneous abortion. *Teratology* 1991;43:460–467.

272. Flynn TJ, Scialli AR, Gibson RR. Cultured organogenesis-staged rat embryos as biomarkers for nutritional factors in human reproductive failure. *Teratology* 1991;43:468–474.

273. Horton WE, Jr, Sadler TW. Mitochondrial alterations in embryos exposed to ß-hydroxybutyrate in whole embryo culture. *Anat Rec* 1985;213:94–101.

274. Turbow MM, Chamberlain JG. Direct effects of 6-aminonicotinamide on the developing rat embryo *in vitro* and *in vivo*. *Teratology* 1968;1:103–108.

275. Kitchin KT, Ebron MT, Svendsgaard D. *In vitro* study of embryotoxic and dysmorphogenic effects of mercuric chloride and methylmercury chloride in the rat. *Food Chem Toxicol* 1984;22:31–37.

276. Klein NW, Vogler MA, Chatot C, Pierro LJ. The use of cultured rat embryos to evaluate the teratogenic activity of serum: cadmium and cyclophosphamide. *Teratology* 1980;21:199–208.

277. Warner CW, Sadler T, Tulis S, Smith M. Zinc amelioration of cadmium induced teratogenesis *in vitro*. *Teratology* 1984;30:47–53.

278. Usami M, Ohno Y. Teratogenic effects of selenium compounds on cultured postimplantation rat embryos. *Teratog Carcinog Mutagen* 1996;16(1):27–36.

279. Klug S, Collins M, Nagao T, Merker HJ, Neubert D. Effect of lithium on rat embryos in culture: growth, development, compartmental distribution and lack of a protective effect of inositol. *Arch Toxicol* 1992;66(10):719–728.

280. Brown NA, Goulding EH, Fabro S. Ethanol embryotoxicity: direct effects on mammalian embryos *in vitro*. *Science* 1979;206:573–575.

281. Campbell MA, Fantel AG. Teratogenicity of acetyldehyde *in vitro*: relevance to the fetal alcohol syndrome. *Life Sci* 1983;32:2,641–2,647.

282. Schmid BP, Attenon P, Cicurel L. *In vitro* exposure of rat embryos to caffeine and its metabolites. *Teratology* 1986;33:57A.

283. Marret S, Gressens P, Van-Maele-Fabry G, Picard J, Evrard P. Caffeine-induced disturbances of early neurogenesis in whole mouse embryo cultures. *Brain Res*

1997;777:213–316.

284. Greenaway JC, Shepard TH, Fantel AG, Juchau MR. Sodium salicylate teratogenicity *in vitro. Teratology* 1982;26:167–171.

285. Greenaway JC, Bark DN, Juchau MR. Embryotoxic effects of salicylates: role of biotransformation. *Toxicol Appl Pharmacol* 1984;74:141–149.

286. Cicurel L, Schmid B. *In vitro* teratogenicity of acetylsalicylic acid on rat embryos: studies with venous culture conditions. *Methods Findings Exp Clin Pharmacol* 1986;8:227–232.

287. McGarrity C, Samani NJ, Beck F. The *in vivo* and *in vitro* action of sodium salicylate on rat embryos. *J Anat* 1978;127:646–649.

288. Greenaway JC, Fantel AG, Shepard TN, Juchau MR. The *in vitro* teratogenicity of cyclophosphamide in rat embryos. *Teratology* 1982;25:335–343.

289. Kitchin KT, Schmid BP, Sanyal MK. Teratogenicity of cyclophosphamide in a coupled microsomal activating/embryo culture system. *Biochem Pharmacol* 1981;30:59–64.

290. Mirkes PE, Greenaway JC, Shepard TH. A kinetic analysis of rat embryo response to cyclophosphamide exposure *in vitro. Teratology* 1983;28:249–256.

291. Mirkes PE, Greenaway JC, Hilton J, Brundrett R. Morphological and biochemical aspects of monofunctional phosphoramide mustard teratogenicity in rat embryos cultured *in vitro. Teratology* 1985;32:241–249.

292. Kochhar DM. Assessment of teratogenic response in cultured postimplantation mouse embryos: effects of hydroxyurea. In: Neubert D, Merker HJ, eds. *New Approaches to the Evaluation of Abnormal Embryonic Development.* Stuttgart: Thieme-Edition, 1975:250–277.

293. Mirkes PK, Greenaway JC. Teratogenicity of chlorambucil in rat embryos *in vitro. Teratology* 1982;26:135–143.

294. Beyer BK, Juchau MR. Cytochrome P-450 dependent embryotoxicity of estradiol 17. *Teratology* 1986;33:45C.

295. Beyer BK, Greenaway JC, Juchau MR. DES-induced embryotoxicity *in vitro. Teratology* 1985;31:45A.

296. Bechter R, Terlouw GDC, Tsuchiya M, Tsuchiya T, Kistler A. Teratogenicity of arotinoids (retinoids) in the rat whole embryo culture. *Arch Toxicol* 1992;66(3):193–197.

297. Goulding EN, Pratt RM. Isotretinoin teratogenicity in mouse whole embryo culture. *J Craniofacial Genet Dev Biol* 1986;6:99–112.

298. Morriss GM, Steele CE. Comparison of the effects of retinol and retinoic acid on postimplantation rat embryos *in vitro. Teratology* 1977;15:109–120.

299. Webster WS, Johnston MC, Lammer EJ, Sulik KK. Isotretinoin embryopathy and the cranial neural crest: an *in vivo* and *in vitro* study. *J Craniofacial Gen Dev Biol* 1986;6:211–222.

300. Kitchin KT, Ebron MT. Further development of rodent whole embryo culture: solvent toxicity and water insoluble compound delivery system. *Toxicology* 1984;30:45–57.

301. Kitchin KT, Ebron MT. Studies of saccharin and cyclohexylamine in a coupled microsomal activating/embryo culture system. *Food Chem Toxicol* 1983;21:537–541.

302. Schmid BP. Monitoring of organ formation in rat embryos after *in vitro* exposure to azathioprine, mercaptopurine, methotrexate or cyclosporin A. *Toxicology* 1984;31:9–21.

303. Svoboda KK, O'Shea KS. Optic vesicle defects induced by vincristine sulfate: an *in vivo* and *in vitro* study in the mouse embryo. *Teratology* 1984;29:223–239.

304. Naruse I, Shoji R. Whole embryo culture as a primary screening system for teratogens––effects of vinblastine on rat embryos *in vitro*. *Proc Jpn Acad* 1986;62B:31–34.

305. Greenaway JC, Fantel AG. Enhancement of rifampin teratogenicity in cultured rat embryos. *Toxicol Appl Pharmacol* 1983;69:81–88.

306. Fujinaga M, Park HW, Shepard TH, Mirkes PE, Baden JM. Staurosporine does not prevent adrenergic-induced situs inversus, but causes a unique syndrome of defects in rat embryos grown in culture. *Teratology* 1994;50(4):261–274.

307. Yamamoto Y, Yamamoto K, Hayase T, Fukui Y, Shiota K. Effects of amphetamine on rat embryos developing *in vitro*. *Reprod Toxicol* 1998;12(2):133–137.

308. Bruckner A, Lee YJ, O'Shea KS, Henneberry RC. Teratogenic effects of valproic acid and diphenylhydantoin on mouse embryos in culture. *Teratology* 1983;27:29–42.

309. Tesh JM, Newall DR. Teratogenic effects of thalidomide on rat embryos *in vitro*. *Teratology* 1985;31:68A.

310. Itoh A, Matsumoto N. Organ-specific susceptibility to clastogenic effect of urethane, a trial of application of whole embryo culture to testing system for clastogen. *J Toxicol Sci* 1984;9:175–192.

311. Kao J, Brown NA, Schmid B, Goulding EN, Fabro S. Teratogenicity of valproic acid: *in vivo* and *in vitro* teratogenicity. *Teratog Carcinog Mutagen* 1981;1:367–382.

312. Andrews JE, Ebron McCoy ME, Bojic U, Nau H, Kavlock RJ. Validation of an *in vitro* teratology system using chiral substances: stereoselective teratogenicity of 4-yn-valproic acid in cultured mouse embryos. *Toxicol Appl Pharmacol* 1995;132:310–316.

313. Crockroft DL, New DAT. Abnormalities induced in cultured rat embryos by hyperthermia. *Teratology* 1978;17:277–284.

314. Hirsekorn JM. The effects of hyperthermia on the developing central nervous system in the mouse embryo. *Anat Rec* 1980;196:79A.

315. Mirkes PK. Effects of acute exposure to elevated temperatures on rat embryo growth and development *in vitro*. *Teratology* 1985;32:259–266.

316. Kimmel GL, Cuff JM, Kimmel CA, Heredia DJ, Tudor N, Silverman PM. Embryonic development *in vitro* following short-duration exposure to heat. *Teratology* 1993;47(3):243–251.

317. Mirkes PE, Little SA, Cornel L, Welsh MJ, Laney TN, Wright FH. Induction of heat shock protein 27 in rat embryos exposed to hyperthermia. *Mol Reprod Dev* 1996;45(3):276–284.

318. Mirkes PE, Cornel LM, Park HW, Cunningham ML. Induction of thermotolerance in early postimplantation rat embryos is associated with increased resistance to hyperthermia-induced apoptosis. *Teratology* 1997;56(3):210–219.

319. Fantel AG, Greenaway JC, Shepard TH, Juchau MR, Selleck SB. The teratogenicity of cytochalasin D and its inhibition by drug metabolism. *Teratology* 1981;23:223–231.

320. Geissler FT, Faustman-Watts E. Teratogenicity of aflatoxin B1 role of biotransformation.

Toxicologist 1985;5:187.

321. Geissler F, Faustman EM. Developmental toxicity of aflatoxin B1 in the rodent embryo *in vitro*: contribution of exogenous biotransformation systems to toxicity. *Teratology* 1987;37:101– 111.

322. Geissler FT, Eaton D, Faustman-Watts E. Investigation of the role of endogenous embryonic biotransformation in mycotoxin-induced dysmorphogenesis. *Teratology* 1986;33:45A.

323. Flynn TJ, Stack ME, Troy AL, Chirtel SJ. Assessment of the embryotoxic potential of the total hydrolysis product of fumonisin B1 using cultured organogenesis-staged rat embryos. *Food Chem Toxicol* 1997;35(12):1,135–1,141.

324. Faustman E, Little S, Kirby Z, Mirkes P. Detection of DNA damage in rodent embryos exposed to alkylating agents *in vitro*: correlations with developmental toxicity. *Toxicologist* 1986;6:96.

325. Faustman E, Kirby Z, Gage D, Varnum M. *In vitro* developmental toxicity of five direct-acting alkylating agents in rodent embryos: structure-activity patterns. *Teratology* 1989;40:199–210.

326. Faustman-Watts EM, Greenaway JC, Namkung MJ, Fantel AG, Juchau MR. Teratogenicity *in vitro* of two deacetylated metabolites of N-hydroxy-2-acetylaminofluorene. *Toxicol Appl Pharmacol* 1984;76:161–171.

327. Faustman-Watts EM, Yang HYL, Namkung MJ, Greenaway JC, Fantel AG, Juchau MR. Mutagenic, cytotoxic and teratogenic effects of 2-acetylaminofluorene and reactive metabolites *in vitro*. *Teratog Carcinog Mutagen* 1984;4:273–283.

328. Faustman-Watts EM, Fiachelli CM, Juchau MR. Carbon monoxide inhibits monooxygenation by the conceptus and embryotoxic effects of proteratogens *in vitro*. *Toxicol Appl Pharmacol* 1986;83:590–595.

329. Faustman-Watts E, Namkung M, Juchau M. Modulation of the embryotoxicity *in vitro* of reactive metabolites of 2-acetylaminofluorene by reduced glutathione, ascorbate and via sulfation. *Toxicol Appl Pharmacol* 1986;86:400–410.

330. Rawlings SJ, Shuker DEG, Webb M, Brown NA. The teratogenic potential of alkoxy acids in postimplantation rat embryo culture: structure-activity relationships. *Toxicol Lett* 1985;28:49–58.

331. Greenaway JC, Fantel AG, Juchau MR. On the capacity of nitroheterocyclic compounds to elicit an unusual axial asymmetry in cultured rat embryos. *Toxicol Appl Pharmacol* 1986;82:307–315.

332. Brown-Woodman PDC, Hayes L, Huq F, Herlihy C, Picker K, Webster W. *In vitro* assessment of the effect of halogenated hydrocarbons: chloroform, dichloromethane, and dibromoethane on embryonic development of the rat. *Teratology* 1998;57:321–333.

333. Brown-Woodman PD, Webster WS, Picker K, Huq F. *In vitro* assessment of individual and interactive effects of aromatic hydrocarbons on embryonic development of the rat. *Reprod Toxicol* 1994;8(2):121–135.

334. Sidney Hunter E, Rogers EH, Schmid JE, Richard A. Comparative effects of haloacetic acids in whole embryo culture. *Teratology* 1996;54:57–64.

335. Kavlock RJ, Oglesby LA, Hall LL, et al. *In vivo* and *in vitro* structure-dosimetry–activity relationships of substituted phenols in developmental toxicity assays. *Reprod Toxicol* 1991;5:255–258.

336. Mirkes PE, Doggett B, Cornel L. Induction of a heat shock response (HSP 72) in rat embryos exposed to selected chemical teratogens. *Teratology* 1994;49(2):135–142.

337. Wilk AL, Greenberg JH, Horigan EA, Pratt RM, Martin GR. Detection of teratogenic compounds using differentiating embryonic cells. *In Vitro* 1980;16(4):269–276.

338. Paulsen DF, Langille RM, Dress V, Solursh M. Selective stimulation of *in vitro* limb-bud chondrogenesis by retinoic acid. *Differentiation* 1988;39(2):123–130.

339. Watanabe T, Iwase T. Developmental and dysmorphogenic effects of glufosinate ammonium on mouse embryos in culture. *Teratog Carcinog Mutagen* 1996;16(6):287–299.

340. Iwase T, Arishima K, Ohyama N, Inazawa K, Iwase Y, Ikeda Y, Shirai M, Yamamoto M, Somiya H, Eguchi Y. *In vitro* study of teratogenic effects of caffeine on cultured rat embryos and embryonic cells. *J Vet Med Sci* 1994;56(3):619–621.

341. Gregotti CF, Kirby Z, Manzo L, Costa LG, Faustman EM. Effects of styrene oxide on differentiation and viability of rodent embryo. *Toxicol Appl Pharmacol* 1994;128:25–35.

342. Cicurel L, Schmid BP. Postimplantation embryo culture for the assessment of the teratogenic potential and potency of compounds. *Experientia* 1988;44(10):833–840.

343. Sadler TW, Warner CW, Tulis SA, Smith MK, Doerger J. Factors determining the *in vitro* response of rodent embryos to teratogens. *Concepts Toxicol* 1985;3:36–45.

344. Welsch F. Short-term methods of assessing developmental toxicity hazard. *Issues Rev Toxicol* 1990;5:115–153.

345. Bass R, Bochert G, Merker HJ, Neubert D. Some aspects of teratogenesis and mutagenesis in mammalian embryos. *J Toxicol Environ Health* 1977;2:1,353–1,374.

346. Kochhar DM. Embryonic limb bud organ culture in assessment of teratogenicity of environmental agents. *Teratog Carcinog Mutagen* 1982;2:303–312.

347. Neubert D, Lessmollmann U, Hinz N, Dillmann I, Fuchs G. Interference of 6-mercaptopurine riboside, 6 methylmercaptopurine riboside and azathioprine with the morphogenetic differentiation of mouse extremities *in vivo* and in organ culture. *Naunyn Schmiedebergs Arch Pharmacol* 1977;298:93–105.

348. Manson JM, Simons R. *In vitro* metabolism of cyclophosphamide in limb bud culture. *Teratology* 1979;19:149–158.

349. Sadler TW, Kochhar DM. Chlorambucil-induced cell death in embryonic mouse limb buds. *Toxicol Appl Pharmacol* 1976;37:237–256.

350. Neubert D, Tapken S, Baumann I. Influence of potential thalidomide metabolites and hydrolysis products on limb development in organ culture and on the activity of proline hydroxylase—further data on our hypothesis on the thalidomide embryopathy. In: Neubert D, Merker HJ, Nau N, Langman J, eds. *Role of Pharmacokinetics in Prenatal and Perinatal Toxicology*. Stuttgart: Georg Thieme, 1978:359–382.

351. Welsch F, Baumann I, Neubert D. Effects of methyl-parathion and methyl-paraoxon on

morphogenetic differentiation of mouse limb buds in organ culture. In: Neubert D, Merker HJ, Nau H, Langman I, eds. *Role of Pharmacokinetics in Prenatal and Perinatal Toxicology*. Stuttgart: Georg Thieme, 1978:351–358.

352. Flint OP. The effects of sodium salicylate, cytosine arabinoside, and eserine sulphate on rat limb buds in culture. In: Merker HJ, Nau H. Neubert D, eds. *Teratology of the Limbs* New York: Walter deGruyter, 1980:325–338.

353. Schreiner CM, Zimmerman EF, Wee EL, Scott WJ, Jr. Caffeine effects on cyclic AMP levels in the mouse embryonic limb and palate *in vitro*. *Teratology* 1986;34:21–27.

354. Kawanishi N, Falion JT, Holmes LB. Effects of acetazolamide on limb chondrogenesis *in vitro* of the day 10.5 mouse embryo. *Teratology* 1985;32:28B.

355. Shury T, Rousseaux G, Keiner J, Politis M. Interactive effects of cadmium and retinoic acid on mouse limb bud development *in vitro*. *Bull Environ Contam Toxicol* 1994;53:570–576.

356. Kastner M, Blankenburg G, Neubert D. Isolated and reconstituted monooxygenases as supplement for organ culture. *Teratology* 1985;32:24A–25A.

357. Shiota K, Kosazuma T, Klug S, Neubert D. Development of the fetal mouse palate in suspension organ culture. *Acta Anat* 1990;137:59–64.

358. Buckalew AR, Abbott BD. A new procedure for serum-free palatal culture. *Teratology* 1991; 43:461.

359. Abbott BD. Responses of embryonic palates to selected teratogens in serum-free organ culture. *Teratology* 1991;43:461.

360. Abbott BD, Schmid JE, Brown JG, Wood CR, White RD, Buckalew AR, Held GA. RT-PCR quantification of AHR, ARNT, GR, and CYP1A1 mRNA in craniofacial tissues of embryonic mice exposed to 2,3,7,8-tetrachlorodibenzo-p-dioxin and hydrocortisone. *Toxicol Sci* 1999;47(1):76–85.

361. Abbott BD, Held GA, Wood CR, Buckalew AR, Brown JG, Schmid J. AhR, ARNT, and CYP1A1 mRNA quantitation in cultured human embryonic palates exposed to TCDD and comparison with mouse palate *in vivo* and in culture. *Toxicol Sci* 1999;47(1):62–75.

362. Abbott BD, Probst MR, Perdew GH, Buckalew AR. AH receptor, ARNT, glucocorticoid receptor, EGF receptor, EGF, TGF alpha, TGF beta 1, TGF beta 2, and TGF beta 3 expression in human embryonic palate, and effects of 2,3,7,8-tetrachlorodibenzo-p-dioxin (TCDD). *Teratology* 1998;58(2):30–43.

363. Abbott BD. Review of the interaction between TCDD and glucocorticoids in embryonic palate. *Toxicology* 1995;105(2–3):365–373.

364. Couture LA, Abbott BD, Birnbaum LS. A critical review of the developmental toxicity and teratogenicity of 2,3,7,8-tetrachlorodibenzo-p-dioxin: recent advances toward understanding the mechanism. *Teratology* 1990;42(6):619–627.

365. Abbott BD, Logsdon TR, Wilke TS. Effects of methanol on embryonic mouse palate in serum-free organ culture. *Teratology* 1994;49(2):122–134.

366. Abbott BD, Lau C, Buckalew AR, Logsdon TR, Setzer W, Zucker RM, Elstein KH, Kavlock RJ. Effects of 5–fluorouracil on embryonic rat palate *in vitro*: fusion in the ab-

sence of proliferation. *Teratology* 1993;47(6):541–554.

367. Lee EL, Taylor P. Endocrine disruptors signal the need for receptor models and mechanisms to inform policy. *Cell* 1998;93:157–163.

368. European Centre for the Validation of Alternative Methods (ECVAM). The use of biokinetics and *in vitro* methods in toxicological risk evaluation. *ATLA* 1995;24(4):473–498.

369. Anderson ME. Combining *in vitro* alternatives and physiologically-based computer modeling will improve quantitative health risk assessments. In: Goldberg AM, van Zutphen LFM, eds. *The World Congress on Alternatives and Animal Use in the Life Sciences: Education, Research, Testing*. New York: Mary Ann Lieber, 1995.

370. Nau H, Scott WJ. *Pharmacokinetics in Teratogenesis*. Boca Raton, FL: CRC Press, 1987.

371. O'Flaherty EJ, Scott W, Schreiner C, Beliles RP. A physiologically based kinetic model of rat and mouse gestation: disposition of a weak acid. *Toxicol Appl Pharmacol* 1992;112(2):245–256.

372. O'Flaherty EJ, Nau H, McCandless D, Beliles RP, Schreiner CM, Scott WJ, Jr. Physiologically based pharmacokinetics of methoxyacetic acid: dose-effect considerations in C57BL/6 mice. *Teratology* 1995;52(2):78–89.

373. Fisher JW, Whittaker TA, Taylor DH, Clewell HJ III, Andersen ME. Physiologically based pharmacokinetic modeling of the pregnant rat: a multiroute exposure model for trichloroethylene and its metabolite, trichloroacetic acid. *Toxicol Appl Pharmacol* 1989;99(3):395–414.

374. Gabrielsson JL, Johansson P, Bondesson U, Karlsson M, Paalzow LK. Analysis of pethidine disposition in the pregnant rat by means of a physiological flow model. *J Pharmacokinet Biopharm* 1986;14(4):381–395.

374. Neubert D. Toxicity studies with cellular models of differentiation. *Xenobiotica* 1985;15:649–660.

375. Johnson EM, Christian MS. When is a teratology study not an evaluation of teratogenicity? *J Am Coll Toxicol* 1984;3:431–434.

376. Neubert D, Blankenburg C, Lewandowski C, Klug S. Misinterpretations of results and creation of "artifacts" in studies on developmental toxicity using systems simpler than *in vitro* systems. *Prog Clin Biol Res* 1985;171:241–266.

377. Faustman EM. Short-term tests for teratogens. *Mutation Res* 1988;205:355–384.

378. Fabro S, Brown NA, Scialli AR. Alternative tests for teratogens. *Reprod Toxicol: Med Lett* 1984;3:21–24.

379. Johnson EM. A review of advances in prescreening for teratogenic hazards. *Prog Drug Res* 1985;29:121–154.

380. Homburger F, Marquis J. Third international conference on safety evaluation and regulation and joint American-Swiss seminar on alternative embryotoxicity and teratogenicity tests. *J Am Coll Toxicol* 1985;4:185–191.

381. Kimmel GL, Smith K, Kochhar DM, Pratt RM. Proceedings of the consensus workshop on *in vitro* teratogenesis testing. *Teratog Carcinog Mutagen* 1981;2:i–v.

382. Neubert D, Merker HJ, eds. *Culture Techniques: Applicability for Studies in Prenatal*

Differentiation and Toxicity. New York: Walter de Gruyter, 1981.

383. Picard JJ. Proceedings, *in vitro* culture of post-implantation rodent embryos. *Reprod Toxicol* 1991;5:221–222.

384. Schwetz B. *Interpretation of Segment II Developmental Toxicity Studies: Identification of Criteria for Short-Term Test Validation*. Research Triangle Park, NC: NIEHS, 1991.

385. Balls JB, Blaabauer BJ, Fentem JG, Bruner L, Combes RD, Ekwall B, Fielder RJ, Guillouzo A, Lewis RW, Lovell DP, Reinhardt CA, Repetto G, Sladowski D, Spielmann H, Zucco F. Practical aspects of the validation of toxicity test procedures—The report and recommendations of ECVAM workshop 5. *ATLA* 1995;23:129–147.

386. Brown NA, Spielmann H, Bechter R, Flint OP, Freeman SJ, Jelinek RJ, Koch E, Nau H, Newall, DR, Palmer AK, Renault JY, Repetto MF, Vogel R, Wiger R. Screening chemicals for reproductive toxicity—the current alternatives. The report and recommendations of ECVAM workshop 12. *ATLA* 1995;23:868–882.

9

Neurotoxicology
In Vitro

Shayne Cox Gad

Gad Consulting Services, Raleigh, North Carolina

In vitro systems have taken their place as major tools in neuroscience. Since the mid-1970s, the number of specific preparations and the problems that can be addressed have multiplied, so that today, essentially, a complete array of neurobiologicsystems at the molecular, cellular, and system level is represented in the tissue culture arena. Summarized here are some examples of neurobiologic studies presently in progress as possible uses in neurotoxicologic applications. A number of systems will be presented with regard to toxicologic utility, and problems and prospects for neurotoxicology screening and analytic work will be discussed. Specifically, the difficult correlation between *in vivo* and *in vitro* observations is considered.

The central issue in the utilization of *in vitro* systems is the need to clearly differentiate selective neurotoxicants from those agents generally toxic to cells or other classes of *in vitro* preparations. This issue must be explicitly addressed in the evaluation of each potentially useful system. Such studies must specifically provide the means to answer one or more of the following questions to be of use in actually assessing neurotoxic risk to humans [75,77,78]:

Is the neurotoxic effect direct or secondary to toxic damage in nonneural tissues?
Is the neurotoxic effect transient or permanent?
Is the neurotoxic effect caused by functional or morphologic changes?
Does the neurotoxic effect show a dose-response relationship?

Furthermore, whenever possible, the metabolism of the compound under investigation in the model chosen for the study should resemble that in humans. Only when these questions have been answered can a meaningful risk assessment be performed.

9.1 BIOCHEMICAL AND CELLULAR MECHANISMS OF TOXICITY

Much of the physiology and metabolism of the normal central nervous system (CNS) has not been fully elucidated. Thus, although the biochemical and cellular mechanisms of some neurotoxins are known, the exact mechanisms by which many toxins cause damage to the nervous system are not completely understood.

Considered at a cellular level, toxins may cause injury by a direct action on the neuron, or alternatively, the neuronal injury may be indirect, subsequent to toxic damage to other cell types or to a change in the environment around the neuron. Direct-acting toxins may cause damage by interfering with any of the components of the neuron, so that the mechanisms of toxicity are diverse. The range of actions by which toxins indirectly cause neurotoxic damage is also extensive: The toxic effect may be in nonneuronal cells of the nervous system itself, or in cells remote from the nervous system [76].

The biochemical and cellular mechanisms of neurotoxicity may be classified in a number of ways, but the simplest is according to their toxic actions on the nervous system. However, it is important to remember that some neurotoxins may cause changes in more than one of these categories [79]:

Interference with aerobic metabolism

Interference with protein synthesis

Interference with intermediate metabolism

Changes to cellular membranes

Interference with neurotransmission

Disturbances of axonal transport

Damage to nonneuronal cells

Damage to capillaries

Neurocarcinogenesis

Developmental neurotoxicity

In evaluating the actual validity and utility of *in vitro* systems, then, one should be able to differentiate the effects of known neurotoxins on established target sites of the nervous system. Table 9.1 presents a list of such agents and their sites of action.

9.2 *IN VITRO* SYSTEMS IN NEUROTOXICOLOGY

Striking advances in the methodologies involved in preparing dissociated cell cultures and isolated organ preparations from the CNS have been made. Essentially, any region of the vertebrate and, indeed, mammalian CNS can be maintained successfully for a period of many days, weeks, or months. Selected cell types can be identified by a number of techniques, and isolation of homogeneous populations of

TABLE 9.1. Agents associated with neurotoxicity

Site of action	Neurotoxins	
Neuron	Methyl chloride	Trimethyltin
	Carbon monoxide	MPTP (1-methyl-4-phenyl-1, 2, 3,
	Nitrites	6-tetrahydropyridine)
	Hydrogen cyanide	Aluminum
	Hydrogen sulphide	Kainic acid
	Methylmercury	Domoic acid
	Methyl alcohol	Ricin
Neurons of dorsal root ganglia	Doxorubicin	Cisplatin
Axon	γ-Diketones	Vinca alkaloids
	Acrylamide	Taxol
	β, β'-Iminodipropionitrile	Organophosphates
	Colchicine	Tetrodotoxin
	Local anaesthetics	Dichlorodiphenylethanes
Synapses	Nicotine	Botulinum toxin
	L-Glutamic acid	GABA-inhibiting organochlorines
	Domoic acid	Organophosphates
	Kainic acid	Carbamic esters
	Strychnine	Cocaine
Astrocytes	Ammonia	Methionine sulphoxamine
Oligodendrocytes, Schwann cells, and myelin	Triethyltin	Tellurium
	Hexachlorophene	Lead
	Cuprizone	GABA-transaminase inhibitors
Ependymal cells	Amoscanate	
Vascular endothelium	Lead	Ethyl chloride
	Cadmium	Thiaminases
	Mercury	Mycotoxins

neurons can be accomplished by the separation of identified cells or selective main-tenance of desired cell types. We are approaching the realization of the goal implicit in the dissociated cell culture approach, of being able to study specific cell types in isolation and in controlled combinations, so that interaction among the cells of the nervous system can be analyzed in some detail. The mechanism involved in such cellular interactions can be expected to play a central role in the regulation of ner-vous system development. Morphologic, neurochemical, and physiologic methods, similarly, have shown a remarkable increase in power and precision in the past few years. The conjunction between analytic capabilities and tissue culture preparations has resulted in considerable progress in cellular neuroscience. Patch-clamp analysis of a variety of membrane ionic channels and receptors has been carried out in pe-ripheral and central neurons and glial cells. Excitatory and inhibitory receptors have been characterized pharmacologically, and the relationship between pharmacologic and synaptic receptors has been studied [1,2]. Intracellular changes in calcium ion concentration have been measured using a variety of techniques under a variety of experimental conditions [3]. A number of trophic factors important for development of the nervous system have been identified and purified, and their genes have been cloned [4–5,107] We will deal with some of the techniques that have been successful in preparing cultured neurons and glia to maximize their experimental usefulness, and we will describe some of the experimental methods recently applied to cultured neutral material and give a detailed account of studies illustrating the new approaches to important questions.

9.3 CULTURES OF SPECIFIC CELL TYPES

9.3.1 Glia

In 1980, McCarthy and DeVellis [6] developed a procedure for preparing nearly pure populations of astroglia and oligodendroglia from the postnatal rat cerebral cortex. Their relatively simple and reliable technique has been used in a number of laboratories, with minor variations in subsequent years. The method is based on the principle that the different cell types within the complex tissue of the brain have a number of intrinsic properties that can be used to produce an enriched population under appropriately designed culture conditions. These authors list cellular adhesiveness, culture medium requirements, developmental time course, and growth patterns as particularly useful properties to exploit for purposes of cell separation. Appropriate criteria for judging the success of any cell purification scheme are essential, and these authors used ultrastructural properties, marker enzymes, and pharmacologic responsiveness as a means of identifying the cell types they were trying to isolate. We will describe this method in some detail because it exemplifies a number of considerations pertinent to the general problem of establishing a pure culture of known types.

Postnatal (1–2 days) rats were chosen because it was established easily that with a conventional dissociation procedure and culture medium containing a relatively high (15%) fetal calf serum concentration, essentially no neurons survived. Thus, a preparation was obtained that contained primarily astroglia and oligodendroglia, and the purification problem became that of selectively growing these two types of cells. It was noted that the mixed culture became a strictly layered or stratified structure at a specific time (9–10 days) after establishment of the culture. The morphologic appearance suggested that this stratification, in fact, represented a separation of the two cell types into (1) a bed layer of astrocytes attached to the culture dish and (2) a layer of oligodendroglia growing on top of the astrocytes. The astrocytes were tightly adherent to the surface of the culture flask, with the oligodendrocytes less well attached to the astrocytes, and the authors were able to design a method of mechanically shaking the oligodendrocytes free while leaving the astrocytes still attached to the flask. This differential adhesiveness, combined with the fact that the astrocytes were more active mitotically, allowed (with another cycle of plating, separation, and replating) the preparation of cultures >98% pure for one or the other cell type.

Clearly, it is essential in any purification procedure to have a means of assessing the degree of purification that is obtained. No doubt existed that the two cell populations prepared by the method of McCarthy and DeVellis [6] were different by light-microscopic criteria, but, in addition, fine structural, pharmacologic, and enzyme marker studies established satisfactorily that the cell types had properties characteristic of astrocytes and oligodendrocytes. The presence of numerous intermediate filaments and a paucity of microtubules characterized the astrocyte population, whereas the converse was true for the oligodendrocytes. Astrocytes, but not oligodendrocytes, exhibited characteristic morphologic response to cyclic adenosine monophosphate (cAMP) or brain extracts. Cyclic nucleotide phosphohydrolase (CNPase) appears to be localized in myelin and oligodendrocytes *in vivo*, and appropriately, it could be detected readily in the purified oligodendrocyte cultures, but it was undetectable in the astrocyte preparations. Glycerol phosphate dehydrogenase (GPDH) is an oligodendrocyte marker induced markedly *in vivo* by hydrocortisone. A low level of this enzyme was detected in the astrocyte culture, and this was induced approximately threefold by hydrocortisone. A tenfold higher level was measured in the oligodendrocyte cultures, and this was induced 16-fold further by hydrocortisone. Changes in cAMP levels produced by several pharmacologic agents (α- and β-adrenergic agonists or antagonists, adenosine, and prostaglandin E) were different for the two types of purified cell cultures, in accordance with their presumptive cell type. Subsequent studies with immunocytochemical staining for the astroglial antigen, glial fibrillary acidic protein (GFAP) [7], have corroborated fully these findings.

Thus, a relatively simple method provided a well-validated, fairly large-scale separation of purified populations of two important cellular components of the mammalian CNS. A large number of studies have been done subsequently using these preparations to characterize these cells further and to analyze their interaction with other cells. Some representative studies will be described below.

A step fundamental in the understanding of the function of glial cells was to characterize the receptors on their surface membranes. The β-adrenergic receptors on different classes of cells from the CNS in cell cultures have been studied quantitatively using a combination of the immunocytochemical and autoradiographic techniques [7]. Most of these receptors were located on astroglia having a flat polygonal morphology. Fibroblasts and process-bearing GFAP-positive cells had a much lower level of β-adrenergic receptors and neurons, and oligodendroglia had essentially none. The importance of culture conditions and care in documenting the types of cells present in the cultures in interpreting receptor-binding data is emphasized in these studies. Meningeal fibroblasts express $β_2$-, whereas astrocytes express $β_1$, adrenergic receptors, and the relative dominance of the cultures by these cell types must be ascertained (e.g., by combined immunocytochemistry for fibronectin and GFAP). Nominally minor changes in culture methodology can affect the mix of cell types that one obtains.

Purified glial cultures have been used effectively to study the second messenger systems to which glial receptors are coupled. Perhaps, not surprisingly, both cyclic nucleotide and phospholipid systems can be regulated in these cells by a broad variety of catecholaminergic, cholinergic, and peptidergic agonists [8]. Altered levels of cAMP resulted from catecholamine and from petidergic stimulation, and this was accompanied by an altered state of phosphorylation of GFAP and glial filaments [9]. A transition from a polygonal to a process-bearing morphology occurred with these treatments, but further analysis indicated that no causal role of intermediate filament protein phosphorylation could be established with regard to the morphologic change.

Astroglia cells have proven to be highly useful "feeder layers" for low-density neuronal cultures and, as noted below, they synthesize and secrete powerful neurotrophic materials. The glia respond to a number of agonists with large changes in intracellular calcium [10,11], and these changes appear to propagate in waves through populations of the glia. Thus, it is clear that, rather than being passive support cells, glial cells play an active dynamic role in nervous system development and function.

9.3.2 Neurons

The situation regarding glial lineage, purification, and identification is complicated, and this is even more true for neurons; in addition, the strategies for dealing with neuronal identification and separation are correspondingly complex.

Dodd and Jessell [12] have taken an immunocytochemical approach to the identification of subsets of spinal sensory neurons from dorsal root ganglia (DRG). They have used antibodies directed against intracellular peptides (substance P and somatostatin) and a sensory-neuron–specific enzyme (fluoride-resistant acid phosphatase). These three markers serve to distinguish three nonoverlapping populations of small DRG neurons that have a distinctive laminar projection pattern in the dorsal horn of the spinal cord. In addition, a number of complex surface-membrane carbohydrate-containing structures have been shown to identify subsets of sensory ganglion neurons and their distinctive axonal projection patterns within the spinal cord. The working assumption is that these surface glycoconjugates may be involved in the establishment of the specific connections characteristic of different functional types of sensory neurons. At least some aspects of this surface-membrane carbohydrate structure specificity is retained in culture; however, some antibodies that labeled adult DFG neurons failed to label cultured neurons, and in general, antibodies against the carbohydrates labeled a lower proportion of cultured neurons than did neurons in mature DRGs. These anticarbohydrate antibodies provide powerful tools for identifying, and potentially isolating, functionally meaningful populations of DRG neurons, and at least some of these markers molecules are expressed *in vitro*.

Immunohistochemical methods have been used extensively in identifying different types of central neurons as well. This technique can be used in conjunction with autoradiographic studies of transmitter uptake. Inhibitory neurons using gamma-aminobutyric acid (GABA) have been identified in this way by Neale et al. [13]. A high degree of concordance was found between neurons that take up GABA and those that express glutamic acid decarboxylase (GAD), an enzyme involved in the production of GABA. Many other molecular neuronal phenotypes can be identified and are essential in analyzing experiments done with the complex mixtures of cell types occurring in typical dissociated cell cultures from the CNS [14].

A series of experiments illustrating the utility of some of these methodologies has been done by Brenneman et al. [15,16]. These workers are interested in the role of electrical activity in neurons in regulating development of the nervous system. Extensive experimentation *in vivo* established that electrical activity does exert a critical developmental influence on neuronal survival and synaptic connections. The cell culture methodology is advantageous in investigating the mechanics involved in this activity-development coupling.

It was straightforward to establish that, *in vitro*, electrical activity does indeed have a major impact on neuronal survival; blockade of electrical activity with the specific voltage-dependent sodium channel blocker, tetrodotoxin (TTX), results in a substantial decrement in the number of surviving neurons. This sensitivity to action potential blockade occurs over a restricted developmental time window (from about 1 week to 3 weeks *in vitro* for cultures prepared form approximately 2-week-old mouse embryos). Furthermore, some cell types do not survive TTX exposure.

The hypothesis was tested that neuropeptides might be involved in the activity-dependent regulation of neuronal survival. The experiment was to add putative trophically active peptides to cultures blocked with TTX to see if these peptides could reverse the effects of TTX. One peptide, vasoactive intestinal peptide (VIP), proved to be effective in this regard for extraordinarily low concentrations (10^{-10} M, or even less). Using a variety of techniques, it was then possible to show the following:

1. VIP-containing neurons exist in the cultures; 1–2% of the neurons showed VIP-like immunoreactivity.

2. VIP is released in an activity-dependent manner, because radioimmunoassay demonstrated VIP in the culture medium from control cultures, but no VIP was detectable from TTX-blocked cultures.

3. VIP does not appear to act directly on neurons. Cultures grown in a defined medium containing no fetal bovine or horse serum had a relatively enriched number of neurons with a relatively lower number of glial cells. Neurons in these cultures exposed to TTX are not rescued by VIP.

4. Glial cells possess VIP receptors.

5. Glial cells produce a trophic material promoting neuronal survival, and the secretion of this material by glial is increased substantially by VIP treatment at concentrations like those described above.

A number of other closely related peptides (e.g., secretin) do not have the survival-promoting activity of VIP [17]. Antibodies to VIP or VIP-blockading peptide fragments (VIP10-28) have a deleterious effect on neuronal survival indistinguishable from TTX.

Taken together, these findings led to a developmental schema of the following sort. Electrically active, VIP-containing neurons release VIP, which interacts with VIP receptors of glial cells. These activated glial cells elaborate a proteinaceous factor acting on neurons as a trophic material promoting their survival. An attractive feature of this scheme is that it would seem to serve as a paradigm for much more extensive and general neuron–glial–neuron interactions that might be involved in developmental regulation and maintenance of neuronal integrity. As noted above, glial cells have an extraordinarily rich complement of receptors on their surface, giving them potential responsiveness to a broad array of neurotransmitters.

These developmental studies involved the use of a relatively complex culture system with several types of neurons and background cells. Identification of specific cell types was essential for these studies, and the ability to grow pure glial preparations and enriched neuronal populations was a key feature.

The development of a powerful antagonist for VIP has allowed a test of whether VIP plays a role in the development of the nervous system *in vivo* [18]. When the antagonist is injected into rat pups or into the ventricle of older animals, substantial behavioral deficits and neuronal dysgeneses were produced. Such whole-animal validation of results obtained with tissue culture preparations is a crucial step in mechanistic understanding of developmental or neurotoxicologic processes. Other examples in which this has been accomplished in developmental neuroscience include cholinergic induction in the autonomic nervous system [5,19,20] and regulation of the motoneuron number in the spinal cord [21–23].

Preparations of central neuronal cultures have been developed that, although not providing purified cell populations, represent useful experimental preparations for a variety of biochemical, morphologic, and physiologic studies. These studies include cell cultures prepared from (1) the ventral or dorsal horn of the spinal cord, (2) the hippocampus, (3) the cerebral cortex, (4) granule cells from the cerebellum, (5) the brain stem, and (6) the basal nuclei from the forebrain (see ref. 24 for an excellent review).

An effective means of identifying certain populations of neurons has been developed and is being used in several laboratories, both for identification and for separation and purification of the neurons [24–26]. The method relies on knowing the axonal projection pattern of the population of interest. Marker molecules are injected into the region to which the axons project, are taken up by the axonal endings, and are transported retrograde to the cell bodies of the neurons. Insofar as a population of neurons in a given brain region are the only ones in that region projecting to the injection site, this method serves to identify that population of neurons. This method has been used to label spinal motoneurons, because only the spinal cord neurons whose axons terminated in muscle would be motoneurons and labeled by materials injected into skeletal muscle. Similarly, however, cerebral cortical neurons projecting to the pyramidal tract or to a subcortical structure, the superior colliculus, have been labeled by suitably injected retrograde-transported marker molecules [27].

9.3.3 Use in Physiologic Studies

Dissociated cell cultures of the CNS have been particularly useful in allowing rigorous biophysical characterization of membrane mechanisms involved in synaptic and pharmacologic responses of neurons. Neurons from nearly every region of the mammalian central neurons have been available for a level of analysis far beyond that which can be obtained *in vivo*. Recent progress in our understanding of excitatory amino acid (EAA) responses exemplifies this work; an excellent comprehensive review is available [1].

A large body of physiologic and pharmacologic experimentation *in vivo* indicated that at least two and possibly three different amino acid receptors were involved in fast excitatory responses of central neurons. Considerable uncertainty has been attached to the interpretation of the membrane mechanisms involved in the different responses, however. In particular, it was unclear whether the response to glutamate involved an increase or decrease of membrane conductance and what ions contributed to the response. Tissue culture methods allowed the application of voltage-clamp techniques to the analysis of the problem, in conjunction with the neuronal membrane. Known concentrations of the receptor agonists and antagonists are crucial for these experiments.

Glutamate has been shown to be a mixed excitatory agonist, activating receptors of both the N-methyl-D-aspartate (NMDA) and non-NMDA (kainate or quisqualate or K/Q preferring) type. Activation of K/Q-preferring receptor results in a voltage-independent increase in membrane conductance to monovalent cations; the reversal potential for this response is about 0 mV. Divalent cations have little effect on these responses. The response to NMDA receptor activation, by contrast, is markedly voltage-dependent in physiologic solutions, and a region of negative slope conductance occurs between about -60 and -20 mV. This negative slope region is dependent on the presence of Mg^{2+} ions in the external medium; in 0 Mg^{2+} solutions, the response to NMDA is nearly voltage independent. The reversal potential of the NMDA response is also about 0 mV in physiologic solutions. The conductance increase elicited by MNDA-type agonists has been shown to involve monovalent cations and Ca^{2+} ions. In fact, the conductance increase of Ca^{2+} is large. Direct demonstration of an increase in cytoplasmic Ca^{2+} as a result of NMDA receptor activation has been possible using Ca^{2+}-sensitive dyes and optical detection methods [28,29].

Numerous experiments have shown that non-NMDA receptors were involved in excitatory interactions both *in vivo* and *in vitro*, and patch-recording methods have demonstrated clearly a major contribution of NMDA receptors in excitatory synaptic potentials in both hippocampal and spinal cord cultures [30]. The NMDA component of synaptic currents has a much longer time course than does the non-NMDA component. Persistent activation of the NMDA receptor is probably responsible for this prolonged synaptic current. Raising external Ca^{2+} from 1 to 20 mM results in a shift in the reversal potential of the NMDA-mediated component of the synaptic current from 0 mV to +10 mV, whereas the non-NMDA reversal potential is unaffected. This shift indicated that, as with the pharmacologic responses, synaptically activated NMDA conductance involves channels through which Ca^{2+} ions as well as monovalent cations can move. In the presence of external Mg^{2+} ions, this conductance displays the same strong voltage dependence as do the pharmacologic NMDA responses.

The dissociated neuronal cell culture has been used in a variety of pharmacologic studies directed at understanding the mechanism of action of anticonvulsant and other neuroactive compounds [31]. Recent advances in patch-clamp and single-channel recording techniques give excellent promise of understanding the molecular basis of the activity of such agents on a variety of CNS cell types.

The tissue culture methodology has been useful in analyzing the presynaptic transmitter release mechanism. In particular, the relationship between the morphology and physiology of excitatory synaptic connections has been studied [32,33]. The statistical properties of individual synaptic connections have been measured in conjunction with presynaptic and postsynaptic injection of fluorescent dyes and horseradish peroxidase to identify the synaptic structure involved in the synaptic activity. The physiologic studies defined the number of functional release elements subserving a given connection. Each release element can release no more than one quantum of transmitter, and the probability that release of a quantum would occur after a presynaptic action potential has a value somewhere between zero and one. The number of release elements and their probability of release essentially define the physiologic transmitter release apparatus. It was found that in some cases the number of anatomical synaptic boutons corresponded closely to the number of functional release elements. In a substantial proportion of cases, however, the number of boutons was much larger than was the number of release elements, suggesting that up to one-half or two-thirds of boutons were not functional. This result implies that a substantial reserve of functionally inactive synapses may be available for recruitment under appropriate circumstances.

Optical techniques, in conjunction with a voltage- or ion-sensitive dyes, provide a direct, relatively noninvasive and sensitive method for following different neuronal activities. A dye designed by Grybkiewicz et al. [34], Fura-2, has provided excellent quantitative measurement of cytosolic Ca^{2+} in single-rate cerebellar granule cells in the outgrowth zone of explant cultures [35]. The output of a charge-coupled device (CCD) camera is recorded on a 320 x 512 pixel array, with exposure times of 0.25–0.5 seconds adequate to obtain each image. Emission at 500 nM, with excitation of 340 versus 380 nM, is compared with obtain Ca_{2+} -dependent emission. A formula exists for relating the 340/380 emission ration to the free Ca_{2+} in the area corresponding to a given period, so that the processed data give a picture with $[Ca_{2+}]$ coded in color. High-potassium depolarization of the granule cells was accompanied by an increase in cytosolic Ca^{2+}; this effect was more pronounced in cells that had been maintained *in vitro* for longer periods. Furthermore, a pronounced accentuation of the rise in cytosolic calcium under depolarization was observed with repeated application of the high-K^+ solution. In intermediate stage cells (longer than 12 days *in vitro*), application of 25-mM K^+

raised cytosolic calcium from 100 mM to approximately 2 mM. Spike inactivation by TTX diminished by about one-half the calcium response to high-K+ solutions, implying that spike activity was partly responsible for the Ca^{2+} rise. Nifedipine, an organic blocker of some types of voltage-sensitive calcium, also blocked about 50% of the calcium response to high-K^+ solutions. Applications of 10-M GABA produced a consistent increase in cytosolic calcium, although this increase (a doubling or so) is much less than that produced by a high K^+. Small, inconsistent calcium responses were produced by glutamate in younger cultures, but rarely in older cultures; more consistent responses were seen with kainate application [3].

9.4 TISSUE CULTURE METHODOLOGY

Recently, several presentations of neural tissue culture methodologies have appeared with both general discussion and detailed procedures (see ref. 24). We will make some comments on some preparations with which we are familiar, emphasizing those features we feel are relevant to the design and utilization of an *in vitro* neurotoxicologic screen.

The overriding imperatives for toxicologic testing using *in vitro* model systems are the three experimental R's: relevance, reliability, and reproducibility. In addressing these imperatives for dissociated neural cultures, the ideal has been best achieved for reproducibility and reliability. Our experience with primary dissociated systems has indicated that the following variables are important in optimizing these systems for both analytical purposes and reproducibility: plating density, culture age, monitoring of electrical activity, cellular diversity, nutrient media, duration of test period, schedule of medium changes, and gestational age of tissue. Among the most important culture variables is plating density, requiring a sufficient number of cells to allow survival and yet few enough cells to retain cellular resolution. A threshold exists of neuronal cell number that must be achieved to permit the survival of neurons, which probably is contingent on having enough background support cells and sufficient synaptic contact with other neurons. In the case of the dissociated spinal cord and hippocampal systems, low-density neuronal cultures can be achieved with a confluent layer of background cells with a high degree of reliability and reproducibility.

The number and type of support glial cells present in a test system are of major importance in several respects: The glia may themselves be the target of the test substance, thus producing neuronal damage indirectly by interfering with glial-derived support; in addition, the glia provide a cellular matrix upon which neurons can grow. These background cells can be manipulated in several ways to optimize a neuronal test system for toxicologic screening. Of practical importance, by "seeding" neurons onto a confluent layer of astrocytes (typically from cerebral cortex), one can obtain an excellent dispersion of nonaggregated neurons at high plating efficiency. Indeed, our experience with neuronal cultures from hippocampus and cerebral cortex has been that this seeding strategy provides cultures with greater reproducibility, increased longevity, and an impressive cellular resolution that is so important for many of the neuronal surface- and image-based assays. Another important aspect of the background cells is control of their cellular division to prevent astrocyte overgrowth and the proliferation of microglia. Typically, a 1- to 4-day treatment (depending on the culture) with uridine plus fluorodeoxyuridine provides adequate inhibition of the nonneuronal cells. In all systems, we have worked with, this antimitotic treatment increased the longevity of the preparation and substantially improved the reliability of the test system.

Included in our list of important considerations for optimizing primary neuronal test systems is the age of the cultures when the test compound is added. The type and number of cellular processes that will be vulnerable to a test compound will vary depending on what age is chosen for the treatment period. For example, if the test compound is added at the time of plating or seeding, neuronal survival could be affected by substances interfering with the attachment of the neurons to their background matrix. Because of this complication, we generally allow cultures to develop for a week before adding test compounds. This period allows for the neurons to migrate and establish a network of interacting, synaptically connected cells. Depending on the goals of the testing, one can limit the period to one in which the neurons are developing instead of one in which the neurons have matured. In general, this maturation requires 3–4 weeks *in vitro*. In screening compounds for their potential effects on developing systems, the concept of a "critical period" comes into consideration. A critical period is a finite stage of development during which the cultured cells are vulnerable to toxic effects of the test substance. Most often, this period occurs during a time when the system is undergoing rapid growth and differentiation. Indeed, this period varies with the type of preparation and brain arca chosen, and thus, this interval needs to be empirically determined for each test system. In the dissociated spinal cord/DRG preparation, this period exists for approximately 2 weeks, from day 7 to day 21 *in vitro*.

Another important variable in obtaining reproducible and reliable test cultures is the manipulation and control of the nutrient medium in which the cells grow. As mentioned previously, altering the growth medium and varying the number of complete changes of medium can manipulate the cellular composition of the primary cultures. In general, to obtain test cultures that are highly enriched in neurons versus nonneuronal cells, a serum-free, defined medium should be used. After plating the cells in a medium containing 10% fetal calf serum, we typically place the cultures into a serum-free medium within 24 hours. Such cultures should be plated on poly-L-lysine or another suitable matrix protein, such as collagen. Often, a "sandwich" of matrix proteins is of value in optimizing cell adherence and stability in these preparations. Because few nonneuronal cells exist in such preparations, the matrix protein becomes an important variable in maintaining cellular adherence and stability during the assay procedures. As previously discussed, such "pure" neuronal cultures probably are not the best initial screen for toxicity. Rather, a preparation comprising a mixed population of cells, including neurons and glia, is more appropriate, so that indirect- as well as direct-acting agents will be detected. Other important medium-related variables include the number of complete medium changes and the age of the culture when these changes are made. The reason for this emphasis is the potential influence of conditioning substances released by cells over time in culture. The vulnerability of neurons can be dramatically affected by the absence or presence of conditioning factors. For instance, TTX, a neurotoxin blocking voltage-dependent sodium channels, produces a 30–50% loss of neurons when added to cultures after a complete exchange of medium, whereas the same treatment in cultures that have not received a complete change of medium actually prevents neuronal death normally occurring in control cultures. Thus, an opposite pharmacologic effect can be obtained depending on the manipulation of conditioning substances during medium exchanges. Thus, changes of medium should be made purposefully and under tight control to avoid inconsistencies during the screening process.

Many of the primary neuronal culture systems exhibit spontaneous electrical activity that can vary from an occasional postsynaptic potential to complex patterns of electrical bursting activity. In the case of developing neuronal cultures, the ability of a test compound to decrease this activity can itself result in neuronal cell death. Thus, the tested substance may not have an intrinsic toxic interaction with cellular metabolism, yet still produce deleterious action through its alteration of the ionic milieu secondary to activity changes.

The most basic experiment for screening potentially toxic substances is the dose-response. Typically, we screen substances at log concentration intervals over five orders of magnitude. The number of replications can be kept to a minimum (two to three) at this stage of screening. Although the broad range of concentrations chosen for the screen may appear excessive, we have found that by using this approach we have discovered atypical toxic effects that a more limited range would have missed. For example, our experiments with the envelope protein from the human immunodeficiency virus (HIV) have shown that toxicity is observed primarily at low concentrations (≤ 1 pM) of the purified protein, with higher amounts producing significantly less or no toxicity [51]. The envelope protein studies serve to illustrate that in the case of peptide/protein test substances, higher concentrations do not necessarily produce greater toxicity. Whereas for most substances the dose-effect curve will be proportional, one should be aware of response properties that may be unique to a given class of substances.

9.5 ASSAYS OF TOXICITY: CELL CULTURE MODELS

The intuitive strategy for screening compounds for their neurotoxic potential might be to devise a series of tests proceeding from a sensitive, albeit nonspecific measure of global cellular structure and function to assays directed at specific neuronal phenotypes or pointed at a specific biochemical pathway, which may explicate a neurochemical mechanism. Within this strategy, proceeding from the general to specific, one has the option of employing either morphologic or biochemically oriented measures to assess the level of toxicity in cultured cells. Although primary neuronal cultures provide for a degree of complexity that may increase the relevancy to effects *in vivo*, the morphologic and neurochemical heterogeneity of the neurons comprising the primary systems pose a problem in finding assays effective in assessing representative amounts of neurotoxic damage. However, it is our opinion that the advantage of having the interacting cell types present in the test system far outweighs any difficulty in estimating damage in such heterogeneous test systems. Indeed, large-scale studies conducted to determine the utility of assays using established cell lines have shown that they are not highly predictive of teratogenic potential [36], hence, the need for relevant interacting cell types in the test system.

The *prima facie* approach in neurotoxic assessment is gross morphologic evaluation for any abnormal appearance of neurons, including vacuolization, degeneration of axons or dendrites, lysis of cell bodies, and arborization shrinkage. In some types of preparations (e.g., dissociated spinal cord/DRG cultures), a need also exists to immunocytochemically identify neurons from glia because of their similar appearance. For this purpose, antisera- to neuron-specific enolase is a good choice for immunocytochemical verification of neurons [37]. Antisera to neurofilament proteins is also used to identify neurons, but this often results in a heavily stained neuropile, making somal body identification difficult in high-density cultures. We have used neuronal cell counts as the most direct assessment of neuronal survival in chronic (3–5 days) test paradigms [38]. Counts are conducted from coded dishes on predetermined coordinate locations and, of course, without knowledge of the treatment group. With this method, one can evaluate neurotoxic effects on the number of morphologically distinct neurons (e.g., bipolar versus multipolar) or immunocytochemically identified neuronal phenotypes. Because these direct counting assays are time consuming and laborious, we usually restrict their use to a confirmatory role of the more indirect screening assays enumerated below. With the advent of sophisticated computer image analysis, it is now possible to count neurons and estimate neurite length by automated techniques [39,40]. Although such systems can provide a rapid, quantitative method for obtaining morphometric parameters, the major disadvantages are cost and the specialized technical knowledge required for accurate image analysis.

For quantitative screening purposes employing biochemically oriented assays, several alternates are available that should be considered. In the most general category, the direct measurement of cellular protein or nucleic acid is commonly used. These, in general, are not sensitive, and in complex systems of dissociated neural tissue, the amount of protein associated with neurons may be an insignificant amount of the total culture protein. In preparations enriched for neurons in comparison to nonneuronal cells, the protein assay could be used with greater sensitivity and utility [41]. Another general cytotoxic measure often employed is the release of the soluble enzyme lactate dehydrogenase (LDH) [42]. Whereas this assay has no cellular specificity, it can be used advantageously in screening compounds for general cytotoxicity. The LDH assay can be automated and scaled down to achieve a rapid and quantitative measure of cytotoxicity [43]. In addition, quantification of neurotoxic and neurotrophic effects has been reported using fluorescein diacetate, a dye taken up by living cells. For this assay, the total amount of fluorescein produced from the dye is measured in cell lysates. [44]. This method was found to be proportional to the number of cells counted under fluorescence microscopy.

In some cases, an assay more specifically directed at neurons is required. Options we have used for this purpose include iodinated tetanus toxin fixation, tritiated ouabain binding, and an assay of neurotransmitter-released enzymes, such as choline acetyltransferase (cholinergic neurons), GAD (GABAergic neurons), or tyrosine hydroxylase (catecholaminergic neurons) [45,46]. All of the above assays have interpretive disadvantages but possess the obvious benefit of speed and precise quantitation. In this brief discussion of these neuron-directed, biochemical methodologies, we should point out interpretive caveats inherent in the assay as well as indicate some of their potential applications. Radiolabeled tetanus toxin binds to neurons with high affinity and can be used to estimate neuronal surface area [47]. This method is rapid and easy to perform but has several limitations. Tetanus toxin does not bind well to newly dissociated neurons for as long as 4 days after plating. Thus, the utility of this assay is confined to neuronal cultures that have developed in cultures for about a week. In addition, tetanus toxin can bind to type II astrocytes, thereby limiting the conclusions that can be drawn. Tetanus toxin also can bind to dead neurons, and thus, sufficient time must be allowed for cell lysis to occur to observe neuronal deficits by this method. More recently, ouabain binding has been employed to estimate neuronal damage [48]. Ouabain binds with high affinity to the Na^+, K^+ -ATPase, which is enriched on neurons in comparison to nonneuronal cells. In addition, the binding of cardiac glycosides to various areas of rat brain have suggested that the high-affinity binding of ouabain selectively labels the neuronal form of the Na^+, K^+ -ATPase [49]. Membrane or whole-cell preparations may be used in the ouabain assay. Good correlations have been found between neuronal cell number and ouabain binding. Excitotoxin-mediated decrements have also been detected with high precision and reproducibility [49].

9.6 NEUROTOXICOLOGY

A basic distinction that should be kept in mind in neurotoxicology (as with other areas of toxicology) has to do with the design of a screening system for agents of unknown effect as compared with analytic systems for studying mechanisms that might be responsible for known toxic effects or related to structures having a high index of toxicologic suspicion. In the latter case, experimental approaches analogous to those involved in mechanistic neurobiologic studies generally seem appropriate; that is, appropriate target neuronal or glial cells populations must be identified. These populations may be highly selective (i.e., only oligodendrocytes or cholinergic neurons) in some cases, whereas any neuronal, glial, or indeed any cell type may be usable in others. Possible involvement of specific molecular or cell biologic processes can then be investigated regarding their direct or secondary vulnerability to the toxic agent in question.

The design of an adequate screening system is perhaps conceptually more difficult. An extreme position that only the behaving human is an adequate test object is strictly correct, but not helpful. We will explore the characteristics of *in vitro* systems that may be useful in identifying compounds with neurotoxic potential, keeping in mind the necessity of avoiding both errors of omission (false-negative) and errors of commission (false-positive). For a screening system, the former sin would appear to be the most serious in that, logically, an initial screen has utility if it removes from further consideration those compounds deemed not to be toxic. Therefore, a negative result on the screen needs to be well validated as strongly indicating that a compound will not have adverse effects on the intact nervous system. In general, studies that would justify using *in vitro* models in this strong, screening fashion are not available. It is beyond the scope of this review to provide a strategy for establishing such a role for *in vitro* systems; we believe it is possible to develop such a strategy, and if this is not done, the systems will be far less useful than would be desirable. Positive indications of toxicity require further investigation; if this reveals no significant effects, the initial false indicators will have resulted in "unnecessary" study of the compound, but no injury to the potential human population of exposed individuals. An excessively high proportion of false positives, however, renders the initial screen of little utility.

As has been noted previously, the nervous system *in vitro* comes in three or four different general versions varying in complexity and the degree to which they retain structural and functional properties of the parent tissue. Detailed descriptions of these different types of cultures are available [24]: (1) Continuously dividing tumor or transformed cell lines form the simplest preparation. These lines can be induced to stop dividing and exhibit a variety of neuronal or glial phenotypes, including the formation of synapses with target cells. (2) Primary dissociated cell cultures are formed by growing a single-cell suspension from some central or peripheral neural structure on the surface of a culture dish. The single-cell suspension or the cultures themselves can be processed in various ways to produce pure neuronal or glial populations or even (with various marking and sorting methods) relatively pure populations of a given cell type, such as spinal motoneurons or a particular type of glia. (3) These cell suspensions can be manipulated to produce reaggregated cultures that may reconstitute various aspects of the parent tissue. (4) Small pieces or slices of neural tissue can be cultured, giving preparation preserving a considerable degree of homology with the parent brain structure. Different pieces of the brain can be positioned in the culture dish so that they establish appropriate synaptic connections with one another (spinal cord with muscle, retina with thalamus, etc.).

We will argue, as have others, [50] that (1) for an initial screening instrument, the more complex multicomponent systems are appropriate and (2) for the assay of toxic effect, more general indices of development and integrity are to be preferred.

A large number of assays are now available for evaluating the development and function of these preparations. These assays range in specificity from those reporting the activity of single molecules peculiar to an individual cell type (enzymes, receptors, specific antigens) to global indicators, such as cell number, total proteins or RNA, general surface membrane markers, and so on. Intermediate markers are available for classes of cells (neurons versus glia; oligoglia versus astroglia).

9.6.1 The Use of More Highly Organized, Multicomponent Systems as a Screening Tool

Because the utility of a screen depends on it being responsive to as broad a range of toxicologic agents and mechanisms as possible, obviously for a preparation to signal an agent with a given target molecule (i.e., mechanism of action), that molecule must be present in the test system. The mixed or complex systems incorporate a wider range of cell types and, hence, should be responsive to a broader range of agents. Furthermore, some agents may act on mechanisms involved in the interaction between different types of cells. In such a case, no single-cell–type preparation would reveal a potentially significant toxic effect. An example may be useful in illustrating this situation. One of the distressing aspects of infection with HIV experienced by many patients with acquired immunodeficiency syndrome (AIDS) is dementia. Histopathology has revealed that a loss of neurons occurs in the cerebral cortex of some people infected with HIV, and experiments in cell cultures have demonstrated that gp120, an HIV coat protein shed from the virus and present in the blood of the infected individual, is capable of producing neuronal death [51]. However, this result is true in hippocampal cultures only if both neurons and glia cells are present, and the gp120 protein does not produce neuronal killing when the glia population has been reduced or eliminated by appropriate culture conditions. One parsimonious interpretation is that the gp120 acts primarily on the astrocyte and that the neuronal death is a secondary effect of this primary action. Of course, this process may not be the only action of the gp120, and indeed the death of retinal ganglion cells produced by gp120 may well be because of a direct effect of the peptide on those neurons [52].

The "excitotoxic" action of the amino acid alpha-amino-beta-methylamine propionic acid (BMAA) is thought to be indirect [50] because of the structural features of the BMAA, its relatively low potency, and the rather slow onset of excitant action after its administration.

9.7 ISOLATED TISSUE ASSAYS

A series of isolated tissue preparation bioassays, conducted with appropriate standards, can be used to determine if the material acts pharmacologically directly on neural receptor sites or transmission properties. Though a classic pharmacologist normally performs these bioassays, a good technician can be trained to conduct them. The required equipment consists of a Magnus (or similar type) tissue bath [80–82], a physiograph or kymograph, force transducer, glassware, a stimulator, and bench spectrophotometer. The assays used in the screening battery are listed in Table 9.2, along with the original reference describing each preparation and assay. The assays are performed as per the original author's descriptions with only minor modifications, except that control standards (as listed in Table 9.2) are always used. Only those assays appropriate for the neurologic/muscular alterations observed in the screen are used. Note that all of these assays are intact organ preparations, not minced tissue preparations, as others [83] have recommended for biochemical assays.

The first modification in each assay is that, when available, both positive and negative standard controls (pharmacologic agonists and antagonists, respectively) are employed. Before the preparation is used to assay the test material, the issue preparation is exposed to the agonist to ensure that the preparation is functional and to provide a baseline dose-response curve against which the activity of the test material can be quantitatively compared. After the test material has been assayed (if a dose-response curve has been generated), one can determine whether the antagonist will selectively block the activity of the test material. If so, specific activity at that receptor can be considered as established. In this assay sequence, it must be kept in mind that a test material may act to either stimulate or depress activity, and therefore, the roles of the standard agonists and antagonists may be reversed.

Commonly overlooked when performing these assays is the possibility of metabolism to an active form that can be assessed in this *in vitro* model. The test material should be tested in both original and "metabolized" forms. The metabolized form is prepared by incubating a 5% solution (in aerated Tyrodes) or other appropriate physiologic salt solution with strips of suitably prepared test species liver for 30 minutes. A filtered supernatant is then collected from this incubation and tested for activity. Suitable metabolic blanks should also be tested.

9.7.1 Electrophysiology Methods

A number of electrophysiologic techniques are available that can be used to detect or assess neurotoxicity. These techniques can be divided into two broad general categories: those focused on CNS function and those focused on peripheral nervous system function [96].

First, however, the function of the individual components of the nervous sys-

TABLE 9.2. Isolated tissue pharmacologic assays

Assay system	Endpoint	Standards (agonist/antagonist)	References
Rat ileum	General activity	None (side-spectrum assay for intrinsic activity)	84
Guinea pig *vas deferens*	Muscarinic nicotinic	Methacholine/atrophine Methacholine/ hexamethonium	85
	or Muscarinic	Methacholine/atropine	
Rat serosal strip	Nicotinic	Methacholine/ hexamethonium	86
Rat *vas deferens*	Alpha adrenergic	Norepinephrine/ phenoxybenz-amine	87
Rat uterus	Beta adrenergic	Epinephrine/propranol	88
Rat uterus	Kinin receptors	Bradykinin/none	89
Guinea pig tracheal chain	Dopaminergic	Dopamine/none	84
Rat serosal strips	Tryptaminergic	5-Hydroxytryptamine (serotonin)/dibenzyline or lysergicacid dibromide	90
Guinea pig tracheal chain	Histaminergic	Histamine/benadryl	91, 92
Guinea pig ileum (electrically stimulated)	Endorphan receptors	Methenkephaline/none	93
Red blood cell hemolysis	Membrane stabilization	Chlorpromazine (not a receptor-mediated activity)	94
Frog rectus abdominis	Membrane depolarization	Decamethonium iodide (not a receptor-mediated activity)	95

tem, how they are connected together, and how they operate as a complete system, should be briefly overviewed.

Data collection and communication in the nervous system occurs by means of graded potentials, action potentials, and synaptic coupling of neurons. These electrical potentials may be recorded and analyzed at two different levels, depending on the electrical coupling arrangements: individual cell (that is, intracellular and extracellular) or multiple cell [e.g., electroencephalogram (EEG), evoked potentials (EPs), slow potentials]. These potentials may be recorded in specific central or peripheral nervous system areas (e.g., visual cortex, hippocampus, sensory and motor nerves, muscle spindles) during various behavioral states or in *in vitro* preparations (e.g., nerve-muscle, retinal photoreceptor, brain slice).

9.7.2 Neurochemical and Biochemical Assays

Though some elegant methods are now available to study the biochemistry of the brain and nervous system, none has yet discovered any generalized marker chemicals, which will serve as reliable indicators or early warnings of neurotoxic actions or potential actions. However, some useful methods exist. Before looking at these, one should understand the basic problems involved.

Normal biochemical events surrounding the maintenance and functions of the nervous system center on energy metabolism, biosynthesis of macromolecules, and neurotransmitter synthesis, storage, release, uptake, and degradation. Measurement of these events is complicated by the sequestered nature of the components of the nervous system and the transient and liable nature of the moieties involved. Use of measurements of alterations in these functions as indicators of neurotoxicity is further complicated by our lack of a complete understanding of the normal operation of these systems and by the multitude of day-to-day occurrences (such as diurnal cycle, diet, temperature, age, sex, and endocrine status) constantly modulating the baseline system. For detailed discussions of these difficulties, the reader is advised to see Damstra and Bondy [97,98].

Two specific markers may be measured to evaluate the occurrence of specific neurotoxic events. These markers are neurotoxic esterase (NTE; the inhibition of which is a marker for organophosphate-induced delayed neuropathy) [99] and β-galactosidase (which is a marker for Wallerian degeneration of nerves). Johnson and Lotti [100–102] have established that inhibition of 70–90% of normal levels of NTE in hens 36 hours after being dosed with a test compound is correlated with the development some 15 days later of ataxia and the other classic physiologic signs of delayed neuropathy. Johnson's 1977 [101] article clearly describes the actual assay procedure.

β-galactosidase is associated not with a single class of compounds, but rather with a particular expression of neurotoxicity—Wallerian degeneration. In nerves undergoing Wallerian degeneration after nerve section, the activity of -galactosidase increases by over 1,000%. Evidence also exists that this enzyme is increased in the peripheral nerves and ganglia of rats suffering from certain toxic neuropathies. This assay, therefore, can be used as a biochemical method for detecting neurotoxic effects of compounds [103].

β-galactosidase is a constituent of lysosomes whose function is to split β-galactosides; for example, it will convert lactose into galactose and glucose. The assay method below uses an artificial substrate, 4-methylumbelliferyl β-D-galactopyranoside (MUG). At an acid pH, β-galactosidase will split galactose from this compound to leave a product fluorescing in alkaline solution.

In summary, animals exposed to or dosed with the chemical are given a necropsy, and peripheral nervous tissue is collected. The β-galactosidase activity of peripheral nervous tissue homogenates is determined by incubating 0.2 ml of 1% weight/volume (w/v) homogenates with 1×10^{-3} M methylumbelliferyl β-galactoside in 1M glycine buffer, pH 3.0 for 1 hour at 37°C. The enzyme releases methylumbelliferone, which can be measured fluorometrically in alkaline solution (excitation wavelength 325–380 nm, emission wavelength 450 nm) [104].

Progress in the more generalized methodologies of evaluating alterations in neurotransmitter levels has not been as conclusive and is reviewed by Bondy [105, 106], and specific methodologies are presented by Ho and Hoskins [108].

9.8 THE USE OF GLOBAL MEASURES AS ASSAYS FOR TOXICOLOGIC DAMAGE

Even extremely specific agents may have striking effects on general indicators of neuronal or glial well being while leaving some more specific measures unaffected. An example is the voltage-sensitive sodium channel blocker, TTX, as noted above. Incubation of fetal CNS cultures with 0.5-M TTX for a period of a few days results in a 30–50% decrease in neuronal number and of neuronal surface membrane, as measured by tetanus toxin binding [15]. The activity of GAD is, however, unchanged. Thus, although some neuronal populations are vulnerable to the effects of sodium channel blockade, others are not. If only selected subpopulations are affected, of course, this results in the signal indicative of toxic effect, which will be less than that of the culture as a whole. *The culture system must be extremely reliable and well characterized so that relatively small quantitative changes can be interpreted with confidence.* This rule requires that the large number of variables determining the development and maintenance of the culture will be controlled.

Of course, if structure-activity data or other information concerning a potential neurotoxicant are available and suggest a specific target or neuropathogenetic mechanism, a more targeted test system and assays would be appropriate.

9.9 SPECIFIC NEUROTOXICOLOGIC STUDIES

We will make no attempt to review the large literature on *in vitro* neurotoxicologic testing (see refs. 53 and 54). Rather, we will take three examples illustrating some of the problems and the promise of the field and discuss them in some detail. These examples include (1) studies of the possible pathogenetic potential of various anticonvulsant medications; (2) the excitotoxic responses exemplified by motor system damage produced in lathyrism, by nonprotein amino acids from the pathogenetic plant *Lathyrus sativus* and by domoic acid and from some species of mussels; and (3) issues related to heavy metal intoxication and neurotoxicology.

9.9.1 Phenobarbital, Phenytoin, and Other Anticonvulsant Agents

For many years and up until the 1970s, children with a febrile convulsion were treated for up to 5 years with phenobarbital. Dosage levels were such that blood levels were above 65 mol (15 g/ml) and up to 130 mol (30 g/ml). Concern existed as to the therapeutic need for treatment and effects on brain development in children, and in the 1970s and early 1980s, a number of experimental studies in rats and mice suggested that phenobarbital administered prenatally or postnatally indeed impaired brain growth (see, e.g., ref. 55). Several studies using *in vitro* methods have now been done to assess the neurotoxicity of phenobarbital and a number of other anticonvulsant medications.

In 1981, Bergey et al. [56] examined the effect of phenobarbital on developing dissociated cultures of mouse spinal cord. In 1983, Swaiman et al. [57] described similar experiments using phenytoin and mouse cortical cultures. A number of morphologic and neurochemical markers were used to quantify the effects of various concentrations of the anticonvulsant agents on both neurons and glial cells. A more comprehensive pair of studies in 1985 [58,59] examined phenytoin, phenobarbital, carbamazepine, valproic acid, diazepam, and ethosuximide for their effect on mouse cerebral cortical culture, with the aim of determining the relative neurotoxicity of these different anticonvulsant agents. Parameters measured included the number of surviving neurons, total protein, tetanus toxin fixation (an indication of total neuronal surface membrane), high-affinity uptake of β-aminobutyric acid and β-alanine, choline acetyltransferase activity, and specific and clonazepam-displaceable benzodiazepam binding. Ethosuximide and carbamazepine had minimal toxic effects. Valproate, diazepam, and ethosuximide were examined for their effect on mouse cerebral cortical culture, with the aim of determining the relative neurotoxicity of these different anticonvulsant agents. Parameters measured included the number of surviving neurons, total protein, tetanus toxin fixation (an indication of total neuronal surface membrane), high-affinity uptake of β-aminobutyric acid and β-alanine, choline acetyltransferase activity, and specific and clonazepam-displaceable benzodiazepam binding. Ethosuximide and carbamazepine had minimal toxic effects, valproate and diazepam had modest effects, and phenobarbital and phenytoin were definitely detrimental to neuronal survival and development. Serrano et al. [60] performed a detailed morphologic study of phenobarbital effects on neuronal development and concluded that phenobarbital produced a reaction in neuronal survival, and that in surviving neurons, dendritic branching pattern and length were reduced in a dose-dependent manner.

The question of appropriate dosage is discussed in some detail in these papers and constitutes a substantial problem of interpretation. Because the anticonvulsants are lipophilic molecules, much of the agents in serum are bound in some form or another. The question of whether to use total concentration of only the free form of the drug must be considered. Serrano et al. pointed out that in an equilibrium situation, brain levels may be closer to the total concentration in the serum than to the free concentration and that this may be particularly true in neonatal or young animals in which immaturity of the blood–brain barrier and reduced binding by plasma proteins may be important considerations.

Similar studies in 1990 by Regan et al. [61] used a shorter exposure period and incorporated (in addition to primary mouse cerebral cortical culture) the use of neural (neuro-2A) and glial (C6) cell lines to test for anticonvulsant effects on cell proliferation. The cytotoxic effect of phenytoin was confirmed in these studies, and a specific effect on the mitotic rate in the cells lines by valproate and the benzodiazepines was noted.

One of the interesting features of the Serrano et al. study was the effects of phenobarbital on cell number and process length, and complexity was more pronounced for neurons treated from day 14 to week 6 of culture than for neurons treated from day 2 to week 6, despite the fact that the latter cultures were treated for a longer total time.

Do these *in vitro* studies have a parallel in *in vivo* or clinical situations? A number of experimental studies in rodents demonstrated both structural and behavioral effects of phenobarbital given to neonates or to pregnant animals (effects on offspring). A randomized clinical study published by Farwell et al. [62] in 1990 compared the IQ of 217 children having experienced at least one febrile seizure and randomly assigned to a treatment group or to a placebo control group. The children in the treatment group received 4–5-mg/kg body weight of phenobarbital. This group had blood levels of phenobarbital between 15 and 30 g/ml during the course of the study. The treatment group had an average IQ score 8.4 points below that of the control group after 2 years of treatment, and at 6 months after cessation of treatment, the IQ of the treated children was 5.2 points lower than that of the control placebo group. Thus, it appears that phenobarbital treatment may well be accompanied by some decrement in tested intellectual function. In addition, this study showed a lack of clinical efficacy in that no difference existed between treatment group and placebo control in the number of seizures experienced.

9.9.2   Excitotoxins

Since the early descriptions by Olney and Sharpe [63] of neurotoxic damage to neural tissue by EAAs, particularly glutamate, a fairly complete scheme has emerged for understanding this pathogenic process. Receptors for EAAs exist in neurons and glia, and in fact, these receptors are responsible for the normal excitatory synaptic interactions occurring among neurons and are essential features of brain function. At least three or four pharmacologically and physiologically distinguishable species of these receptors exist, and molecular biologic techniques have demonstrated several subspecies of receptor subunit polypeptides. Importantly, a number of agonists and specific competitive and noncompetitive blockers of these receptors have been developed that are extremely useful in testing whether a given neurotoxic result may be caused by or involve a component of EAA excitotoxicity (see ref. 1 for review).

It seems likely excitotoxicity may contribute to the actions of a broad range of neurotoxins, and consideration of the cellular mechanisms involved in excitotoxic damage to the nervous system may be useful. Work from a number of laboratories has contributed to the scheme summarized by Choi [64].

Glutamate is present in high (mM) concentrations in neural tissue, particularly in nerve terminals. It is released from excitatory terminals during normal physiologic activity to mediate synaptic transmission between these terminals and post-synaptic receptor cells. The EAA receptors on these neurons are of at least two broad varieties termed NMDA and non-NMDA or K/Q receptor. The NMDA receptor is characterized by its voltage-dependence, its permeability to calcium and sodium, its blockability by Mg^{2+} ions, and its requirement for low levels of glycine for activation. The non-NMDA receptors exhibit none of these characteristics. The conductance change (channel openings) associated with synaptic activation of NMDA receptors is of considerably longer duration than that associated with non-NMDA receptor activation. The excitotoxicity hypothesis states that excessive uncontrolled activation of these EAA receptors produces a cascade of events resulting in brain damage.

A number of observations strongly suggest that an increase in intracellular calcium ion concentration is a key step in the excitotoxic process.

Cell biologic investigations over the past 2 decades have revealed the central role that intracellular $[Ca^{2+}]_i$ plays in the regulation of cellular processes. Protein kinase, proteases, phospholipases, and xanthine oxidase are all affected, and in various ways, these may contribute to cell damage. Proteases such as calpain I degrade cellular structural proteins, the phospholipases can break down cell membrane constituents, and xanthine oxidase and lipid breakdown products generate potentially destructive superoxide radicals.

Depolarization of neurons by the EAAs may initiate a cycle in which depolarization-induced release of EAAs produces further depolarization, EAA buildup, and pathologic consequences.

It should be noted that these cell biologic effectors are extremely general and impinged on by a number of factors. Lebel and Bondy [65] have emphasized the potential importance of oxygen radicals as mediators of neurotoxicity and point out that six of eight common neurotoxic agents (including such diverse components as methyl mercury, toluene, and methamphetamines) increase cerebral oxygen radical formation. Such a "final common pathway," as these authors term it, would provide a basis for synergistic interactions between different types of neurotoxicants.

The neurotoxic effect of glutamate (and other EAAs) has been demonstrated by direct injection in vivo and by a number of in vivo experiments involving anoxia, ischemia, or physical trauma. The importance of EAAs in producing brain damage in these various models has been shown by the sparing effect that specific EAA receptor blockers produce. That is, NMDA receptor blockers are "neuroprotective in a variety of hypoxia paradigms" [59].

Two examples have been investigated in which ingestion of EAAs may be causal for neurodegenerative diseases. The most clear-cut example is that of lathyrism, caused by the toxin β-N-oxaly-lamino-L-alanine (BOAA) from the chickpea [50,66]. The onset of symptoms occurs after prolonged periods of consumption of the agent—in contrast to the acute symptomatology seen with contaminated mussels [67,68], in which domoic acid is the EAA responsible for the neurotoxicity. A potentially important link has been suggested between ingestion of flour made from seeds of the cycad plant and development of amyotrophic lateral sclerosis-Parkinsonism-dementia among people in the Western Pacific islands, particularly Guam [69]. The suspect agent, BMAA, is a weak and atypical EAA, and its mechanism of action is unclear; indeed, its involvement in the Guam syndrome is challenged [70]. The possibility, however, that such environmental agents may contribute through an excitotoxic mechanism to chronic neurodegenerative disease is being considered most seriously. It is clear that although *in vitro* systems cannot be complete models for clinical conditions of such a long time course, they provide excellent material for evaluating the excitotoxic potential of any suspect compounds.

9.9.3 Heavy Metals

Mercury, tin, and lead exemplify environmentally pervasive compounds that have a clear neurotoxic effect. Catastrophic acute effects have been conclusively shown at levels occurring in real-life situations, as exemplified by the widespread mercury poisoning that occurred as a result of pollution of Minamata Bay in Japan [54,71]. Acute lead poisoning is also an all-too-common clinical picture, and *in vitro* models have been useful in establishing possible molecular and cellular mechanisms underlying such toxicity [72]. In those experiments, cells of the glial cell line C6 were treated with various concentrations of lead, and the effect on two enzymes and on the induction of those enzymes by cortisol was noted. The two enzymes were GPDH and LDH. Basal levels of the two enzymes were unaffected by lead doses up to the 1-mM range. The induction of GPDH, however, was inhibited in a dose-dependent manner by lead, and at 10^{-4} M, the block of induction was 40–50%. The experiments showed that the failure of GPDH induction by cortisol under lead treatment was caused by its block of synthesis of the enzyme; GPDH degradation was unaffected by lead treatment. The effect of lead seems to be specific, with respect to GPDH induction; no effect is seen on basal enzyme levels, on norepinephrine induction of cAMP, on cell viability, or on protein synthesis generally. The authors conclude that lead acts at some point between cortical binding in the nucleus of the C_6 cells and translation of GPDH mRNA.

GPDH is a specific marker for oligodendrocytes and myelin-forming cells generally, and GPDH is thought to be involved in myelination. One of the symptoms of lead intoxication is a deficit in peripheral and central myelination. The demonstrated effects of lead on GPDH may well provide a molecular basis for some of the symptomatology of lead intoxication.

The discussion of lead as an environmental neurotoxicant has shifted considerably in recent years, because the possibility has been raised by epidemiologic studies that low levels of lead ingestion and blood concentration 10–100 times lower, for instance, than are those explored in the study described above, over prolonged periods, may have a deleterious effect on intellectual performance [73]. Intense controversy surrounds this issue, and it poses one version of the neurotoxicology problem. For a compound known to be neurotoxic at high doses and for which total elimination may be difficult, or at least expensive, can "acceptable" levels be satisfactorily established? This problem, of course, is a common and difficult one in other areas of toxicology, notably, with respect to carcinogenic agents. How might *in vitro* methods contribute in this regard? Two possible strategies might be considered: (1) It is possible to maintain cultured preparations for relatively prolonged periods. Both dissociated cell culture preparations and organotypic slice preparations can be maintained for several months. Although not approximating the lifetime exposure that might be involved with human populations, this may nevertheless allow longer time-course pathogenetic processes to reveal themselves with low-dose exposures for neurotoxic agents. (2) Quantitative and precise indicators of toxic actions are provided by the many assays available for evaluating the *in vitro* system. Particularly if these assays are directed at the cell biologic and molecular entities directly affected by the neurotoxic agents, great sensitivity may be attained. Thus, even quantitatively minor effects produced by low levels of the agents may be unequivocally demonstrated in these systems at relatively short (days to weeks) periods. Although each situation would have to be evaluated carefully, such modest but reliable indications of damage might be useful in the context of evaluating the levels at which thresholds should be set. The epidemiologic data provide the definitive basis for such threshold setting, but this sort of data is extremely difficult to obtain and evaluate and is costly. Exploring the degree to which data from *in vitro* test systems might be helpful in this regard would be well worthwhile.

9.10 PROBLEMS

1. A prominent difficulty or limitation of the *in vitro* system is that it does not deal at all with issues of metabolism and toxicokinetics. Many environmental agents may not themselves be neurotoxic, but when metabolized in the liver or elsewhere, the products generated may be so. This result may be offset in part by testing independently the known metabolites of neurotoxicants when they are established. Hybrid culture schemes have also been proposed combining potential biotransforming cellular components (such as liver or kidney cells) in some sort of coculture with the test neural tissue. The presumption would be that such cocultures might provide an efficient combination of high-resolution *in vitro* test systems with some component, at least, of whole-animal biotransformations of potential neurotoxicants.

2. Most primary neural cultures of either dissociated cells or organotypic explants are composed of nondividing neurons with, in many cases, a largely static population of glial cells. The effect of some neurotoxic agents may be on neuroblast division and differentiation, so that teratologic, development abnormalities are dominant. As noted

earlier, dividing cell lines with potential for neural expression may be useful in evaluating such agents. Such cell lines are useful in probing for molecular and cellular sites of action as well (see above for discussion of lead oligodendrocyte toxicity). A large number of cell lines are available that can be differentiated in various ways by different culture manipulation. Extremely interesting studies on viral transformation of neural and glial cells are creating material with rich potential for neurotoxicologic evaluation.

9.11 PROSPECTS FOR *IN VITRO–IN VIVO* APPROACHES

Considerable recognition has been given to the potential importance of *in vitro* approaches to neurotoxic evaluation of the large number of essential untested chemicals in use (60,000–70,000) and being added to the inventory (1,000–1,500 per year) [74]. After considering pros and cons of *in vitro* testing, a report by the Office of Technology Assessment (OTA) in 1990 [54] concluded the following: "Nevertheless all test systems have limitations, and there is general agreement that the many advantages of *in vitro* testing present a strong incentive for continued development and increased utilization." Reservations about the role of *in vitro* testing are summarized, however, by Tilson [74]: "Clearly before *in vitro* techniques were adopted to problems of hazard detection in neurotoxicology, research will be needed to devise a strategy to develop, refine *and validate* these procedures."

References

1. Mayer ML, Westbrook GL. The physiology of excitatory amino acids in the vertebrate central nervous system. *Prog Neurobiol* 1987;28:197–276.

2. Mayer ML, Vyklicky L, Jr, Patneau DK. Glutamate receptors in cultures of mouse hippocampus studied with fast application of agonists, modulators and drugs. In: Ben-Ari Y, ed. *Excitatory Amino Acids and Neuronal Plasticity*. New York: Plenum Press, 1990:3–11.

3. Kater SB, Mattson MP, Cohan CS, Connor JA. Calcium regulation of the neuronal growth cone. *Trends Neurosci* 1988;11:315–321.

4. Barde Y-A. Trophic factors and neuronal survival. *Neuron* 1989;2:1,525–1,534.

5. Patterson PH. Environmental determination of autonomic neurotransmitter function. *Annu Rev Neurosci* 1987;1:1–17.

6. McCarthy KD, DeVellis J. Preparation of separate astroglial and oligodendroglial cell cultures from rat cerebral tissue. *J Cell Biol* 1980;85:890–896.

7. Trimmer PA, Evans T, Smith MM, Harden TK, McCarthy KD. Combination of immunocytochemistry and radioligand receptor assay to identify β-adrenergic receptor subtypes on astroglia *in vitro*. *J Neurosci* 1984;4:1,598–1,607.

8. Evans T, McCarthy KD, Harden TK. Regulation of cyclic AMP accumulation by peptide hormone receptors in immunocytochemically defined astroglial cells. *J Neurochem* 1984;43:131–139.

9. McCarthy KD, Prime J, Harmon T, Pollenz R. Receptor-mediated phosphorylation of astroglial intermediate filament proteins in cultured astroglia. *J Neurochem* 1985;44:723–735.

10. Cornell-Bell AH, Finkbeiner SM, Cooper MS, Smith SJ. Glutamate induces calcium waves in cultured astrocytes: long range glial signaling. *Science* 1990;247:470–473.

11. Fatatis A, Russell JT. Spontaneous changes in intracellular calcium concentration in type I astrocytes from rat cerebral cortex in primary culture. *Glia* 199;26:84–971.

12. Dodd J, Jessell TM. Lactoseries carbohydrates specify subsets of dorsal root ganglion neurons projecting to the superficial dorsal horn of rat spinal cord. *J Neurosci* 1985;5:3,278–3,281.

13. Neale EA, Oertel WG, Bowers LM, Weise VK. Glutamate decarboxylase immunoreactivity and -[3H]aminobutyric acid accumulation within the same neurons in dissociated cell cultures of cerebral cortex. *J Neurosci* 1983;2:376–382.

14. Neale EA, Matthew E, Zimmerman EA, Nelson PG. Substance P-like immunoreactivity in neurons in dissociated cell cultures of mammalian spinal cord and dorsal root ganglia. *J Neurosci* 1982;2:169–177.

15. Brenneman DE, Nelson PG. Neuronal development in culture. Role of electrical activity. In: Bottenstein JE, Sato G, eds. *Cell Culture in the Neurosciences*. New York: Plenum Press, 1985:299–316.

16. Brenneman DE, Neale EA, Foster GA, d'Autremont SW, Westbrook GL. Nonneronal cells mediate neurotropic action of vasoactive intestinal peptide. *J Cell Biol* 1987;104:1,603–1,610.

17. Brenneman DE, Foster GE. Structural specificity of peptides influencing neuronal survival during development. *Peptides* 1987;8:687–694.

18. Hill JM, Gozes I, Hill JL, Fridkin M, Brenneman DE. Vasoactive intestinal peptide antagonist retards the development of neonatal behavior in the rat. *Peptides* 1991;12:187–192.

19. Landis SC, Keefe D. Evidence for neurotransmitter plasticity *in vivo*: developmental changes in properties of cholinergic sympathetic neurons. *Dev Biol* 1983;98:349–372.

20. Schotzinger RS, Landis SC. Cholinergic phenotype developed by noradrenergic sympathetic neurons after innervation of a noval cholinergic target *in vivo*. *Nature* 1988;335:637–639.

21. Giller EL, Jr, Neale JH, Bullock PN, Schrier BK, Nelson PG. Choline acetyltransferase activity of spinal cord cultures increased by co-culture with muscle and by muscle conditioned medium. *J Cell Biol* 1977;74:16–29.

22. McManaman JL, Crawford FG, Steward SS, Appel SH. Purification of a skeletal muscle polypeptide which stimulates choline acetyltransferase activity in cultured spinal cord neurons. *J Biol Chem* 1988;263:5,890–5,897.

23. Houenou LJ, McManaman JL, Prevette D, Oppenheim RW. Regulation of putative muscle-derived neurotrophic factors by muscle activity and innervation: *in vivo* and *in vitro* studies. *J Neurosci* 1991;11:2,893–2,837.

24. Banker G, Goslin K. *Culturing Nerve Cells*. Cambridge, MA: MIT Press, 1991.

25. Schaffner AE, St. John PA, Barker JL. Fluorescence-activated cell sorting of embryonic mouse and rat motoneurons and their long-term survival *in vitro*. *J Neurosci* 1987;7:3,088–3,104.

26. O'Brien RJ, Fischbach GD. Isolation of embryonic chick motoneurons and their survival *in vitro*. *J Neurosci* 1986;6:3,265–3,269.

27. Huettner JE, Baughman RW. Primary culture of identified neurons from the visual cortex of postnatal rats. *J Neurosci* 1986;6:3,044–3,049.

28. MacDermott AB, Mayer ML, Westbrook GL, Smith SJ, Barker JL. NMDA-receptor activation increases cytoplasmic calcium concentration in cultured spinal cord neurones. *Nature* 1986;321:519–522.

29. Mayer ML, MacDermott AB, Westbrook GL, Smith SJ, Barker JL. Agonist and voltage-gated calcium entry in cultured mouse spinal cord neurons under voltage clamp measured with arsenazo III. *J Neurosci* 1987;7:3,230–3,244.

30. Forsythe ID, Westbrook GL. Slow excitatory postsynaptic currents mediated by *N*-methyl-D-aspartate receptors on cultured mouse central neurons. *J Physiol* 1988;396:505–533.

31. Barker JL, Ransom BR. Pentobarbitone pharmacology of mammalian central neurones grown in tissue culture. *J Physiol (Lond)* 1987;280:355–372.

32. Neale EA, Nelson PG, Macdonald RL, Christian CN, Bowers LM. Synaptic interactions between mammalian central neurons in cell culture. III. Morphophysiological correlates of quantal synaptic transmission. *J Neurophysiol* 1983;49:1,459–1,468.

33. Pun RYK, Neale EA, Guthrie PB, Nelson PG. Active and inactive central synapses in cell culture. *J Neurophysiol* 1986;1,242–1,256.

34. Grybkiewicz G, Poenie M, Tsien RY. A new generation of Ca^{2+} indicators with greatly improved fluorescence properties. *J Biol Chem* 1985;260:3,440–3,450.

35. Hockberger PE, Tseng H-Y, Connor JA. Immunocytochemical and electrophysiological differentiation of rat cerebellar granule cells in explant cultures. *J Neurosci* 1987;7:1,370–1,375.

36. Steele VE, Morrissey RE, Elmore EL, Gurganus-Rocha D, Wilkinson BP, Curren RD, Schmetter BS, Louie AT, Lamb JC, Yang LL. Evaluation of two *in vitro* assays to screen for potential developmental toxicants. *Fundam Appl Toxicol* 1988;11:673–684.

37. Schmechel D, Marangos PJ, Zis AP, Brightman M, Goodwin FK. Brain enolases as specific markers of neuronal and glial cells. *Science* 1978;199:313–315.

38. Brenneman DE, Buzy JM, Ruff MR, Pert CB. Peptide T sequences prevent neuronal cell death produced by the envelope protein (gp120) of the human immunodeficiency virus. *Drug Dev Res* 1988;15:361–369.

39. Clements JD, Buzy JM. Automated image analysis for counting unstained cultured neurones. *J Neurosci Methods* 1991;36:1–8.

40. Matsumoto T, Oshima K, Miyamoto A, Sakurai M, Goto M, Hayashi S. Image analysis of CNS neurotrophic factor effects on neuronal survival and neurite outgrowth. *J Neurosci Methods* 1990;31:153–162.

41. Hayashi M, Tanii H, Horiguchi M, Hashimoto K. Cytotoxic effects of acrylamide and its related compounds assessed by protein content, LDH activity and cumulative glucose consumption of neuron-rich cultures in a chemically defined medium. *Arch Toxicol* 1989;63:308–313.

42. Koh J, Choi DW. Quantitative determination of glutamate mediated cortical neuronal injury in cell culture by lactate dehydrogenase efflux assay. *J Neurosci Methods* 1990;20:83–90.

43. Klingman JG, Hartley DM, Choi DW. Automated determination of excitatory amino acid neurotoxicity in cortical culture. *J Neurosci Methods* 1990;31:47–51.

44. Didier M, Heaulme M, Soutrie P, Bockaert J, Pin JP. Rapid, sensitive and simple method for quantification of both neurotoxic and neurotrophic effects of NMDA on cultured cerebellar granule cells. *J Neurosci Res* 1990;27:25–35.

45. Brenneman DE. Role of electrical activity and trophic factors during cholinergic development in dissociated cultures. *Can J Physiol Pharmacol* 1986;64:356–362.

46. Brenneman DE, Neale EA, Habig WH, Bowers LM, Nelson PG. Developmental and neurochemical specificity of neuronal deficits produced by electrical impulse blockade in dissociated spinal cord cultures. *Dev Brain Res* 1983;9:13–27.

47. Dimpfel W, Huang RTC, Habermann E. Gangliosides in nervous tissue cultures and binding of ^{125}I-labeled tetanus toxin, a neuronal marker. *J Neurochem* 1977;29:329–334.

48. Markwell MAK, Sheng HZ, Brenneman DE, Paul SM. A rapid method to quantify neurons in mixed cultures based on the specific binding of [3H]ouabain to neuronal Na+, K+ -ATPase. *Brain Res* 1991;538:1–8.

49. Hauger R, Luu MD, Goodwin FK, Paul SM. Characterization of [3H]ouabain binding in the rat central nervous system. *J Neurochem* 1985;33:1,709–1,715.

50. Spencer PS, Ross SM, Nunn PB, Roy DN, Seelig M. Detection and characterization of plant-derived amino acid motorsystem toxin in mouse CNS cultures. In: Shahar A, Goldstein AM, eds. *Model Systems in Neurotoxicologic Alternative Approaches to Animal Testing*. New York: Alan R. Liss, 1987:349–361.

51. Brenneman DE, Westbrook GL, Fitzgerald SC, Ennist DL, Elkins KL, Ruff MR, Pert CB. Neuronal cell killing by the envelope protein of HIV and its prevention by vasoactive intestinal peptide. *Nature* 1988;335:639–642.

52. Lipton SA. HIV-related neurotoxicity. *Brain Pathol* 1991;1:193–199.

53. Shahar A, Goldstein AM, eds. *Model Systems in Neurotoxicologic Alternative Approaches to Animal Testing*. New York: Alan R Liss, 1987.

54. Congress of the United States, Office of Technology Assessment. *Neurotoxicity. Identifying and Controlling Poisons of the Nervous System. New Developments in Neuroscience*. Washington, DC: U.S. Government Printing Office, 1990. OTA Publication BA-436.

55. Yanai J, Bergman A. Neuronal deficits after neonatal exposure to phenobarbital. *Exp Neurol* 1981;73:199–208.

56. Bergey GK, Swaiman KF, Schrier BK, Fitzgerald S, Nelson PG. Adverse effects of phenobarbital on morphological and biochemical development of fetal mouse spinal cord neurons in culture. *Annu Neurol* 1981;9:584–589.

57. Swaiman KF, Neale EA, Schrier BK, Nelson PG. Toxic effect of phenytoin on developing cortical neurons in culture. *Annu Neurol* 1983;13:48–52.

58. Neale EA, Sher PK, Graubard BI, Habig WH, Fitzgerald SC, Nelson PG. Differential toxicity of chronic exposure to phenytoin, phenobarbital, or carbamazepine in cerebral cortical cell cultures. *Pediatr Neurol* 1985;1:143–150.

59. Sher PK, Neale EA, Graubard BI, Habig WH, Fitzgerald SC, Nelson PG. Differential neurochemical effects of chronic exposure of cerebral cortical cell culture to valproic acid, diazepam, or ethosuximide. *Pediatr Neurol* 1985;1:232–237.

60. Serrano EF, Kunis DM, Ransom BR. Effects of chronic phenobarbital exposure on cultured mouse spinal cord neurons. *Annu Neurol* 1988;24:429–438.

61. Regan CM, Gorman AMC, Larsson OM, Maguire C, Martin ML, Schousboe A, Williams DC. *In vitro* screening for anticonvulsant-induced teratogenesis in neural primary cultures and cell lines. *Int J Neurosci Methods* 1990;8:143–150.

62. Farwell R, Jr, Lee YS, Hirtz DG, Sulzbacher SI, Ellenberg JH, Nelson KB. Phenobarbital for fibril seizures—effects on intelligence and on seizure recurrence. *N Eng J Med* 1990;322:364–369.

63. Olney JW, Sharpe LG. Brain lesions in an infant rhesus monkey treated with monosodium glutamate. *Science* 1969;166:386–388.

64. Choi DW. Glutamate neurotoxicity and diseases of the nervous system. *Neuron* 1988;1:623–639.

65. Lebel CP, Bondy SC. Oxygen radicals' common mediators of neurotoxicity. *Neurotoxicol Teratol* 1991;13:341–346.

66. Spencer PS, Allen RG, Kisby GE, Ludolph AC, Excitotoxic disorders. *Science* 1990;245:144–148.

67. Perl TN, Bedard L, Kosatsky T, Hockin JC, Todd ECD, Remis RS. An outbreak of toxic encephalopathy caused by eating mussels contaminated with domoic acid. *N Engl J Med* 1990;322:1,775–1,780.

68. Teitelbaum JS, Zatorre RJ, Carpenter S, Gendron D, Evans AC, Gjedde A, Cashman NR. Neurologic sequel of domoic acid intoxication due to the ingestion of contaminated mussels. *N Engl J Med* 1990;322:1,781–1,787.

69. Spencer PS, Nunn PB, Hugon J, Ludolph AC, Ross SM, Roy DN, Robertson RC. Guam amyotrophic lateral sclerosis Parkinsonian dementia linked to a plant excitant neurotoxin. *Science* 1987;237:517–522.

70. Duncan MW, Steele JC, Kopin IJ, Markey SP. 2-amino-3(methylamino)-propanoic acid (BMAA) in cycad flour: an unlikely cause of amyotrophic lateral sclerosis and Parkinsonism-dementia of Guam. *Neurology* 1990;40:767–772.

71. Chang LW. Mercury. In: Spencer PS, Schaumberg HH, eds. *Clinical and Experimental Neurotoxicology.* Baltimore, MD: Williams & Wilkins, 1980:508–509.

72. DeVellis J, McGinn JF, Cole R. Selective effects of lead on the hormonal regulation of glial cell proliferation in cell culture. In: Shahar A, Goldstein AM, eds. *Model Systems in Neurotoxicologic Alternative Approaches to Animal-Testing.* New York: Alan R Liss, 1987;217–227.

73. Bellinger D, Leviton A, Waternaux C, Needleman H, Rabinowitz M. Longitudinal analyses of prenatal and postnatal lead exposure and early cognitive development. *N Engl J Med* 1987;316:1,037–1,043.

74. Tilson HA. Neurotoxicology in the 1990s. *Neurotoxicol Teratol* 1990;12:293–300.

75. Williams SP, O'Brien S, Whitmore K, Purcell WM, Cookson MR, Mead C, Pentreath VW, Atterwill CK. An *in vitro* neurotoxicity testing scheme: evaluation of cytotoxicity determinations in neural and non-neural cells. *In Vitro Toxicol* 1996;9:83–92.

76. Andres MI, Repetto G, Sanz P, Repetto M. Determination of phosphofructokinase on enolase activities in cultured mouse neuroblastoma cells: application to the *in vitro* detection of neurotoxic effects. *ATLA* 1995;23:63–71.

77. Schmuck G, Schluter G. An *in vitro* model for toxicological investigations of environmental neurotoxins in primary neuronal cell cultures. *Toxicol Indust Health* 1996;12:683–696.

78. Binding N, Madeja M, Murshaff U, Neidt U, Altrup U, Speckmann E-J, Witting U. Prediction of neurotoxic potency of hazardous substances with a modular *in vitro* test battery. *Toxicol Lett* 1996;88:115–120.

79. Campbell IC, Abdulla EM. Strategic approaches to *in vitro* neurotoxicity. In: Chang LW, Skipper W, eds. *Neurotoxicology: Approaches and Methods.* San Diego: Academic Press, 1995:495–506.

80. Turner RA. *Screening Methods in Pharmacology,* Vols. I and II. New York: Academic, 1965:42–47, 60–68, 27–128.

81. Offermeier J, Ariens EJ. Serotonin I. Receptors involved in its action. *Arch Int Pharmacodyn Ther* 1966;164:92–215.

82. Nodine JH, Siegler PE, eds. *Animal and Clinical Techniques in Drug Evaluation.* Chicago: Year Book Medical Publishers, 1964.

83. Bondy SC. Rapid screening of neurotoxic agents by *in vivo* means. In: Gryder RM, Frankos VH, eds. *Effects of Food and Drugs on the Development and Function of the Nervous System: Methods for Predicting Toxicity.* Washington, DC: Office of Health Affairs, FDA, 1979:133–143.

84. Domer FR. *Animal Experiments in Pharmacological Analysis.* Springfield, IL: Charles C. Thomas, 1971:98, 115, 155, 164, 220.

85. Leach GDH. Estimation of drug antagonisms in the isolated gui pig *vas deferens. J Pharm Pharmacol* 1956;8:501–503.

86. Khayyal MT, Tolba NM, El-Hawary MB, El-Wahed SA. A sensitive method for the bioassay of acetylcholine. *Eur J Pharmacol* 1974;25:287–290.

87. Rossum JM van. Different types of sympathomimetic β-receptors. *J Pharm Pharmacol* 1965;17:202–205.

88. Levy B, Tozzi S. The adrenergic receptive mechanism of the rat uterus. *J Pharmacol Exp Ther* 1963;142:178–180.

89. Gecse A, Zsilinsky E, Szekeres L. Bradykinin antagonism. In: Sicuteri F, Back N, Haberland G, eds. *Kinins; Pharmacodynamics and Biological Roles.* New York: Plenum Press, 1976:5–13.

90. Lin RCY, Yeoh TS. An improvement of Vane's stomach strip preparation for the assay of 5-hydroxytryptamine. *J Pharm Pharmacol* 1965;17:524–525.

91. Castillo JC, DeBeer EJ. The guinea pig tracheal chain as an assay for histamine agonists. *Fed Proc* 1947;6:315.

92. Castillo JC, DeBeer EJ. The tracheal chain. *J Pharmacol Exp Ther* 1947b;90:104.

93. Cox BM, Opheim KE, Teschemach H, Goldstein A. A peptide-like substance from pituitary that acts like morphine 2. Purification and properties. *Life Sci* 1975;16:1,777–1,782.

94. Seeman P, Weinstein J. Erythrocyte membrane stabilization by tranquilizers and antihistamines. *Biochem Pharmacol* 1966;15:1,737–1,752.

95. Burns BD, Paton WDM. Depolarization of the motor end-plate by decamethonium and acetylcholine. *J Physiol* 1951;115:41–73.

96. Seppalainen AM. Applications of neurophysiological methods in occupational medicine: a review. *Scand J Work Environ Health* 1975:1:1–14.

97. Damstra T, Bondy SC. The current status and future of biochemical assays for neurotoxicity. In: Spencer PS, Schaumburg HH, eds. Experimental and Clinical Neurotoxicology. Baltimore, MD: Williams and Wilkins, 1980:820–833.

98. Damstra T, Bondy SC. Neurochemcial approaches to the deletion of neurotoxicity. In: Mitchell CL, ed. *Nervous System Toxicology.* New York: Raven Press, 1982:349–373.

99. Fedalei A, Nardone RM. An *in vitro* alternative for testing the effect of organo-phosphates on neurotoxic esterase activity. In: Goldberg AM, ed. *Product Safety Evaluation.* New York: Mary Ann Liebert, 1983:253–269.

100. Johnson MK. The delayed neuropathy caused by some organophosphorus esters: Mechanism & challenge. *Crit Rev Toxicol* 1975;3:289–316.

101. Johnson MK. Improved assay of neurotoxic esterase for screening. Organophosphates for delayed neurotoxicity potential. *Arch Toxicol* 1977;37:113–115.

102. Johnson MK, Lotti M. Delayed neurotoxicity caused by chronic feeding of organophosphates requires a high-point of inhibition of neurotoxic esterase. *Toxicol Lett* 1980;5:99–102.

103. Dewar AJ, Moffett BJ. Biochemical methods for detecting neurotoxicity—a short review. *Pharmacol Ther* 1979;5:545–562.

104. Dewar AJ. Neurotoxicity testing with particular references to biochemical methods. In: Gorrod W, ed. *Testing for Toxicity*. London: Taylor & Francis, 1981:199–217.

105. Bondy SC. Neurotransmitter binding interactions as a screen for neurotoxicity. In: Prasad KN, Vernadakis A, eds. *Mechanisms of Actions of Neurotoxic Substances*. New York: Raven Press, 1982:25–50.

106. Bondy SC. Especial consideration for neurotoxicological research. *Crit Rev Toxicol* 1984; 14(4):381–402.

107. Yamamori T, Fukada K, Aebersold R, Korsching S, Fann M-J, Patterson PH. The cholinergic neuronal differentiation factor from heart cells is identical to leukemia inhibitory factor. *Science* 1989;246:1,412–1,416.

108. Ho IK, Hoskins B. Biochemical methods for neurotoxicological analyses of neuroregulators and cyclic nucleotides. In: Hayes AW, ed. *Principles and Methods of Toxicology*. New York: Raven Press, 1982:375–406.

10

In Vitro Assessment of Nephrotoxicity

Joan B. Tarloff

University of the Sciences in Philadelphia, Pennsylvania

The kidney is a complex and heterogeneous organ, composed of vascular as well as tubular components, and it is frequently a site of injury after exposure to chemicals or during drug treatment. Susceptibility of the kidney to toxicity may be related to any one or more of a combination of factors. First, the kidneys receive a disproportionately high percentage of cardiac output (20% of total cardiac output distributed to organs accounting for less than 1% of body weight), thereby exposing renal tissue to high concentrations of toxicant. Second, potential toxicants may become highly concentrated within the tubular lumen after reabsorption of electrolytes, nutrients, and water by the nephron. Thus, tubular epithelial cells may be exposed to higher concentrations of toxicants than may be found in other tissues. Third, the proximal tubule actively reabsorbs solutes, such as glucose and amino acids, while actively secreting metabolic products, such as organic acids (e.g., urate, mercapturates) and organic bases (e.g., dopamine, creatinine). If a toxicant is reabsorbed or secreted during active transport, that toxicant may be concentrated within proximal tubular cells, causing site-specific injury. Fourth, proximal and distal tubular cells, as well as renomedullary interstitial cells, contain enzymes (e.g., cytochrome P450s, cysteine conjugate 2-lyase, prostaglandin H synthetase) capable of bioactivating xenobiotics. Any one or a combination of these factors may contribute to the development of nephrotoxicity, and each is difficult to evaluate *in vivo* because of the complexity of the kidney. Therefore, investigators have developed *in vitro* methods to delineate more clearly mechanisms involved in nephrotoxicity.

During development of new therapeutic agents or chemicals to which humans or animals may be exposed (e.g., pesticides, herbicides), a compound may be found to produce nephrotoxicity *in vivo*. Because *in vivo* studies are time consuming, expensive, and may require numerous animals to be used, it would be desirable to rapidly and efficiently screen a series of compounds for nephrotoxic potential. Consequently, *in vitro* methods that can assess relative nephrotoxicity have been developed and proven useful in the development of new drugs and chemicals.

Renal physiologists, pharmacologists, and toxicologists use numerous *in vitro* methods. The focus of this chapter is to review several of the *in vitro* methods used by toxicologists, ranging from whole-organ perfusion to isolated cell systems, by outlining the techniques, the specific advantages and limitations, and discussing several examples of studies using each technique.

10.1 ISOLATED PERFUSED KIDNEY

Evaluation of the nephrotoxic potential of xenobiotics involves determining the effects of a toxicant on specific renal functions, such as urinary concentrating and diluting mechanisms, electrolyte reabsorption, and solute (e.g., glucose, amino acids) reabsorption. Traditionally, these renal functions have been assessed using *in vivo* clearance techniques or urinalysis. The isolated perfused kidney (IPK) may be used to assess renal function either after pretreatment of animals with toxicant (*ex vivo*) or during perfusion with a toxicant (*in vitro*). In addition, the IPK is used with increasing frequency to assess the role of the kidney in xenobiotic metabolism and to investigate the mechanisms whereby chemicals are eliminated by the kidneys.

10.1.1 Methodology

An animal, usually a rat or rabbit, is anesthetized and the ureter is cannulated to enable collection of urine. The renal artery is cannulated, usually via the mesenteric artery, and arterial perfusion is established in situ. The kidney is removed from the animal, trimmed of fat, and placed in a perfusion chamber (Fig. 10.1). The kidney is perfused with a blood-free medium (e.g., Krebs–Ringer or Krebs–Henseleit buffers) containing glucose, amino acids, and albumin as an oncotic agent. A recirculating perfusion system is usually employed, and the perfusate is supplemented with metabolic substrates [1]. Perfusion may be set at constant speed or constant pressure [1,2]. Inulin is generally present in the perfusate to determine glomerular filtration rate (GFR) [1]. Other investigators have incorporated in-line features for instantaneous determination of GFR. For example, Cox et al. [3] included cyanocobalamin in the perfusate and incorporated a microflow-through cuvette in which urinary cyanocobalamin was determined colorimetrically. In this manner, GFR can be monitored continuously throughout an experiment, allowing an IPK to be discarded if functional capacity is not at the desired value.

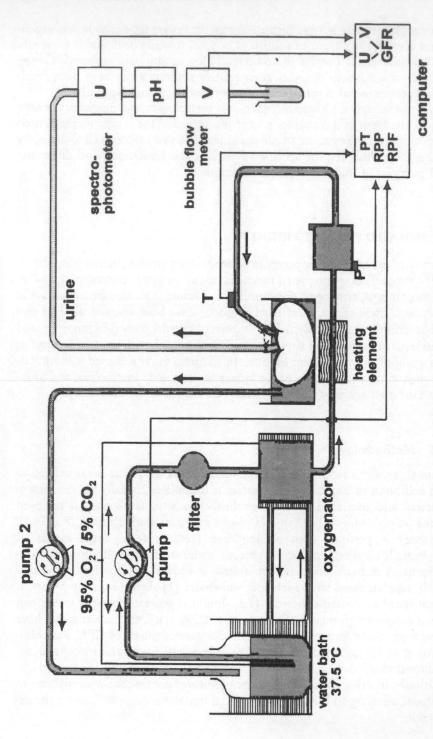

FIG. 10.1. Schematic diagram of apparatus used for isolated rat kidney perfusion. Shaded areas indicate perfusate undergoing rapid recirculation. Arrows indicate the direction of flow through the rapidly recirculating system. Reproduced from Cox et al. [3] and Elsevier Science Publishers with permission.

During development of the IPK, investigators noted that GFR and sodium re-absorption were low compared with *in vivo* observations. In addition, in the IPK, high perfusion flow rates and pressures were required to maintain oxygenation, leading to impaired urinary concentrating ability that was not corrected by exoge-nous antidiuretic hormone [4–6]. These observations led investigators to question the stability and viability of the IPK. Subsequent studies indicated that early in the course of perfusion, the IPK developed an irreversible lesion affecting the thick ascending limb of the loop of Henle [6], and, in some instances, the medullary portion of the proximal straight tubule [7]. Morphologic damage, con-sisting of cytoplasmic flocculation and vacuolization, was apparent in the thick as-cending limb of the loop of Henle as early as 15 minutes after initiating perfusion. Damage progressed to cytoplasmic disruption and nuclear pyknosis and extended to the distal tubule within 60 to 90 minutes of perfusion [6]. Similar cytoplasmic and nuclear degeneration was observed in medullary proximal straight tubules after 100 or 200 minutes of perfusion [7]. The lesion was exacerbated with low oxygen tension in the perfusate and attenuated when erythrocytes or fluorinated hydrocarbons were used [8,9]. However, erythrocytes or fluorinated hydrocarbons are not routinely included in the perfusate of an IPK because of technical prob-lems, including clumping and clotting with erythrocytes and uncertainty about in-herent toxicity of fluorinated hydrocarbons. When oxygen consumption by the IPK was reduced, by inhibiting sodium transport with ouabain or furosemide, morphologic damage to the thick ascending limb was minimized [10]. These ob-servations suggest that relative hypoxia/anoxia contributes importantly to the de-velopment of damage in the medullary proximal tubule and thick ascending limb of the loop of Henle. Further, the IPK may be an unsuitable preparation to inves-tigate toxic responses of the more distal portions of the nephron, such as the thick ascending limb, distal tubule, collecting tubule, and collecting duct. However, function and morphology in glomerular and early proximal tubular structures are well maintained in the IPK [6,7], making this preparation useful in assessment of glomerular and proximal tubular integrity.

10.1.2 Advantages

For certain studies, the IPK presents distinct advantages that cannot be duplicated *in vivo* or by other *in vitro* techniques. Most importantly, the structural and mor-phologic integrity of vascular and tubular components of the kidney is main-tained in the IPK, unlike other *in vitro* techniques in which tubules are dissociated from glomeruli and capillaries. Structural integrity is a particular advantage in ex-amining glomerular function in response to toxicants. Glomerular functional re-sponses cannot be directly assessed with other currently available *in vitro* techniques, and evaluation of direct glomerular damage *in vivo* may be compli-cated by toxicant-induced changes in renal blood flow or cardiovascular function.

Another advantage of the IPK is that this technique uses an artificial perfusate, and the contents and composition of the perfusate may be rigorously defined and controlled. The ability to alter perfusate content is valuable in determining the roles of filtration and tubular transport in xenobiotic accumulation by tubular epithelial cells. The ability to control perfusate composition allows the investigator to manipulate variables, such as degree of protein binding of a xenobiotic, urinary pH, and urinary flow rate [2].

An important advantage of the IPK is that kidneys may be obtained from animals pretreated with toxicant as well as from naive animals, allowing for detailed *in vivo–in vitro* comparisons. In this manner, intrinsic responses of the kidneys may be differentiated from responses because of alterations in renal hemodynamics, cardiovascular function, or extra-renal factors, such as hepatic metabolism or bioactivation.

The IPK offers substantial utility in the study of xenobiotic metabolism and disposition, and it has been used in elucidating mechanisms by which the kidney handles drugs and toxicants. In the IPK, xenobiotics will undergo filtration, reabsorption, or secretion to the extent that those processes occur *in vivo*. Nonrenal factors that may influence xenobiotic disposition *in vivo*, such as extra-renal metabolism and binding to extra-renal tissue, may be circumvented by use of the IPK [2]. For many xenobiotics, metabolism is catalyzed by cytochromes P450, enzymes present in both liver and kidney. It is difficult to quantitate the contribution of renal cytochromes P450 to overall xenobiotic metabolism, because renal P450 content is only about 10% of hepatic P450 content [11,12]. However, renal concentrating mechanisms, active transport systems, and high intrinsic permeability of certain nephron segments may lead to intracellular concentrations of xenobiotics that are much higher in the kidney than they are in other tissues [11,12]. Isolated tissue preparations, such as tubules, cells, or microsomes, may allow xenobiotics to gain access to enzymes catalyzing metabolism, whereas such access may be restricted *in vivo*. Metabolism of a xenobiotic by the IPK is convincing evidence that the xenobiotic can be metabolized by the kidney *in vivo* [13].

The IPK allows evaluation of the relative contribution of filtration (a glomerular function), tubular transport (largely a function of tubular basolateral or luminal membranes), and tubular reabsorption (largely a function of tubular luminal membranes) in xenobiotic accumulation and metabolism by tubular epithelial cells. Glomerular filtration *in vivo* can be reduced or abolished by maneuvers such as ligating the ureter or clamping the renal artery. Both of these procedures will eliminate glomerular filtration, but clamping the renal artery will also eliminate peritubular blood flow, making it impossible to assess the role of basolateral membrane transport in intrarenal accumulation of drugs. Ureteral ligation is a nonphysiological technique relying on increase of intraluminal pressure to eventually stop glomerular filtration. In contrast, glomerular filtration can be reduced in the IPK by raising perfusate albumin concentration so that perfusate colloid osmotic pressure approaches glomerular capillary hydrostatic pressure. In this manner, perfusate flow through the IPK is preserved, including flow through peritubular capillaries, whereas glomerular filtration is abolished. This maneuver allows clear dissociation of glomerular filtration from tubular transport and enables investigators to determine if xenobiotic accumulation in tubular epithelial cells is a consequence of basolateral or luminal transport [13].

10.1.3 Limitations

The IPK has several distinct weaknesses limiting its utility in toxicologic studies. Considerable equipment is required to support the IPK. For example, custom-designed perfusion chambers, oxygenators, pumps, and filters are among a few of the items included in the perfusion circuit (Fig. 10.1). In addition, as with many *in vitro* techniques, the IPK is not a technically simple procedure. Great care must be taken in cannulating the renal artery to not produce ischemia [1]. A significant limitation of the IPK is that renal function tends to decline over time: GFR and sodium reabsorption are stable for only 2 hours or so [1]. If nephrotoxicity requires a period of several hours or longer to develop, the IPK may be unsuitable for *in vitro* monitoring of the progression of toxicity.

The morphologic lesion in the thick ascending limb of the loop of Henle as well as the inability to concentrate urine in the IPK, despite the presence of antidiuretic hormone, are significant limitations, and they make the IPK unsuitable for investigations of toxicants injuring distal nephron structures (e.g., thick ascending limb, distal tubule, collecting tubule, and collecting duct). In particular, the concentrating defect may dilute luminal concentrations of some toxicants so that these compounds may not achieve sufficiently high concentrations in tubular epithelial cells to produce toxic responses [13].

Finally, functional assessments in the IPK do not allow identification of the site of nephrotoxic injury. For example, declines in GFR or increases in urine output are integrated responses of the whole kidney and cannot be ascribed to a single mechanism. Thus, the IPK continues the clearance "blackbox" approach of comparing input with output, except that extra-renal mechanisms, such as changes in renal hemodynamics or cardiovascular function, may be excluded from consideration.

10.2 USE OF THE ISOLATED PERFUSED KIDNEY IN RENAL TOXICOLOGY

10.2.1 Acetaminophen Nephrotoxicity and Metabolism in the Isolated Perfused Kidney

Acetaminophen overdosage is characterized primarily by hepatic necrosis. In addition, some patients may develop acute proximal tubular necrosis in the presence or absence of hepatotoxicity after acetaminophen overdosage. In rats, acetaminophen-induced hepatotoxicity occurs after cytochrome P450-dependent formation of a reactive quinoneimine intermediate [14,15]. In contrast, the precise pathways responsible for acute nephrotoxicity after acetaminophen overdosage in rats are unclear [16–18]. The liver is the primary site of acetaminophen metabolism [14]. The role of the kidney, if any, in the metabolism of acetaminophen is uncertain. Therefore, investigators have used the IPK to evaluate the ability of the kidney to metabolize acetaminophen and to correlate acetaminophen metabolism with nephrotoxicity.

In the IPK, at toxicologically relevant concentrations of acetaminophen (1–3 mM), fractional excretion of acetaminophen was about 25%, indicating that approximately 75% of filtered acetaminophen was reabsorbed by the tubular epithelium [13,19]. Probenecid (0.1 mM) failed to alter acetaminophen fractional excretion in the IPK, indicating that secretion was not involved in the renal handling of acetaminophen [20]. Acetaminophen metabolism by the kidney was suggested by the identification of glucuronide, sulfate, cysteine conjugates, and mercapturic acid metabolites of acetaminophen in urine, but not in the perfusate, of the IPK [13,19]. All of these metabolic pathways were saturable with differing characteristic maxima [13], analogous to saturation of hepatic metabolism of acetaminophen [15]. Thus, all major metabolites of acetaminophen formed in the liver are also formed in the kidney, although at considerably lower rates and in lower amounts than they are in the liver. Renal metabolism is unlikely to contribute importantly to overall acetaminophen elimination. However, intrarenal bioactivation of acetaminophen by cytochrome P450-dependent pathways, such as occurs in the liver to produce hepatotoxicity, cannot be excluded as a possible factor in acetaminophen nephrotoxicity.

In examining acetaminophen metabolism in the IPK, investigators have tried to identify renal functions impaired by acetaminophen. Acetaminophen-induced nephrotoxicity requires about 24 hours to develop in vivo, and the IPK has not been particularly useful in investigating acetaminophen toxicity in vitro. For example, concentrations of acetaminophen in excess of 10 mM were required to produce diuresis and natriuresis in the IPK [13,20]. In contrast, other investigators have observed no effects of 3×10^{-8}-M to 3×10^{-5}-M acetaminophen on GFR, urine output, or sodium excretion [19]. The only effect in IPKs perfused with acetaminophen was a 50% reduction of intracellular reduced glutathione (GSH) after 2 hours of drug treatment in vitro [19]. Acetaminophen-induced GSH depletion was potentiated in kidneys from rats pretreated with polybrominated biphenyls, inducers of cytochromes P450, and attenuated in kidneys from rats pretreated with piperonyl butoxide, an inhibitor of cytochromes P450 [19], suggesting a correlation between oxidative metabolism of acetaminophen and GSH depletion. However, recovery of sulfur-containing metabolites of acetaminophen could not quantitatively account for GSH depletion observed in these IPKs, leading to the suggestion that a portion of the GSH depletion seen with acetaminophen may be caused by interaction of a reactive acetaminophen intermediate with enzymes responsible for GSH synthesis [19]. Alternatively, oxidative stress may be a component of acetaminophen-induced nephrotoxicity, similar to a mechanism proposed in acetaminophen-induced hepatotoxicity [21,22].

Thus, the IPK has been useful in defining the renal metabolism of acetaminophen [13,19]. In addition, acetaminophen-induced GSH depletion in the IPK; the magnitude and time course of GSH depletion *in vitro* was similar to that observed *in vivo* [16,19]. However, the time limitations of the IPK do not allow full expression of acetaminophen-induced nephrotoxicity, and the role of intrarenal metabolism in acetaminophen nephrotoxicity remains uncertain.

10.2.2 Cisplatin Nephrotoxicity in the Isolated Perfused Kidney

A significant limitation in cisplatin therapy is the development of nephrotoxicity, characterized by reductions of renal plasma flow and GFR as well as tubular dysfunction [23–25]. However, *in vivo* clearance studies have not dissociated the effects of cisplatin on GFR from that on renal hemodynamics, or tubular or glomerular damage. The IPK is an excellent technique to allow differentiation among effects on renal hemodynamics, glomerular filtration, and tubular function, and it has been used to investigate cisplatin nephrotoxicity.

When rats were pretreated with cisplatin and kidneys perfused 48 hours later, GFR and sodium and glucose reabsorption were significantly reduced [26]. Renal perfusion flow was similar in kidneys from naive or cisplatin-pretreated rats [26]. Thus, cisplatin produced alterations in glomerular and tubular functionals without apparent alterations of renal hemodynamics *ex vivo*. When 0.5-mM cisplatin was included in the perfusate, IPKs from naive rats showed a time-dependent decline in GFR and sodium and potassium reabsorption in the absence of changes in renal perfusate flow (Fig. 10.2) [26]. The earliest manifestations of functional damage were reductions in GFR and sodium reabsorption within 30–40 minutes of perfusion (Fig. 10.2). Potassium reabsorption was not significantly reduced until 90–100 minutes of perfusion, whereas glucose reabsorption was unaltered during perfusion with cisplatin [26]. In addition, clearances of para-aminohippurate (PAH) and tetraethylammonium (TEA) were markedly reduced during perfusion with cisplatin, when compared with those of controls [26].

In nonfiltering kidneys, inclusion of 0.5-mM cisplatin in the perfusate reduced renal perfusion pressure in a time-dependent manner [26]. PAH and TEA clearances were also markedly reduced in the nonfiltering kidney perfused with cisplatin [26]. Further, reductions in PAH and TEA clearances were not caused by reductions in perfusion pressure, because control kidneys perfused at flow rates comparable with those of the cisplatin-perfused kidneys maintained nearly normal cortical accumulation of PAH and TEA [26].

Thus, cisplatin treatment *in vivo* or *in vitro* markedly decreased GFR and tubular function. These effects were not mediated by changes in renal hemodynamics [26]. However, because both glomerular and tubular functions were reduced by cisplatin, the primary site of cisplatin nephrotoxicity cannot be identified by these studies. The cisplatin-induced reduction in GFR may be related to changes in the

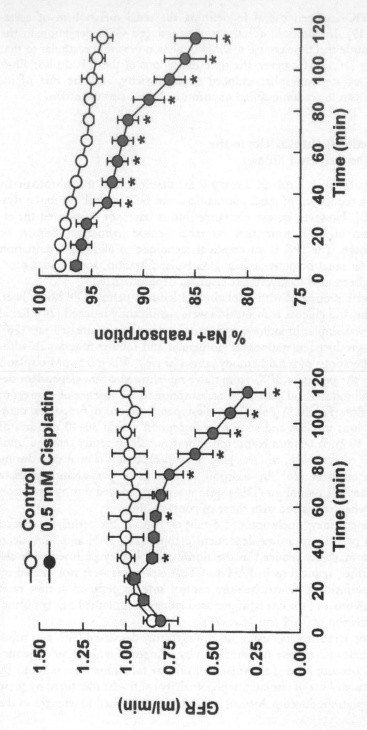

FIG. 10.2. Effects of cisplatin on glomerular filtration rate (GFR) (left panel) and sodium reabsorption (as percent of filtered load) (right panel) in the isolated perfused kidney (IPK). Asterisks indicate significant differences from control ($p < 0.05$). Reproduced from Miura et al. [26] and Elsevier Scientific Publishers with permission.

ultrafiltration coefficient of the glomerulus or events subsequent to tubular damage, for example, tubular back-leak. Data from the nonfiltering IPK indicate that filtration is not a prerequisite for cisplatin nephrotoxicity and that transport or diffusion across the tubular basolateral membrane is sufficient for cisplatin to induce a tubular injury [26].

10.3   RENAL SLICES

Renal cortical slices were developed initially to investigate proximal tubular transport of organic anions, such as PAH, and organic cations, such as TEA or N-methyl-nicotinamide (NMN). These early studies indicated a reasonably good correlation between *in vivo* secretion and *in vitro* renal cortical slice accumulation of PAH and NMN or TEA [27]. For example, substances that stimulated organic anion secretion *in vivo* (e.g., acetate or lactate) also stimulated renal cortical slice accumulation of these ions *in vitro*. Similarly, substances that decreased renal tubular secretion of organic ions *in vivo* (e.g., 2,4-dinitrophenol) also decreased renal cortical slice accumulation *in vitro*. In addition, xenobiotics, such as the non-nutritive sweetener, saccharin, or drugs, such as sulotroban and cimetidine, have been shown to be actively secreted via organic ion transport *in vivo* and are accumulated by renal cortical slices *in vitro* via similar organic ion transporters [28–30]. Thus, renal cortical slices have been established as a valuable technique to evaluate tubular transport of xenobiotics. More recently, the renal cortical slice technique has been used to evaluate the biochemical and functional responses to toxicants after either *in vivo* treatment or exposure to a nephrotoxicant *in vitro*.

10.3.1   Methodology

Several procedures are available for the preparation of renal slices. The oldest technique for the preparation of renal slices is a free-hand method in which slices of 0.2 to 0.5 mm in thickness can be prepared routinely. A Stadie–Riggs microtome also can be used, resulting in greater uniformity in the thickness of tissue slices. Once prepared, 50–100 mg of tissue is transferred to flasks containing oxygenated buffer and the flasks are placed in a metabolic shaker. Frequently, the Cross–Taggart medium [31] is used, although a Krebs–Ringer bicarbonate buffer also can be used. To maximize the renal cortical slice accumulation of organic anions, incubation medium is often supplemented with 10-mM lactate or acetate. Usually, tissue slices are shaken at either 25°C or 37°C in the presence of 100% oxygen or 95% O_2/5% CO_2, depending on the choice of buffered medium. Unless the flasks are stoppered, a continual flow of gas is needed. After an appropriate amount of time, slices are removed from the incubation flask, blotted, and weighed. Slices are then processed for determination of the biochemical endpoints of interest. For example, to assess the effects of a xenobiotic on PAH and TEA transport in renal cortical slices, renal cortical slices are incubated in a medium supplemented with PAH and TEA. Subsequently, slices are homogenized and assayed for PAH and TEA content; an aliquot of the incubation media is similarly processed. Data obtained from such an experiment are expressed by the slice/medium (S/M) concentration ratio of PAH and TEA. In general, S/M ratios exceeding unity are indicative of active transport. Renal cortical slices also have been used either *ex vivo* or *in vitro* to monitor the effects of a toxicant on gluconeogenesis, formation of peroxidative products such as malondialdehyde (MDA), enzyme [i.e., lactate dehydrogenase (LDH), alkaline phosphatase, maltase] leakage, oxygen consumption, covalent binding, and intracellular content of adenosine triphosphate (ATP) and GSH [32,33]. In addition, metabolism of a xenobiotic by renal cortical slices can be assessed [34].

One limitation of the renal cortical slice method is that although cell–cell contact is maintained, site selective injury cannot be assessed morphologically because of difficulties in identifying cell types in this particular preparation. To circumvent this problem, positional renal slices have been developed, allowing identification of renal cell types by their anatomical location within the slice [35,36]. Briefly, kidneys from rats or rabbits are removed, decapsulated, and cylindrical cores (6–8 mm in diameter) are made through the kidney along its cortical-papillary axis. Cores are obtained from the kidney by gently pushing the kidney against a cork borer mounted to a variable-speed drill motor rotating at approximately 300 rpm. Cores are then sliced perpendicular to this axis using a mechanical slicer. Up to 100 cortical slices (300 µm) from a single rabbit can be collected. Incubations are carried out in vessels designed to support several slices on a porous surface submerged beneath a constantly circulating aerated medium. Slices are incubated in a buffered medium gassed with 95% O_2/5% CO_2 at room temperature. Similar to renal cortical slices, positional slices may be used either *ex vivo* or *in vitro* to monitor the effects of a toxicant on biochemical (intracellular potassium, DNA and ATP content, oxygen consumption) and functional (organic ion transport) indices. In addition, positional slices may be evaluated by light or electron microscopy to identify the exact site of damage. Thus, positional renal slices can be used to evaluate and compare effects of xenobiotics on renal tubular biochemistry and function versus morphology. Ruegg et al. have demonstrated that viability (intracellular potassium/DNA) and structural integrity of positional slices can be maintained for at least 30 hours [35]. Rat renal slices incubated in the dynamic organ culture system maintained normal ATP content with less than 8% LDH leakage during a 24-hour culture period [36].

10.3.2 Advantages

Renal slices offer several advantages over other techniques. The use of renal slices is relatively straightforward and inexpensive. Slices prepared from animals challenged with a toxicant can be used to provide more sensitive indices of chemically induced damage than do standard estimates, such as blood urea nitrogen (BUN) or serum creatinine concentrations. Indeed, studies from several laboratories have demonstrated that the *in vitro* renal cortical slice accumulation of PAH and TEA are among the most sensitive and versatile indicators of proximal tubular injury when compared with other renal functional tests, such as urinalyses or serum biochemistry (BUN, creatinine) [37–39]. It is possible to assess a variety of biochemical and functional endpoints in a single sample. For example, organic ion transport, oxygen consumption, enzyme leakage, and lipid peroxidation can be measured in slices from a single incubation vessel, thus providing an economical means of assessing nephrotoxic potential. Positional renal slices offer the added advantage of assessing site-selective morphologic injury. In addition to *ex vivo* evaluation, slices can be prepared from naive animals and exposed to toxicants *in vitro*. Thus, the *in vivo* and *in vitro* effects of a toxicant can be compared in similar tissue preparations, and such a comparison allows assessment of direct (intrarenal) versus indirect (extrarenal) effects. Evaluation of the temporal sequence of biochemical and functional changes produced by nephrotoxicants *in vitro* can provide valuable information on mechanisms of nephrotoxicity.

Renal slices can be prepared from a number of mammalian species, including mouse, rat, rabbit, hamster, guinea pig, dog, monkey, and human, and within any given species, strain, sex, and age-related differences can be assessed. This wide range of populations from which renal slices can be prepared, coupled with the ease of preparation, facilitates species, strain, sex, and age-related comparisons of toxicant-induced renal damage both *ex vivo* and *in vitro* and ultimately may provide information on more appropriate animal models for toxicologic evaluation.

10.3.3 Limitations

Renal cortical slices contain a heterogeneous population of tubules, and it is difficult to identify the site of injury. In addition, the lumens of the tubules in a slice are collapsed; hence, the basolateral membrane may be preferentially exposed to toxicants in slice preparations. Thus, renal slices may not be appropriate in evaluating the nephrotoxic potential of drugs or chemicals normally entering the proximal tubular cell via the luminal membrane. Using standard incubation conditions, renal cortical slice preparations have a limited viability of approximately 3 hours, although viability may be extended from 8 to 30 hours using specialized incubation methods [35,36,40]. Limited viability may pose some problems when determining the relevance of data gathered from short-term incubations, particularly when toxicity *in vivo* may take days to become manifest.

An additional limitation of renal cortical slices is the potential for generating false positives in assessing nephrotoxicity *in vitro*. Smith [33] investigated the *in vitro* nephrotoxic effects of a wide variety of chemicals, including mercuric chloride, cephaloridine, gentamicin, carbon tetrachloride, hexachlorobutadiene, potassium dichromate, and 4-ipomeanol. Although the nephrotoxic potential of most of these chemicals and drugs *in vitro* correlated well with their known nephrotoxic potential *in vivo*, 4-ipomeanol, which is not nephrotoxic *in vivo*, also produced nephrotoxic effects *in vitro*, albeit at relatively high concentrations. Thus, the use of renal cortical slices to screen unknown chemicals and drugs for nephrotoxicity may not be appropriate. Rather, the greatest utility of this technique relates to screening of a known nephrotoxic class of compounds or investigating biochemical mechanisms of a nephrotoxicant.

10.4 USE OF SLICES IN RENAL TOXICOLOGY

10.4.1 Biochemical Mechanisms of Nephrotoxicity: Cephalosporin Antibiotics

Cephaloridine is a broad-spectrum cephalosporin antibiotic producing dose-related nephrotoxicity when administered in large doses to laboratory animals [41]. Cephaloridine nephrotoxicity *in vivo* has been fairly well characterized, and several studies have indicated that renal cortical slice function *ex vivo* is an exquisitely sensitive indicator of nephrotoxicity. For example, cephaloridine-induced increases in BUN concentrations occurred in rabbits receiving dosages of 150 mg/kg or greater, whereas decreased renal cortical slice accumulation of PAH and TEA and gluconeogenesis were observed at dosages as low as 50–100 mg/kg [42]. Similarly, in cephaloridine-treated F344 rats, renal cortical slice accumulation of PAH and TEA and gluconeogenic capacity were decreased as early as 1 hour after administration of 2000 mg/kg of cephaloridine, a time preceding any detectable change in BUN or serum creatinine concentrations [43]. Importantly, the early biochemical effects of cephaloridine observed after *in vivo* treatment (i.e., organic ion transport, gluconeogenesis, intracellular GSH depletion, lipid peroxidation) have been reproduced in renal cortical slices harvested from naive animals and exposed to cephaloridine *in vitro* [32]. Thus, the similar nephrotoxic effects of cephaloridine observed in renal cortical slices after either *ex vivo* or *in vitro* treatment provided the basis to investigate the mechanisms of cephaloridine nephrotoxicity using renal cortical slices *in vitro*.

In vivo studies have indicated that the incidence and severity of cephaloridine nephrotoxicity were related to the renal cortical concentrations of this drug. For example, both renal cortical concentrations and nephrotoxicity of cephaloridine were greatest in rabbits, intermediate in guinea pigs, and least in rats [44], suggesting a relationship between cortical accumulation and toxicity of this antibiotic. The exact mechanisms mediating the renal cortical accumulation of cephaloridine have been studied in detail in the rabbit. Pretreatment of rabbits with inhibitors of organic anion transport, such as probenecid, decreased the cortex/serum concentration ratio of cephaloridine [44], suggesting that cephaloridine is actively transported into proximal tubular cells by an organic anion transporter. That organic anion transport plays an important role in cephaloridine accumulation, and nephrotoxicity has been suggested by the observation that pretreament of rabbits with probenecid or other competitive inhibitors of organic anion transport, such as PAH or benzylpenicillin, decreased both renal accumulation and nephrotoxicity of cephaloridine [45]. Unlike other cephalosporins, however, cephaloridine is a zwitterion containing a positive charge at the 3 position of the cephem ring and organic cation transport also seems to be important to the renal tubular handling of cephaloridine. Wold and Turnipseed [46] demonstrated that the renal cortical concentrations and nephrotoxicity of cephaloridine were significantly greater after in vivo treatment with an inhibitor of organic cation transport, cyanine. To investigate the mechanisms mediating the effects of cyanine on renal cortical concentrations of cephaloridine, Wold and Turnipseed [46] studied the effects of cyanine on the in vitro uptake and efflux of cephaloridine in renal cortical slices from naive rabbits. Indeed, cyanine did not affect the uptake of cephaloridine, but it did significantly decrease its efflux by renal cortical slices, suggesting that cephaloridine transport from the proximal tubular cell into the tubular fluid was dependent on an organic cation transporter. Thus, the increased severity of cephaloridine nephrotoxicity after cyanine pretreatment was attributed to the ability of cyanine to decrease efflux (excretion) and, hence, increase net renal cortical concentrations of this antibiotic. These studies further demonstrate the utility of renal cortical slices in dissecting the mechanisms of tubular transport of xenobiotics and the potential relationship of tubular transport processes to drug-induced nephrotoxicity.

Although the role of renal tubular transport in cephaloridine nephrotoxicity has been fairly well defined, the exact biochemical mechanisms mediating cephaloridine cytotoxicity are less well understood. Several lines of evidence suggest that lipid peroxidation may play an important role. Most notably, Kuo et al. [47] have reported that conjugated dienes, products of lipid peroxidation, were increased in renal cortical tissue shortly after cephaloridine administration, before the detection of other biochemical or functional evidence of tubular injury. Furthermore, cephaloridine also depleted renal cortical GSH concentrations and increased oxidized glutathione (GSSG) concentrations, consistent with cephaloridine-induced oxidative stress. However, these in vivo observations do not provide direct evidence of a causal link between cephaloridine-induced oxidative stress and nephrotoxicity. To more precisely determine whether cephaloridine-induced oxidative stress is an initiating event mediating cephaloridine nephrotoxicity, the time course of biochemical effects was monitored in renal cortical slices exposed to cephaloridine in vitro. These studies revealed that incubation of renal cortical slices with cephaloridine resulted in a time- and concentration-related increase in lipid peroxidation, reflected by MDA production [32]. Furthermore, the onset of cephaloridine-induced peroxidation preceded cephaloridine-induced inhibition of organic ion transport. More definitive evidence supporting a role for lipid per-

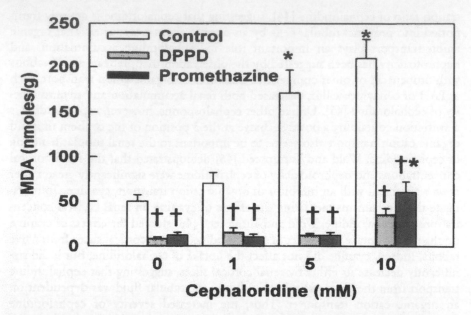

FIG. 10.3. Effects of coincubation with antioxidants on cephaloridine-induced malondialdehyde (MDA) production (left panel) and para-aminohippurate (PAH) uptake (right panel) in renal cortical slices. Asterisks indicate significant differences from values representing no cephaloridine exposure but similar antioxidant treatment. Daggers indicate significant differences from values representing no antioxidant treatment but similar cephaloridine exposure. Reproduced from Goldstein et al. [32] and Academic Press with permission.

oxidation in cephaloridine nephrotoxicity was obtained from studies evaluating the effects of antioxidants on cephaloridine-induced alterations in organic ion transport. Incubation of renal cortical slices with antioxidants, such as promethazine or N,N'-diphenyl-p-phenylenediamine (DPPD), blocked the effects of cephaloridine on both lipid peroxidation (Fig. 10.3) and organic ion transport [32,48], suggesting a cause–effect relationship between cephaloridine-induced lipid peroxidation and inhibition of organic ion transport. Thus, these *in vitro* studies using renal cortical slices have defined more precisely the role of lipid peroxidation in cephaloridine nephrotoxicity.

In addition, *in vitro* biochemical studies of cephaloridine nephrotoxicity indicated that cephaloridine impaired certain metabolic functions of the proximal tubule by mechanisms other than lipid peroxidation. For example, cephaloridine profoundly inhibited pyruvate-supported gluconeogenesis in renal cortical slices *in vitro*, an effect that occurred before the onset of lipid peroxidation [32]. Furthermore, antioxidants, such as promethazine or DPPD, did not block cephaloridine inhibition of gluconeogenesis, suggesting that the effects of cephaloridine on gluconeogenesis were independent of peroxidation. Further, *in vitro* studies were designed to test the hypothesis that cephaloridine-induced inhibition of renal cortical slice gluconeogenesis was related to inhibition of any one or more of the rate-limiting enzymes of gluconeogenesis, specifically, pyruvate carboxylase, phosphoenolpyruvate carboxykinase, fructose diphosphatase, or glucose-6-phosphatase. To test this hypothesis, gluconeogenesis was evaluated using substrates supporting each of the rate-limiting reactions, i.e., pyruvate (pyruvate carboxykinase), oxaloacetate (phosphoenolpyruvate carboxykinase), fructose-1,6-diphosphate (fructose diphophatase), and glucose-6-phosphate (glucose-6-phosphatase) [49]. These studies indicated that cephaloridine-inhibited gluconeogenesis supported by each of these substrates, suggesting either that cephaloridine inhibited each of the rate-limiting enzymes in a nonspecific manner or that cephaloridine inhibited the final step in glucose synthesis, the conversion of glucose-6-phosphate to glucose. Further assessment of the effects of cephaloridine on enzymatic activity in subcellular fractions indicated that cephaloridine specifically inhibited glucose-6-phosphatase activity, suggesting that cephaloridine inhibition of gluconeogenesis was caused by inhibition of the final step in glucose synthesis [49].

In summary, studies using renal cortical slices have yielded valuable information concerning cephaloridine-induced nephrotoxicity. First, *ex vivo* renal cortical slice functions are exquisitely sensitive indicators of cephaloridine nephrotoxicity. Second, the effects of cephaloridine on renal cortical slice functions and biochemistry *in vitro* are consistent with the effects observed after cephaloridine administration *in vivo*. Third, *in vitro* evaluation of the temporal sequence of biochemical effects as well as the effects of antioxidants on cephaloridine nephrotoxicity have more precisely defined the role of lipid peroxidation in cephaloridine nephrotoxicity. Finally, multiple mechanisms of cephaloridine nephrotoxicity exist that are characterized by both peroxidative-dependent and peroxidative-independent effects.

10.4.2 Site-Selective Nephrotoxic Injury

Most nephrotoxicants appear to have their primary effects on discrete segments or regions of the nephron. The proximal tubule, for example, is the primary target for many nephrotoxic antibiotics, antineoplastics, halogenated hydrocarbons, and heavy metals, whereas the glomerulus is the primary target for immune complexes, the loop of Henle/collecting duct for fluoride ions and the medulla/papilla for chronically consumed analgesic mixtures. Although the reasons underlying this site-selective injury are complex, segmental differences in morphology, physiology, and biochemistry appear to play important roles. For example, blood flow to the cortex is disproportionately high compared with other regions of the kidney and thus, blood-borne toxicants likely will be delivered preferentially to the cortical region, rendering cortical structures (predominantly proximal tubules) more vulnerable to chemically induced injury. However, in certain instances, site-selective injury does persist *in vitro* under conditions in which all cell types are exposed to identical extracellular concentrations of a given nephrotoxicant. Thus, under these conditions, site-selective injury cannot be attributed solely to systemic delivery of toxicants but rather may be caused by the biochemical properties of the cell type(s) in question.

Using positional renal slices, work from Gandolfi's laboratory has indicated that nephrotoxicants as diverse as mercuric chloride, potassium dichromate, cisplatin, hexachlorobutadiene, and S-(1,2-dichlorovinyl)-L-cysteine (DCVC) produce the same pattern of selective cellular injury *in vitro* as reported by others after *in vivo* exposure [50–53]. More specifically, both *in vivo* and *in vitro* exposure to mercuric chloride resulted in nephrotoxic lesions to the straight segment (*pars recta*), whereas *in vivo* and *in vitro* exposure to potassium dichromate produced selective injury to the convoluted segment (*pars convoluta*) of the proximal tubule [50]. Wolfgang et al. [52] have used positional renal cortical slices to delineate the temporal sequence of biochemical and histopathologic changes leading to the site-specific injury to the *pars recta* after exposure to DCVC. In these studies, ^{35}S-DCVC uptake by renal cortical slices was shown to be time and concentration dependent. Autoradiography indicated that ^{35}S-DCVC was distributed fairly equally among all proximal tubular segments within the slice, suggesting that the site-selective injury to the *pars recta* could not be attributed solely to a preferential uptake or accumulation of DCVC by the *pars recta*. After DCVC uptake, the initial events related to DCVC toxicity were characterized by covalent binding followed by alterations in intracellular ATP content, oxygen consumption, and leakage of brush-border enzymes. These temporal data further suggested that covalent binding preceded ultrastructural damage to both mitochondria and brush border, both of which appear to be early targets of DCVC toxicity. Light microscopy revealed a sequence of histopathologic changes, beginning with a selective lesion to the *pars recta* followed by lesions to all other proximal tubular segments. Thus, these studies demonstrated that the site-selective injury of DCVC to the *pars recta* was preceded by covalent binding, followed by effects on mitochondria and brush border.

10.5 TUBULE SUSPENSIONS

Several methods have been used to investigate biochemical responses of renal epithelial cells to potential toxicants. In large part, these methodologies have been adapted from physiologic studies examining renal metabolism and transport properties. In general, *in vitro* preparations used for biochemical studies include semipurified portions of tubular epithelium (tubule suspensions) or individual tubular epithelial cells (cell suspensions, cell cultures).

10.5.1 Methodology

Usually, tubules are harvested from rabbits, although tubule suspensions have been prepared from rats and mice [54–61]. Several methods have been published describing preparation of tubule suspensions [62–64].

Many investigators use *in situ* or *ex vivo* perfusion with collagenase to isolate tubules from supporting tissue. After anesthesia, the renal artery is cannulated either directly [65] or through the aorta [56,61,66]. Kidneys are perfused initially with a collagenase-free solution to remove blood. The perfusate is then switched to a solution containing collagenase, and perfusion continued for several minutes. After perfusion, the kidneys are removed from the animal and tubules are obtained by manual dispersion of kidney tissue. The tubules are separated from cellular debris by a variety of methods, including filtering through gauze or centrifugation through discontinuous gradients of Percoll or Ficoll. Alternatively, kidneys may be perfused with a collagenase solution containing magnetic iron oxide. After perfusion and manual dispersion of the tissue, the tubule suspension is strained through a series of wire sieves and glomeruli are removed magnetically [65].

Other investigators use *in vitro* incubation of kidney tissue with collagenase to disperse tubules. In this method, kidneys may be perfused initially with collagenase-free buffer to remove blood cells. Kidneys are then removed from the animal, placed on ice, and minced into small pieces. The kidneys pieces are incubated in collagenase-containing buffer for 30 to 60 minutes at 37°C. The resulting tubule suspension is washed and centrifuged several times to remove cellular debris, and then further purified by centrifugation through a Percoll or Ficoll gradient [56,67–69].

10.5.2 Advantages

Tubule suspensions have several distinct advantages over other *in vitro* methods, such as kidney slices, cell suspensions, or cultured cells. A significant advantage of tubule suspensions is that the nephron segments maintain morphologic integrity and display metabolic characteristics and transport properties similar to those observed *in vivo*. For example, proximal tubules in suspension actively accumulated á-methyl-glucose approximately 20-fold, and this active accumulation was reduced by phloretin or phloridzin, inhibitors of sugar transport [68]. Tubule suspensions responded appropriately to the addition of hormones: proximal tubules produced cyclic adenosine monophosphate (cAMP) in response to parathyroid hormone (PTH) but not ADH, whereas distal tubules produced cAMP in response to ADH but not PTH [67,70].

Another advantage of this technique is that relatively pure suspensions containing the nephron segment of interest may be prepared by careful choice of starting material or purification method. For example, by limiting the starting tissue to the outer cortex, investigators have prepared suspensions containing greater than 90% proximal tubules [64,67]. By limiting dispersion to medullary tissue, other investigators have prepared suspensions containing greater than 90% medullary thick ascending limb of Henle (TALH) [71]. Responses of specific nephron segments to potential toxicants may be examined *in vitro* using these purified tubule suspensions.

Tubule suspensions maintain viability for at least several hours [59,60] or longer [68], a longer lifespan than is observed for the IPK. Thus, toxicants may be tested in tubule suspensions over a period of several hours or more. As with any *in vitro* technique, use of tubule suspensions allows the investigator to precisely control the incubation medium and substrate concentrations. Toxicants may be included for the duration of the incubation period or removed after some specified time. In evaluation of proximal tubular function, transport substrates, such as PAH, TEA, or K-methyl-glucose, may be included to assess tubular integrity after exposure to a potential toxicant.

An additional advantage of this technique is that tubule suspensions yield a relatively large amount of tissue, particularly if a large animal, such as a rabbit, is used or if kidneys from several rats or mice are pooled. Thus, characterization of concentration- and time-response curves or structure-activity relationships (SARs) may be evaluated in a relatively homogeneous preparation.

Tubule suspensions offer the ability to directly or indirectly probe intracellular sites of toxicity. For example, oxygen consumption, cAMP production, hormone responsiveness, transport capabilities, substrate-supported gluconeogenesis, and leakage of intracellular enzymes are among the numerous responses to toxicants that may be measured in tubule suspensions. Because the starting material, i.e., the nephron segment contained in the suspension, is relatively well defined, the responses elicited by toxicant exposure *in vitro* should resemble the characteristic responses of that nephron segment after toxicant exposure *in vivo*. Lack of correlation between responses obtained with tubule suspensions *in vitro* compared with nephrotoxic responses observed *in vivo* would suggest that extra-renal factors are involved in the nephrotoxicity of a compound.

10.5.3 Limitations

Tubule suspensions have several limitations that must be kept in mind when using this technique. When kidneys are perfused *in situ* or *ex vivo*, a risk of producing anoxia or ischemia exists. When magnetic iron is used to separate glomeruli from tubules, clumping and inadequate perfusion may occur with these iron-containing perfusates [65]. By whatever method tubules are prepared, initial steps involve limited collagenase digestion to free tubules from the supporting extracellular matrix. Collagenase digestion may compromise the integrity of tubule membranes, making these membranes more leaky and allowing diffusion to occur more readily than it does in intact tubules. Although tubule suspensions are purified, purification is not complete and trace contamination by other tubular segments or vascular tissue will probably occur to some degree. Therefore, the preparation needs to be carefully evaluated, preferably by microscopic methods, to assess purity. In most cases, suspensions will be enriched greater than 90% in the nephron segment of interest, so that contamination by other nephron segments does not confound interpretation of data.

Compared with the IPK or renal slices, considerably more time is required to make a tubule suspension. Equipment required to support tubule suspensions is not as extensive as with the IPK, but somewhat more equipment is needed than it is with kidney slices. In addition, cost of collagenase and other chemicals required to prepare tubule suspensions makes the technique more costly than are renal slices or the IPK.

Other disadvantages include the finite lifespan of any preparation, which must be rigorously defined by the investigator before toxicity testing or biochemical studies, and the small amount of tissue contained in tubule incubations, compared with the IPK or renal slices. Typically, renal slices use 50–100 mg of tissue per incubation, whereas tubule suspensions use 10–50 mg of tissue for each incubation. Thus, assays need to be downsized or modified to detect small amounts of compounds. Often, investigators will use radiolabeled compounds or custom-made equipment to measure experimental endpoints, further driving up the cost of an experiment.

Finally, limited information is available concerning *in vivo* and *in vitro* correlations for tubule suspensions. As with the IPK and renal slices, because some toxicities are slow to develop, the relatively short lifespan of tubule preparations may make them unsuitable for some studies.

With renal slices and the IPK, tissue may be obtained from animals pretreated with toxicants. The ability to obtain tubule suspensions from animals pretreated with toxicants has not been rigorously investigated or reported, although one group has obtained tubules from rats at various times after renal arterial occlusion and reflow [72]. These investigators obtained viable proximal tubule suspensions from rats subjected to 45 minutes of renal arterial occlusion followed by 15 minutes, 2 hours, or 24 hours of reflow. Tubules were prepared according to a method that yielded primarily proximal convoluted tubules. Viability was monitored by ouabain-sensitive and nystatin-stimulated oxygen consumption, and these indices were reduced compared with control rats only after 15 minutes or 2 hours of reflow [72]. Viability returned to control values within 24 hours of reflow [72]. In addition, the authors provided morphologic evidence of tubular damage after ischemic insult and 15 minutes reflow; however, they failed to comment on the progression or lack of progression of tubular injury at longer reflow times [72]. In contrast, *in vivo* renal arterial occlusion (25 minutes) and 24-hour reflow in rats produced marked proximal tubular injury, primarily affecting proximal straight tubules [73]. Longer periods of ischemia (45–60 minutes) with 24-hour reflow produced extensive damage to both proximal convoluted and proximal straight tubules [74]. Based on these *in vivo* studies, one would expect some degree of damage to be detectable in proximal convoluted tubules harvested from rats subjected to 45-minute ischemia and 24-hour reflow, as used by Gaudio et al. [72]. Alternatively, if 45-minute ischemia and reflow produced damage only in proximal straight tubules, Gaudio et al. isolated tubules that were undamaged by ischemia and reflow, because their procedure harvested primarily proximal convoluted tubules. It is difficult to reconcile the discrepancies between these *in vivo* and *in vitro* studies [71–74]. One possible explanation may be that severely damaged tubules may be difficult or impossible to obtain by conventional procedures and the tubule suspension prepared from rats subjected to *in vivo* ischemia represents uninjured or minimally injured tubules. Thus, it is unclear whether damaged tubules may be obtained after toxicant exposure *in vivo*. Certainly, more work is required to validate this technique for use following pretreatment regimens.

10.6 USE OF RENAL TUBULE SUSPENSIONS IN TOXICOLOGY

10.6.1 Comparison of Tubular Segments: Proximal Versus Distal Tubules

The kidney is a heterogeneous organ, comprising discrete and unique portions. Nowhere is this complexity more evident than it is in the tubule. Each renal tubule contains at least eight segments, differentiated by morphology, localization, and function. In the past, it has been difficult to purify different tubule segments, with the exception of proximal tubules, in quantities necessary to determine intracellular contents, enzymatic components, and other biochemical characteristics. Tubule suspensions have helped overcome some of these difficulties by allowing differential preparation of fractions enriched in particular nephron segments, for example, proximal versus distal segments.

By centrifugation of partially digested kidney cortex through a Percoll gradient, several groups have prepared suspensions enriched in proximal or distal tubules [56,70]. These suspensions, containing greater than 90% proximal or distal tubules, allowed comparison of biochemical characteristics of these tubular segments. Oxygen consumption and ATP content were similar in proximal and distal tubule suspensions [70]. Alkaline phosphatase, 3-glutamyl-transpeptidase, fructose-1,6-diphosphatase, and glucose-6-phosphatase activities were significantly higher in proximal tubules, whereas hexokinase and kallikrein activities were significantly greater in distal tubules [70]. Gluconeogenesis in the distal tubule suspensions was less than 10% of that in proximal tubule suspensions. Cytochrome P450 content and PAH uptake in proximal tubules were significantly greater than they were in distal tubules. Nonprotein sulfhydryl content (primarily GSH) was similar in proximal and distal tubules. Although xenobiotic metabolism is most commonly associated with proximal rather than distal tubules, distal tubular en-

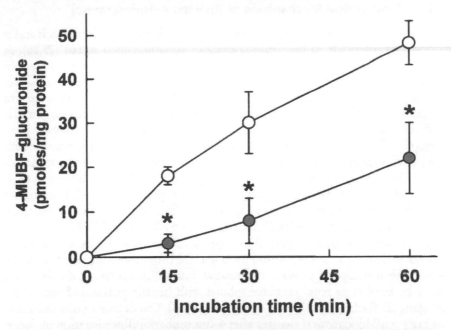

FIG. 10.4. Time course of glucuronidation of 4-methylumbelliferone (4-MUBF) to 4-MUBF-glucuronide by proximal □ and distal □ tubule fractions. Asterisks indicate significant differences between proximal and distal tubular suspensions ($p < 0.05$). Reproduced from Cojocel et al. (56) and Permagon Press with permission.

zymes catalyzed glucuronidation of 4-methylumbelliferone (Fig. 10.4), although activity in distal tubules was only about 50% of that in proximal tubules [56].

Thus, by use of suspensions containing proximal or distal tubules, direct comparison of these segments was possible and biochemical similarities and differences were identified. As expected, proximal tubule suspensions had greatest activity in those functions previously associated with the proximal tubule, such as PAH uptake, gluconeogenesis, and cytochrome P450-dependent metabolism. These activities were minimal or not detectable in distal tubule suspensions. In contrast, distal tubules had significant glucuronidation capacity, as evidenced by glucuronidation of 4-methylumbelliferone (4-MUBF) (Fig. 10.4). Previously, glucuronidation in the kidney was thought to occur only in the proximal tubule. The significance of distal tubule glucuronidation capacity remains unclear, and further investigations to more fully characterize glucuronidation, as well as other biochemical functions, of the distal tubule would help clarify the role of this nephron segment in xenobiotic metabolism.

10.6.2   Biochemical Mechanisms of Toxicity: 4-Aminophenol

4-Aminophenol (para-aminophenol] produces acute proximal tubular necrosis and is proposed to contribute to the nephrotoxicity of acetaminophen in rats [75,76]. *In vivo*, 4-aminophenol produces functional and morphologic changes indistinguishable from those caused by acetaminophen [75], which lead to the hypothesis that acetaminophen nephrotoxicity in rats was caused by *in situ* deacetylation of acetaminophen to yield 4-aminophenol. However, recent investigations suggest that in addition to intrarenal bioactivation, hepatic metabolism may be involved in 4-aminophenol nephrotoxicity. For example, biliary cannulation or pretreatment with buthionine sulfoximine to deplete GSH partially protected rats from 4-aminophenol-induced nephrotoxicity [77], leading to the suggestion that GSH conjugates of 4-aminophenol were at least partially responsible for nephrotoxicity. GSH conjugates of 4-aminophenol were found in the bile of male Wistar rats [78], and 4-aminophenol GSH conjugates produced nephrotoxicity *in vivo* and *in vitro* [78–81].

The concept that 4-aminophenol may be bioactivated by GSH conjugation in the liver is similar to the mechanism proposed for bromobenzene nephrotoxicity [82,83]. We investigated possible cooperative interactions between the liver and kidney by incubating renal proximal tubules with hepatic postmitochondrial supernatant (S9 fraction) as a bioactivating system [84]. Our *in vivo* studies indicated that early renal biochemical changes after 4-aminophenol administration included depletion of GSH, decreases in oxygen consumption, depressed organic anion (PAH) and cation (TEA) uptake, and decreases in ATP [61]. However, incubation of renal proximal tubules with 1-mM 4-aminophenol for up to 4 hours failed to produce significant changes in thee biochemical indices of toxicity [61]. In contrast, decreases in oxygen consumption and ATP content were readily observed when renal proximal tubules were incubated with 0.5-mM 4-aminophenol in the presence of hepatic S9 (10-mg protein) and GSH (1 mM) (Fig. 10.5). Inclusion of an nicotinamide adenine dinucleotide phosphate (NADPH)-generating system prevented 4-aminophenol-induced changes, suggesting that either cytochrome P450s did not contribute to bioactivation of 4-aminophenol or that NADPH kept 4-aminophenol in a reduced form, preventing autoxidation [84].

Our results suggesting that extrarenal metabolism is required for 4-aminophenol nephrotoxicity are at variance with other investigators. For example, Klos et al. [78] observed that 4-aminophenol produced marked decreases in viability, measured as trypan blue exclusion, in renal proximal tubule cells prepared from Wistar rats. GSH conjugates of 4-aminophenol were also cytotoxic in these cells [78]. Lock et al. [85] observed that 4-aminophenol caused significant LDH leakage, loss of ATP, and decreases in state 3 respiration in rabbit proximal tubules. However, GSH conjugates of 4-aminophenol were ineffective in producing cytotoxicity in these rabbit proximal tubules [85]. At present, the explanation for these discrepancies is unknown. However, our ability to combine hepatic and renal tissue in a single incubation system represents a unique system to investigate bioactivation as a factor in chemical-induced nephrotoxicity.

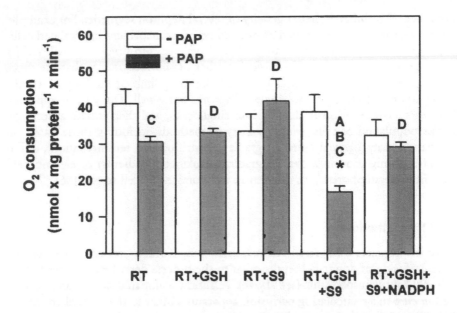

FIG. 10.5. Oxygen consumption in renal tubules incubated with 4-aminophenol in the absence or presence of glutathione (GSH), hepatic postmitochondrial supernatant (S9) fraction, or an nicotinamide adenine dinucleotide phosphate (NADPH)-generating system for 4 hours. Renal tubules (RT, 8–12-mg protein per flask) were incubated for 4 hours in Krebs–Henseleit buffer in the absence or presence of 4-aminophenol (0.5 mM). Some flasks contained GSH (1 mM, RT + GSH), hepatic S9 fraction (10-mg protein, RT + S9), or both (RT + GSH + S9). In addition, some flasks contained an NADPH-generating system (RT + GSH + S9 + NADPH). Values are means ± SE, $n = 4$–11. The asterisk indicates a value significantly different from tubules incubated under the same conditions in the absence of 4-aminophenol. "A" indicates a value significantly different from tubules incubated in buffer alone (RT), "B" indicates a value significantly different from tubules incubated in the presence of GSH alone (RT + GSH), "C" indicates values significantly different from tubules incubated in the presence of hepatic S9 fraction alone (RT + S9), and "D" indicates values significantly different from tubules incubated in the presence of hepatic S9 fraction and GSH (RT + GSH + S9). Reproduced from Shao et al. [84].

10.7 CELL SUSPENSIONS

The utility of hepatocyte preparations led investigators to try similar techniques with kidneys, with varying degrees of success. Whereas the liver contains only a few cell types, the kidney is a complex organ containing numerous cell types. Therefore, methodology had to be developed to obtain fairly homogeneous preparations of the desired cell population. Essentially, two different approaches may be employed. One approach is to prepare tubule fragments, as described above, and further dissociate those fragments into single cells. This method is complicated by inherent heterogeneity of several nephron segments. For example, collecting tubules are composed of two cell types, principal and intercalated cells. Therefore, dissociation of collecting tubule fragments into single cells will yield a heterogeneous population. However, for other nephron segments that are more homogeneous in nature, such as the thick ascending limb of the loop of Henle, dissociation of fragments into single cells works fairly well. The second approach is to dissociate renal tissue directly into single cells, and then isolate a homogeneous population of cells using a variety of methods, including those based on buoyant density (isopycnic-density gradient centrifugation, sedimentation), different rate of migration in a density gradient (isokinetic gradients) or electric field (free-flow electrophoresis), or differences in fluorescence (cell sorter) [86].

10.7.1 Methodology

Most investigators using renal cell suspensions start with *in situ* or *ex vivo* collagenase perfusion of donor kidneys, similar to the process described for preparation of renal tubule suspensions (see above). Perfusion is initiated *in vivo* and continued *in vitro* in a recirculating perfusion apparatus similar to that described for the IPK (Fig. 10.6) [87]. A major difference is that, to prepare cell suspensions, perfusion is continued for a longer period of time (usually 15–30 minutes) than the time required to prepare tubule suspensions (usually 5–10 minutes). After perfusion, further procedures are necessary to isolate individual cells. Generally, separation of proximal and distal tubule cells from the loop of Henle and collecting duct cells is achieved by selecting cortical tissue for processing, while discarding medullary tissue [88]. Alternately, medullary tissue is retained for processing and cortical tissue discarded when suspensions of medullary cells are desired [89].

When cortical tissue is processed, glomeruli may be removed by taking advantage of size differences. For example, the diameter of a rat glomerulus is approximately 100 µm, whereas the diameter of a tubular epithelial cell is considerably less [88]. Thus, sieving tissue digest through a series of screens serves to remove larger tissue fragments, glomeruli, blood vessels, and connective tissue. However, the resulting preparation contains cells of proximal as well as distal tubule origin, and further purification is necessary to obtain a more homogeneous cell suspension. Numerous purification procedures have been described, most involving centrifugation through discontinuous gradients of Ficoll, Percoll, or other polymers [88]. The final preparation is highly enriched in proximal tubule cells, although heterogeneity still exists in that cells are derived from S1 and S2 subtypes [88].

For whatever perfusion and purification method employed, other cell types in the suspension must characterize cells as to nephron segment of origin and extent of contamination. A variety of methods have been employed to characterize cell suspensions, including morphologic examination of cells in suspension (Table 10.1), distribution of marker enzymes (e.g., hexokinase for distal tubule cells, alkaline phosphatase, and 3-glutamyl transpeptidase for proximal tubule cells), and

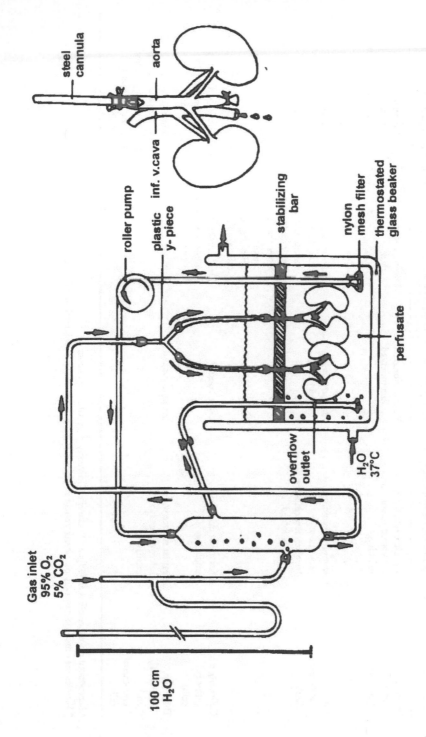

FIG. 10.6. Schematic drawing of apparatus for isolating kidney cells. Arrows indicate direction of perfusate flow through recirculating system. The apparatus allows for simultaneous perfusion of two pairs of kidneys. Right: Illustration of placement of steel cannula into aorta for in situ/ex vivo perfusion. Reproduced from Ormstad et al. [87] and Academic Press with permission.

TABLE 10.1. Structural markers for identification of single isolated cells from rabbit renal medulla[a]

Cells	Plasma membrane	Nucleus	Mitochondria	Cytoplasmic inclusions
Proximal tubule (pars recta)	Long microvilli, thick basolateral membrane infoldings	Round, elongated, often indented	Small, ovoid, dark	Numerous vesicles (mean diameter ~ 0.7 'm), electron-dense material[b]
Thick ascending limb	Few short microvillus projections, basolateral membrane is highly folded	Elongated with indentations	Numerous large, dark, densely packed, filling cell, mostly round, larger than in proximal and collecting tubule cells, visible	Vesicles mostly smaller than are mitochondria[b]
Collecting duct (principal cells)	A few short microvilli, infoldings	Large, round	Dark, small, less numerous than in proximal tubule	Small, light vesicles (mean diameter ~ 0.3 'm)
Thin loop, interstitial, and endothelial cells	Irregular surface with microvillous-like extrusions	Irregular shape	Very few	Some vesicles, cytoplasm is only a small ring around the nucleus.

a Reproduced from Eveloff et al. [89] and Rockefeller University Press with permission.
b Vesicles may include naturally occurring vesicle structures and cross sections of basolateral membranes.

oxygen consumption (inhibited by furosemide in TALH cells, inhibited by amiloride in distal tubule cells) [88–90].

10.7.2 Advantages

In general, cell suspensions share the same advantages as do tubule suspensions. A fairly homogeneous population of the desired cell type may be prepared, and cell population may be exposed to a toxicant over a short period of time. Viability parameters, such as oxygen uptake, LDH content, dye exclusion, and calcium content, may be measured in cell suspensions. Intracellular contents of metabolic substrates, such as adenine nucleotides and GSH, may be monitored, as well as intracellular ion concentrations (e.g., sodium and potassium). All of these parameters may provide valuable information concerning cell viability.

Cell suspensions offer additional advantages over previously discussed *in vitro* methods. In cell suspensions, unstirred layers and anoxic or ischemic damage are minimized because cells are individually and uniformly exposed to medium on all sides. In tubule suspensions, the lumen is open and accessible during separation and purification procedures, but it is not certain that the lumen remains open during subsequent incubations [88]. Thus, during incubation with tubule suspensions, compounds gaining access to tubular epithelial cells by luminal transport or diffusion may not accumulate if tubular lumens do indeed collapse. In contrast, cell suspensions are continually exposed to toxicant throughout incubation, and access of toxicant to intracellular compartments is not a problem. In addition, cell suspensions may be "pulsed" with toxicant. For example, cells may be incubated with toxicant for a short period of time, and then washed and resuspended in incubation medium free of toxicant. In this manner, sequential responses to short toxicant exposures may be studied.

10.7.3 Limitations

As with other *in vitro* techniques, cell suspensions have a limited lifespan of 2–4 hours, similar to tubule suspensions. In addition, a significant amount of time is required to prepare cell suspensions, negating an advantage of renal slices in terms of ease and rapidity of preparation. As mentioned, enzymatic digestion is required to loosen cells from underlying extracellular matrix. Because the incubation or perfusion time required to prepare cells is longer than the time required to prepare tubules, it is possible that membrane integrity may be compromised during digestion and that epithelial cells may lose some specific membrane functions [88].

Another disadvantage of cell suspensions is loss of *in situ* polarity. In suspension, cells are evenly exposed to medium on all sides. Luminal and basolateral attachments are lost, as is luminal and basolateral polarity. Consequently, in renal cell suspensions, more membrane surface area is exposed to toxicants than is exposed in renal slices or tubule suspensions, creating a potential for enhanced diffusion of toxicant into tubular cells.

Another limitation of cell suspensions is the relatively low yield of tissue. Typically, only 25–50 x 10^6 cells are recovered from rat kidneys after digestion and purification [90,91]. Cell yield is usually sufficient to allow experimental maneuvers with appropriate controls. However, analytical techniques may need to be adapted to detect small quantities of substances, and specialized equipment may be necessary to monitor desired parameters.

10.8 USE OF CELL SUSPENSIONS IN RENAL TOXICOLOGY

10.8.1 Comparison of Renal Tubular Cells: Proximal Versus Distal Tubular Cells

Most nephrotoxicants injure renal proximal tubules, although several compounds injuring distal tubules have been identified (amphotericin B, lithium). However, the identification of the proximal tubules as the more common site of chemical-induced injury may reflect a greater ability to detect proximal tubular damage. For example, the proximal tubule has well-defined, unique functions, such as organic anion and cation transport, which make detection of injury relatively easy. The distal tubule has fewer unique functions than does the proximal tubule, making it difficult to identify toxicity in the distal tubule. As noted above, Cojocel et al. [56] used suspensions enriched in proximal or distal tubules to quantitate differences in between these tubular segments. Lash et al. have taken advantage of different densities of proximal and distal tubular cells to prepare purified proximal or distal tubular cell suspensions [90,92,93].

In a series of studies, Lash et al. tested cytotoxic responses of proximal or distal tubular cells to three different classes of compounds: cephalosporin antibiotics, specifically, cephaloridine [90]; oxidizing agents, specifically, *tert*-butyl-hydroperoxide (tBH), menadione, and hydrogen peroxide [92]; and alkylating agents, specifically, methyl vinyl ketone (MVK), allyl alcohol, and N-dimethylnitrosamine (NDMA) [93]. Cephaloridine produces proximal tubular necrosis *in vivo* [94], and cytotoxicity only in proximal tubular cells [90]. Cell viability, measured as trypan blue exclusion, was reduced in a time- and concentration-dependent manner in proximal tubular cells, whereas distal tubular cells incubated with cephaloridine showed no loss of viability over 2 hours [90]. Thus, *in vitro* cell suspensions were in excellent agreement with *in vivo* studies and showed that proximal tubular cells were susceptible to cephaloridine-induced cytotoxicity, whereas distal tubular cells were not.

In contrast to selective susceptibility of proximal tubular cells to cephaloridine, distal tubular cells were much more susceptible than were proximal tubular cells to cytotoxicity induced by oxidizing agents [92]. All three oxidizing agents produced cytotoxicity, measured as LDH leakage, in both proximal and distal tubular cells. However, LDH leakage was significantly greater in distal than in proximal tubular cells (Fig. 10.7) [92]. Oxidative stress is known to alter GSH redox status in cells [95,96]. To elucidate the mechanism(s) underlying greater susceptibility of distal versus proximal tubular cells to oxidative stress, GSH and GSSG were measured in cell suspensions after exposure to tBH. As expected, incubation of either proximal or distal tubular cells with tBH produced loss of GSH and accumulation of GSSG [92]. In addition, incubation of proximal tubular cells in a GSH-containing buffer failed to protect against tBH cytotoxicity, whereas incubation with GSH exerted a protective effect in distal tubular cells [92]. However, exogenous dithiothreitol protected distal tubular cells from tBH-induced cytotoxicity to the same extent as exogenous GSH, suggesting that protection was caused by nonspecific thiol reductant properties rather than by GSH-dependent enzyme-catalyzed reactions [92]. Further studies indicated that GSH status could be dissociated from oxidative injury. For example, preincubation with buthionine sulfoximine, an inhibitor of GSH synthesis, plus acivicin, an inhibitor of GSH degradation, reduced GSH concentration to a similar extent in both proximal and distal tubular cells. However, cytotoxicity of tBH was more pronounced in distal than in proximal tubular cells, despite comparable concentrations of GSH [92]. In addition, increases in cellular GSSG were not uniformly accompanied by increases in LDH leakage. Examination of enzymes involved in GSH status indicated that distal tubular cells had lower activities of GSH peroxidase, GSSG reductase, catalase, and distal tubular-diaphorase than did proximal tubular cells [92]. Thus, distal tubular cells may be more susceptible to oxidative stress than are proximal tubular cells because of a reduced ability of distal tubular cells to detoxify reactive species generated by oxidizing agents.

Both proximal and distal tubular cells were susceptible to cytotoxicity induced by alkylating agents [93]. Distal tubular cells were significantly more susceptible than were proximal tubular cells to MVK-induced LDH leakage and moderately more susceptible to cytotoxicity induced by allyl alcohol. Both MVK and allyl alcohol react with soft nucleophiles, such as GSH and protein sulfhydryl groups. Thus, agents reacting directly with GSH are more cytotoxic to distal versus proximal tubular cells, consistent with decreased ability of distal tubular cells to detoxify reactive intermediates through GSH metabolism (see above). In contrast, NDMA was equally cytotoxic to proximal and distal tubular cells [93]. NDMA undergoes cytochrome P450-dependent metabolism to generate formaldehyde and a methylating intermediate, or it undergoes denitrosation to form formaldehyde and monomethylamine, a metabolically inert species [97]. Similar susceptibility of proximal and distal tubular cells to NDMA cytotoxicity suggests that both cell populations are able to bioactivate NDMA. This observation is somewhat at odds with previous work suggesting that cytochrome P450 is present in the proximal tubule and virtually absent from the distal tubule [11,56]. It is possible that cytochrome P450 content is higher or a specific cytochrome P450 isozyme is

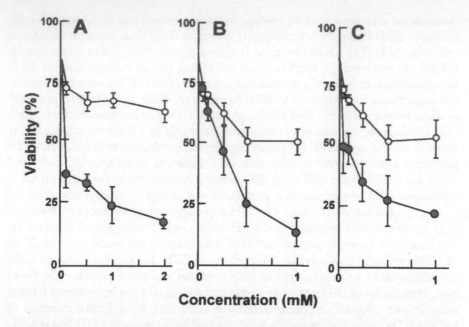

Fig. 10.7. Concentration dependence of cytotoxicity in isolated rat kidney proximal tubule (PT) and distal tubular (DT) cells. Isolated renal PT □ and DT ○ cells were incubated for 1 hour with indicated concentrations of *tert*-butyl hydroperoxide (tBH) (*A*), menadione (*B*), or hydrogen peroxide (*C*). Viability was measured as a fraction of cells that did not leak LDH. Reproduced from Lash and Tokarz [92] and the American Physiological Society with permission.

higher in distal tubules than was previously thought. Cytochrome P450 content of proximal versus distal tubular cells could be reexamined, perhaps using cell suspensions, to resolve this discrepancy.

10.8.2 Biochemical Mechanisms of Cytotoxicity: Halogenated Hydrocarbons

GSH and cysteine-S-conjugates of halogenated hydrocarbons, such as S-(1,2-dichlorovinyl) GSH (DCVG) and DCVC, are potent nephrotoxicants, causing necrosis of renal proximal tubules, increases of BUN concentration, and glucosuria [98–101]. Nephrotoxicity of these compounds depends on metabolic bioactivation: DCVG is metabolized via 3-glutamyltransferase and cysteinylglycine dipeptidase (or aminopeptidase M) to DCVC, and DCVC is further metabolized to pyruvate, ammonia, and a reactive thiovinyl intermediate via cysteine conjugate 2-lyase [100–104]. However, the mechanism whereby the reactive intermediate produces cytotoxicity was not entirely clear. Previous studies suggested that mitochondria are a primary target of DCVC cytotoxicity [105,106], and cell suspensions were employed to more precisely define the events involved in DCVC-induced cytotoxicity.

Incubation of proximal tubular cell suspensions with either DCVG or DCVC caused time- and concentration-dependent loss of viability (measured as trypan blue exclusion and LDH leakage) [107]. DCVG cytotoxicity was attenuated when AT-125, an inhibitor of 3-glutamyltransferase, was included in the incubation medium, and cytotoxicity caused by either DCVG or DCVC was attenuated when aminooxyacetic acid (AOAA), an inhibitor of cysteine conjugate 2-lyase, was included in the incubation medium [107]. Thus, cytotoxicity of both agents in cell suspensions required the same metabolic steps as did nephrotoxicity *in vivo*, supporting the utility of cell suspensions in investigating DCVG and DCVC cytotoxicity.

Incubation of cells with DCVC for 30 minutes caused marked declines in cellular GSH and glutamate concentrations without significantly changing GSSG concentrations [107]. The lack of change in GSSG concentration, as well as absence of thiobarbituric acid (TBA)-reactive material (indicative of lipid peroxidation), suggested that DCVC cytotoxicity did not involve oxidative stress. The decline in glutamate concentration was interpreted to suggest that DCVC induced alterations in cellular energy metabolism.

Subsequent studies monitored the effects of DCVC on three parameters of mitochondrial function: adenine nucleotide status, oxygen consumption, and calcium sequestration. DCVC produced a rapid and pronounced decline in cellular ATP content, coupled with modest increases in cellular adenosine diphosphate (ADP) and adenosine monophosphate (AMP) content. The cellular ATP/ADP ratio and energy charge fell dramatically, suggesting that proximal tubule cells exposed to DCVC would have impaired ability to maintain ATP-dependent functions [107]. DCVC decreased oxygen consumption with succinate as substrate; however, oxygen consumption with glutamate plus malate or ascorbate plus N,N,N',N'-tetramethyl-p-phenylenediamine as electron donors was unaltered by DCVC [107]. Thus, DCVC specifically inhibited succinate oxidation. Both mitochondria and endoplasmic reticulum play major roles in regulation of intracellular calcium homeostasis [108,109]. Incubation of proximal tubule cells with DCVC reduced mitochondrial calcium sequestration by 62% in 2 hours without altering microsomal calcium sequestration [107]. A strong correlation existed between DCVC-induced loss of cell viability and cellular oxygen consumption or mitochondrial calcium sequestration (Fig. 10.8) [107]. Thus, the primary target of DCVC-induced cytotoxicity appears to be mitochondria, leading to alterations in energy charge and calcium sequestration. However, the precise steps leading to cell death remain uncertain.

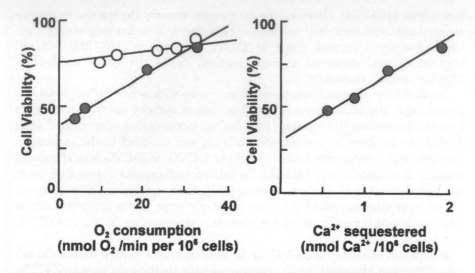

FIG. 10.8. Correlation between effects of S-(1,2-dichlorovinyl) GSH (DCVC) on renal cell viability and mitochondrial function. *A*. Relationship between effects of 1-mM DCVC □ or 1-mM DCVC + 0.1-mM aminooxyacetic acid (AOAA) □ on cell viability and oxygen consumption. *B*. Relationship between effects of 1-mM DCVC on cell viability and mitochondrial calcium sequestration. Reproduced from Lash and Anders [107] and the American Society of Biological Chemists with permission.

10.9 CELL CULTURE

The major disadvantage of previously discussed *in vitro* techniques is the limited lifespan of each preparation. With renal slices, tubule suspensions, or cell suspensions, investigators routinely use high concentrations of toxicant to accelerate the onset of cytotoxic events. However, higher concentrations of toxicants may introduce complications in interpretation. Many toxicities depend on enzyme-catalyzed metabolism for bioactivation or detoxification. Because enzyme-catalyzed processes saturate at finite substrate concentrations, use of high toxicant concentrations may distort normal metabolic pathways, for example, by inappropriately saturating detoxification pathways. It would be more desirable to be able to produce toxicity using concentrations of toxicant similar to those found *in vivo*. Thus, some toxicologists have turned to cell culture systems to investigate mechanisms of cytotoxicity.

Cell culture systems may be divided into established cell lines and primary cell cultures. Established cell lines frequently used in renal toxicology include MDCK cells, considered to be of distal tubular or cortical collecting duct origin and derived from cocker spaniel kidney [110–112]; LLC-PK₁ cells, considered to be of proximal tubular origin and derived from Hampshire pig [113,114]; and OK cells, considered to be of proximal tubular origin and derived from opossum kidney [115,116]. Isolating cells from donor kidneys (usually rat) and allowing these cells to grow and proliferate in defined culture conditions establishes primary cultures [88,117].

10.9.1   Methodology

Cell culture techniques are well established. Cells are grown on plastic dishes or on porous membranes under controlled conditions (CO_2 tension, humidity, temperature, oscillations) in medium containing metabolic substrates and growth factors, such as fetal calf serum. Cultures are allowed to growth to confluency, i.e., to form a monolayer, and then are replated at lower cell density (subculture). Established cell lines may be subcultured infinitely, whereas primary cell cultures may be subcultured a limited number of times before dedifferentiation and loss of specific functions occurs [88].

Primary cultures are initiated by obtaining cells from a donor kidney. In general, the procedure is the same as that used to obtain cell suspensions. Cells are harvested by collagenase digestion, purified as previously discussed, and placed in a Petri dish or other culture system in an appropriate medium. Unfortunately, purification is never complete, and some contamination with cells other than those desired will invariably occur. An additional complication of trace contamination in the initial preparation is that all cells do not grow at the same rate. Therefore, it is possible that a contaminating cell type may grow faster than does the desired cell type, eventually overgrowing and replacing the desired cells in the culture system. In particular, overgrowth of fibroblasts is a problem with proximal tubular cells in culture. Including D-valine and ornithine rather than L-valine and arginine in the culture medium might minimize fibroblast contamination, because proximal tubular cells, but not fibroblasts, can convert D-valine and ornithine into L-valine and arginine, respectively [118,119].

Traditionally, cells in culture are grown on plastic dishes or plates. Cells attach to the plastic surface, forming a basement membrane-like structure. With polarized cells, such as kidney tubule cells, the basolateral surface of the cell is oriented against the plastic support, thereby making the basolateral membrane inaccessible to the incubation medium. The luminal (apical) membrane of kidney cells develops normally in culture, exhibiting microvilli and other characteristic features, and it is accessible to the incubation medium [111–113,120]. When grown on plastic supports, both established cell lines and primary cell cultures develop "domes," areas in which fluid accumulates under the epithelial sheet. Dome formation can be abolished by including an inhibitor of sodium transport, such as ouabain or amiloride, in the incubation medium [111,114], suggesting that dome formation is caused by transepithelial "reabsorption" of solutes and water. Glucose reabsorption in the proximal tubule occurs via a luminal transporter inhibited by phloridzin, and kidney proximal tubular cells grown on plastic supports accumulated á-methylglucose via a phloridzin-sensitive pathway [117]. In contrast, PAH accumulation by proximal tubular cells occurs via a probenecid-sensitive basolateral transport pathway, and PAH was not accumulated by proximal tubular cells in primary culture grown on plastic supports [117], indicating that, with conventional cell culture systems, the basolateral membrane is not accessible to the incubation medium. To overcome this difficulty, investigators have turned to growing cells on porous filters that may be suspended in the incubation medium, as illustrated in Fig. 10.9 [88]. When primary cultures of proximal tubular cells were grown on porous membranes, probenecid-sensitive PAH uptake was observed [117]. In addition, primary cultures of proximal tubular cells grown on porous membranes acidified the medium bathing the apical surface, whereas cells grown on plastic

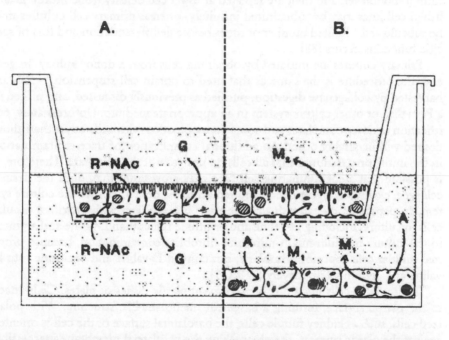

FIG. 10.9. Schematic drawing of proximal tubule cells cultured on a porous membrane, illustrating several possible pathways involved in nephrotoxicity. In panel A, mercapturic acids (R-NAc) applied to the basolateral membrane (lower compartment) may be transported into proximal tubular cells, and then secreted across the luminal membrane into the upper compartment. Glucose (G) applied to the luminal membrane (upper compartment) may be taken up by proximal tubular cells and transported out of cells at the basolateral membrane into the medium (lower compartment). This system allows selective exposure of the luminal and basolateral membranes as well as separate analysis of unidirectional cellular fluxes. In panel B, the porous membrane support system allows the coculture of different cell types on a rigid support system (bottom chamber). In this example, compound A is metabolized by cocultured cells (bottom of the well) to a metabolite (M₁). M₁ is cytotoxic to proximal tubular cells after M₁ is accumulated by transport across the basolateral membrane. Alternatively, M₁ may be further metabolized to M₂, which may be excreted into the medium. Reproduced from Boogaard et al. [88] and Elsevier Scientific Publishers with permission.

dishes did not acidify the incubation medium [121]. Therefore, the support system (plastic versus porous membrane) may be an important consideration when establishing cell cultures, because processes relying on accessibility of the basolateral membrane, such as organic anion accumulation or acidification, may not occur when that membrane is in close contact with the support system.

10.9.2 Advantages

The primary advantage of cell culture over other *in vitro* techniques is that cells may be exposed to toxicant for long periods of time (several days to weeks for established cell lines) at concentrations relevant to those observed *in vivo*. No other *in vitro* technique retains viability for longer than 24 hours. The ability to use pharmacologically or toxicologically relevant concentrations of a compound cannot be overemphasized. Many compounds depend on enzymatic bioactivation or detoxification to exert cytotoxicity. By definition, enzyme-mediated processes may be saturated at sufficiently high concentrations of substrate. If a minor metabolic pathway for a compound involves bioactivation by an enzyme-mediated reaction, such as occurs with acetaminophen in the liver, the traditional method of using high concentrations of substrate with renal slices, tubule suspensions, or cell suspensions will result in production of a greater amount of reactive metabolite than is found *in vivo*. Further, some toxicants may produce injury by multiple, concentration-dependent mechanisms. Exposure of tissue to high concentrations of toxicant for short-term incubations may alter the balance of mechanisms involved in cytotoxicity, such as increased bioactivation when detoxification is saturated. Thus, when investigators use short-term systems relying on high concentrations of toxicants, they should be aware of the potential for differences between *in vitro* and *in vivo* metabolism. Cell culture systems, whether established cell lines or primary cultures, avoid problems with metabolic differences by allowing cells to be exposed to pharmacologically or toxicologically relevant concentrations of toxicant over an extended period of time. In addition, cell cultures may be used for "pulsed" experiments, when toxicant is introduced for a period of time and then removed, allowing cells to fully develop injury or recovery. Although "pulsing" is possible with previously discussed *in vitro* techniques, the short lifespan of other preparations makes it more difficult to monitor expression of cytotoxicity or recovery (repair).

For established cell lines, a unique advantage is that these cells are theoretically immortal and can be reseeded and subcultured repeatedly without change in phenotype or function. In practice, however, established cell lines may undergo some changes over time, so that cell lines need to be characterized periodically by each investigator.

Primary cell cultures, on the other hand, have a finite lifetime and undergo phenotypic and functional changes (see below). Thus, a primary culture established at a different point in time may differ somewhat from a previously established culture.

However, if rigorous care is taken so that starting material and culture conditions are constant for each experiment, primary cell cultures should remain sufficiently homogeneous to allow comparison of data from different cultures.

10.9.3 Limitations

A major limitation of cell culture techniques is the inherent heterogeneity of any preparation. Although heterogeneity may be overcome by using established cell lines, these cell lines have several other limitations. In particular, LLC-PK$_1$ cells, derived from proximal tubule, lack several specific proximal tubular functions, such as organic anion and glucose transport [122]. In addition, LLC-PK$_1$ cells display some characteristics of distal tubule and TALH cells, including vasopressin- and calcitonin-sensitive adenylate cyclases [123,124], and high transepithelial resistance similar to the collecting duct [125–127]. Thus, the use of established cell lines may incorporate some difficulties in interpretation because of the ambiguous origin of those cells.

Primary cultures theoretically avoid the problems of heterogeneity, but in practice, it is nearly impossible to completely purify the starting preparation [88]. However, if culture conditions are optimized as previously discussed, and cultures are carefully and thoroughly characterized as to purity of cells of interest, considerable information may be gained by using primary cultures.

When establishing primary cell cultures, proliferating cells go through a series of dedifferentiations, reverting to a more juvenile cell type, and redifferentiations, acquiring characteristics of the origin cell line [88]. In an ultrastructural study, Koechlin et al. [128] described these processes for rabbit proximal tubular cells grown in culture. Through the first 6 days in culture, proximal tubular cells progressively dedifferentiated, losing brush border, basal interdigitations, and many intracellular organelles. From day 6 through day 20, cells underwent progressive redifferentiation, acquiring microvilli, forming basal interdigitations, and replacing intracellular organelles. However, redifferentiating cells also acquired glycogen granules not seen in proximal tubular cells *in situ*, which Koechlin et al. suggested might be caused by overloading the culture medium with glucose. In addition, the apical microvilli of cells in culture were not as prevalent or well developed as microvilli of cells *in situ*, basal interdigitations remained poorly developed, and mitochondria were less numerous in cultured cells compared with proximal tubular cells *in situ*. From day 20 through 39, cells in culture became overloaded with glycogen granules and other cellular debris, suggesting that cells were undergoing degenerative processes [128]. These data suggest that primary cell cultures do not retain uniform phenotypical characteristics throughout the life of the culture, and timing of experiments relative to days in culture may be a critical variable in studies using cell culture. In addition, cells in culture do not exactly duplicate the appearance of cells *in situ*. The investigator must define the extent to which cells in culture resemble cells *in situ* functionally.

In cultured hepatocytes, investigators have observed a rapid and irreversible decline in cytochrome P450- and glucuronyl transferase-dependent metabolism [129–131]. In a recent investigation, Bruggeman et al. observed that cytochrome P450 was undetectable in microsomes prepared from 3- to 5-day-old cultured proximal tubular cells [132]. Further, GSH concentrations and GSH S-transferase contents (determined by measuring GSH S-transferase subunits by high-performance liquid chromatography) were lower in proximal tubular cell cultured from 1 day through 5 days than they were in freshly isolated proximal tubular cells [132]. Thus, it may be that proximal tubular cells in culture undergo dedifferentiation with loss of enzymes catalyzing metabolism of xenobiotics, similar to the process observed in hepatocytes. To some extent, loss of enzyme activity in cultured proximal tubular cells may be circumvented by including a renal S9 fraction [133]. In addition, investigators are exploring culture systems designed to preserve enzyme activities in cultured hepatocytes [134], and these techniques could easily be extended to cultured renal tubular cells.

Cell culture is a difficult and expensive technique. Considerable equipment is needed to support cell cultures, including sterile facilities for isolating, seeding, and replating cells, incubators for maintaining cells during proliferation, and miscellaneous supplies, solutions, and chemicals, all of which make experiments rather costly. Unfortunately, it is easy to lose a culture to bacterial contamination or fibroblast overgrowth, so that rigorous control of sterility and culture conditions is essential. In addition, a relatively small amount of tissue is contained in a monolayer, so that sensitive methodology is required to detect cytotoxicity [88].

10.10 USE OF CELL CULTURES IN RENAL TOXICOLOGY

10.10.1 Biochemical Mechanisms of Lethal Cell Injury: Halogenated Hydrocarbons

Cysteine conjugates of halogenated hydrocarbons, such as DCVC and DCVG, have been extensively investigated in numerous *in vitro* systems, as previously discussed. However, the intracellular events leading to cell death have been difficult to identify. Studies in proximal tubular cell suspensions indicated that loss of mitochondrial integrity or impaired ability to sequester calcium contributed to DCVC-induced cytotoxicity.

The role of altered calcium distribution in DCVC cytotoxicity was further investigated using cultured LLC-PK$_1$ cells. Intracellular calcium concentration and localization were determined using Fura-2, a calcium-sensitive fluorescent dye. In control cells, intracellular calcium was localized primarily in mitochondria, as shown by rhodamine-123 staining. In contrast, incubation of LLC-PK$_1$ cells for 24 hours with 10^{-4}-M DCVC increased intracellular calcium concentration approximately four-fold while greatly reducing mitochondrial calcium concentration

[135]. Mitochondria remained viable despite calcium loss, as indicated by rho-damine-123 staining properties [135]. Longer incubations with DCVC (up to 96 hours) caused further increases in intracellular calcium concentration, as well as plasma membrane blebbing. Membrane blebs contained high calcium concentrations, whereas calcium was virtually absent from mitochondria. In addition, with longer incubations, mitochondria became nonviable and could no longer be detected by rhodamine-123 staining [135]. Thus, these studies demonstrated that DCVC produced major alterations in intracellular calcium concentration and distribution. Depletion of mitochondrial calcium preceded bleb formation and cell death. In addition, loss of mitochondrial calcium preceded collapse of mitochondrial membrane potential [135]. Vamvakas et al. suggested that calcium release from mitochondria before loss of mitochondrial membrane potential might be related to oxidation and hydrolysis of mitochondrial pyridine nucleotides and ADP-ribosylation of mitochondrial membrane proteins.

Chen et al. tested a variety of cytoprotective agents for their ability to prevent DCVC-induced cytotoxicity in LLC-PK$_1$ cells [136]. Inhibitors of proteolysis (leupeptin, antipain, methylamine), phospholipase (dibucaine, p-bromophenylacyl bromide), calcium channel blockers (nifedipine, verapamil), or a calmodulin antagonist (calmidazolium) were ineffective in preventing LDH leakage after incubation of LLC-PK$_1$ cells with DCVC [136]. In contrast, both AOAA and DPPD were extremely effective in ameliorating DCVC-induced cytotoxicity [136]. AOAA was included as a positive control, because AOAA competitively inhibits cysteine conjugate 2-lyase, the enzyme responsible for bioactivation of DCVC (see above). Several other antioxidants were tested for protective effects against DCVC: Hydrophilic antioxidants (uric acid, ascorbic acid) were ineffective, whereas lipophilic antioxidants (butylated hydroxyanisole, propyl galate, butylated hydroxytoluene, and butylated hydroxyquinone) were effective inhibitors of DCVC toxicity [136]. In contrast to AOAA, which inhibits both DCVC bioactivation and covalent binding, DPPD prevented DCVC-induced cytotoxicity without altering covalent binding of ^{35}S derived from DCVC. Thus, DCVC cytotoxicity may be dissociated from covalent binding of a reactive intermediate of DCVC.

The inability to demonstrate protection against DCVC cytotoxicity by calcium channel blockers (nifedipine, verapamil) observed by Chen et al. [136] appears to be in direct contrast to calcium-dependent DCVC cytotoxicity proposed by Vamvakas et al. [135]. However, Chen et al. noted a biphasic protective effect of DPPD that depended on the concentration of DCVC in the incubation medium. Specifically, DPPD was effective in blocking toxicity at low concentrations of DCVC (25–50 µM) but ineffective at higher DCVC concentrations (250–500 µM) [136]. Thus, the mechanism(s) of DCVC cytotoxicity is(are) concentration dependent. Specifically, lipid peroxidation or oxidative stress may be involved in DCVC cytotoxicity at low DCVC concentrations, whereas antioxidant-insensitive pathways may be involved at higher DCVC concentrations [136]. Concentration-dependent toxicity in cultured or freshly prepared hepatocytes has been described for several compounds, including acetaminophen [137] and carbon tetrachloride [138]. Thus, toxicant concentration may be a critical determinant of the mechanism of cytotoxicity.

The ability of DPPD and other antioxidants to protect cells from DCVC cytotoxicity suggested that lipid peroxidation was a factor in DCVC-induced cell injury. Indeed, TBA-reactive products, indicative of lipid peroxidation, were detectable in LLC-PK$_1$ cells incubated with DCVC before the onset of LDH leakage, suggesting that lipid peroxidation was occurring before, rather than coincident with, cell death. Coincubation of cells with DCVC and DPPD prevented both formation of TBA-positive material and LDH leakage, indicating protection from DCVC cytotoxicity [136]. Deferoxamine, an iron chelator, prevented formation of TBA-positive material and LDH leakage without altering covalent binding of radiolabel derived from DCVC [136], supporting a role for iron-dependent lipid peroxidation in DCVC-induced cell injury. Dithiothreitol, a thiol reducing agent, also protected against DCVC cytotoxicity without altering covalent material of radiolabel derived from DCVC [136]. Thus, DCVC-induced cell death may be caused by oxidative stress. The mechanism of initiation of DCVC-induced oxidative stress remains unclear.

10.10.2 Cellular Accumulation and Metabolism: Aminoglycoside Antibiotics

Aminoglycoside antibiotics, such as gentamicin, are used clinically to treat gram-negative bacterial infections. Nephrotoxicity is a major complication associated with aminoglycoside therapy and has been the subject of numerous investigations [139,140]. However, gentamicin nephrotoxicity is relatively slow to develop. In rats, repeated administration of gentamicin is necessary to produce renal injury [139,140]. Gentamicin nephrotoxicity *in vivo* is characterized by multiple and diffuse abnormalities, including accumulation of myeloid bodies in proximal tubular cells, enzymuria and proteinuria, and decreases in GFR [139,140]. Although some tubular epithelial cells may undergo necrosis and death during continuing administration of gentamicin, other cells may undergo regeneration. These regenerating cells are relatively undifferentiated and immature, and they retain normal morphology despite the presence of gentamicin [141], suggesting that immature cells are not susceptible to gentamicin toxicity.

In vitro methods have been used to investigate the mechanisms leading to gentamicin nephrotoxicity. Alterations in plasma membrane function (e.g., disturbances of phospholipid metabolism, changes in intracellular calcium concentration, alterations in membrane transport properties), alterations in mitochondrial function (e.g., uncoupling of respiration, inhibition of calcium accumulation), as well as alterations in lysosomal membrane (e.g., lysosomal destabilization, diminished phospholipase activity) have been implicated in the development of gentamicin nephrotoxicity [140]. However, the precise mechanisms involved in gentamicin-induced cell injury remain unclear.

Cell culture is an excellent system in which to investigate slowly developing toxicities, such as that caused by gentamicin. However, it should be noted that gentamicin toxicity in cell culture systems generally requires exposure to higher concentrations than do those seen *in vivo* (see below). Presumably, the higher gentamicin concentrations necessary to injure cultured cells may be because of the dedifferentiated or immature state of cultured cells, making them less susceptible to gentamicininduced cytotoxicity than are proximal tubular cells *in vivo*.

It is well established that gentamicin is accumulated within proximal tubular cells via endocytosis across the luminal membrane of proximal tubular cells [139,140]. Cultured LLC-PK$_1$ cells accumulated gentamicin via energy-dependent pathways, inhibited by rotenone and dinitrophenol [142]. Gentamicin accumulation occurred against a concentration gradient, and intracellular gentamicin concentration in LLC-PK$_1$ cells was approximately three times greater than was the gentamicin concentration in the incubation medium [142]. Further, A23187, a calcium ionophore, stimulated gentamicin accumulation, whereas EDTA, a calcium chelator, reduced gentamicin uptake [142]. In addition, gentamicin uptake was sensitive to alterations in calcium concentration of the medium. A23187-stimulated gentamicin uptake was minimal at low external calcium concentrations, whereas gentamicin uptake was maximally stimulated by A23187 at high external calcium concentrations [142]. These studies suggested that LLC-PK$_1$ cells might be a suitable *in vitro* system to investigate gentamicin nephrotoxicity because active gentamicin uptake occurred in these cells. Further, these *in vitro* studies supported the hypothesis that gentamicin accumulation occurred via an endocytotic process by demonstrating that uptake required the presence of calcium ions.

Using LLC-PK$_1$ cells, Schwertz et al. investigated morphologic and functional alterations after incubation with gentamicin [143]. Cells were incubated with concentrations of gentamicin ranging from 0.1 to 2 mM, considerably higher than was the desired therapeutic range of 2–10 µg/ml [139,140], for up to 7 days. Within 4 days of incubation with gentamicin, LLC-PK$_1$ cells developed myeloid bodies characteristic of gentamicin accumulation [143]. However, cell death, indicated by nigrosin permeability, required incubation for at least 7 days with gentamicin, and even then, some cells remained viable despite continued exposure to gentamicin [143]. Incubation of cells for 4 days with up to 2-mM gentamicin did not alter DNA, RNA, protein, or ATP content in LLC-PK$_1$ cells. In contrast, total phospholipid content increased in LLC-PK$_1$ cells incubated with gentamicin, with increased concentrations of phosphatidylinositol and phosphatidylcholine [143]. Gentamicin also altered neutral lipid turnover in LLC-PK$_1$ cells, with increased incorporation of label from [^{3}H]arachidonic acid or [14C]acetate into free fatty acids, monoglyceride, diglyceride, and nonesterified cholesterol, and reduced labeling of triglycerides in cells incubated with gentamicin [143]. Whereas gentamicin might require calcium ions to enter proximal tubular cells, gentamicin-induced alterations in phospholipid metabolism were not dependent on calcium concentration in the incubation medium: Raising calcium concentration in the incubation medium from 0.2 through 0.6 mg/ml did not alter the gentamicin-induced stimulation of phosphatidylinositol and phosphatidylcholine in LLC-PK$_1$ cells [143]. Thus, in LLC-PK$_1$ cells, gentamicin was accumulated and caused phospholipidosis similar to that observed *in vitro*. However, a correlation between phospholipidosis and lethal cell injury could not be established in these studies.

In primary cultures of rabbit proximal tubular cells, incubation with 10–3-M gentamicin for up to 6 days failed to alter cell viability, whereas by day 12 of incubation, cell viability was reduced by about 35% [144]. In addition, gentamicin failed to consistently alter protein and DNA content, or protein and DNA synthesis (measured as [^{14}C]leucine or [^{3}H]thymidine incorporation), until day 12 of incubation [144]. In contrast, gentamicin produced a marked phospholipidosis in cultured rabbit proximal tubular cells as early as day 2 of incubation (Fig. 10.10) [144]. All phospholipids examined, including phosphatidylcholine, phosphatidylserine, and sphingomyelin, were significantly increased after exposure of cells to gentamicin for 6 days (Fig. 10.11) [144]. These studies indicate that one of the earliest functional changes associated with gentamicin toxicity was phospholipidosis affecting all major phospholipids. Gentamicin-induced phospholipidosis was caused, at least in part, by reduced degradation of phospholipids, as indicated by slower turnover of prelabeled phospholipids in cultured cells [144]. In addi-

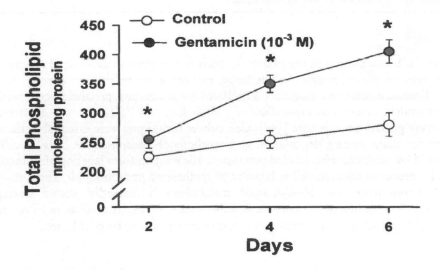

FIG. 10.10. Effect of 10^{-3}-M gentamicin on total phospholipid of cultured rabbit proximal tubular cells as a function of drug exposure. Asterisks indicate significant difference between control and gentamicin groups, $p < 0.01$. Reproduced from Ramsammy et al. [144] and the American Physiological Society with permission.

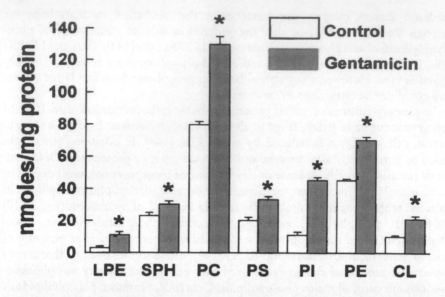

FIG. 10.11. Effect of 10^{-3}-M gentamicin on phospholipid composition of cultured rabbit proximal tubular cells exposed to drug for 6 days. Asterisks indicate significant differences between control and gentamicin groups, * = $p < 0.05$; ** = $p < 0.01$. LPE, lysophosphatidylethanolamine; SPH, sphingomyelin; PC, phosphatidylcholine; PS, phosphatidylserine; PI, phosphatidylinositol; PE, phosphatidylethanolamine; CL, cardiolipin. Reproduced from Ramsammy et al. [144] and the American Physiological Society with permission.

tion, at least a portion of the phosoplipidosis is secondary to more rapid synthesis of selected phospholipids. In particular, incorporation of [3H]myoinositol and [3H]ethanolamine was markedly stimulated by incubating prelabeled cells with gentamicin, whereas incorporation of [3H]choline and [3H]serine was unaltered during gentamicin exposure [144]. Cell culture techniques were essential in these studies characterizing degradation and synthesis of phospholipids because cells could be incubated with labeled precursors, allowing uniform labeling of intracellular precursor pools as well as labeling of synthesized phospholipids. These studies demonstrate that phospholipid metabolism is markedly altered during exposure to gentamicin. However, whether these alterations contribute to or are simply coincident with gentamicin cytotoxicity remains to be established.

10.11 CONCLUDING REMARKS

Numerous *in vitro* techniques have been used to elucidate biochemical and physiologic mechanisms involved in nephrotoxicity. In this chapter, few of the more commonly used *in vitro* methods have been reviewed, emphasizing the distinct advantages and limitations of each method. In addition, studies in which each method has been used have been discussed, with emphasis on the information gained by using *in vitro* or *ex vivo* techniques.

Some *in vitro* techniques are uniquely applicable to study renal function and physiology, such as the IPK, renal slices, and tubule suspensions. Some *in vitro* techniques used in evaluation of nephrotoxicity are more universal and require only a few adaptations or special considerations, such as cell suspensions and cell cultures.

Considerable progress has been made over the last 10–20 years in all aspects of *in vitro* assessment of nephrotoxicity. As in all areas, considerably more research is needed to expand and extend on the fundamental information obtained thus far. In particular, more studies examining *in vivo–in vitro* or *in vivo–ex vitro* correlations in nephrotoxicity are highly desirable. Recently, investigators have recognized the importance of preserving metabolic capabilities, such as cytochromes P450, in cell culture systems, and are devising methods and strategies making cell culture systems resemble more closely the *in vivo* biochemical systems originally present in the cells. Such efforts will greatly extend the utility of cell culture systems in evaluation of nephrotoxicity, as well as in studies aimed at delineating mechanisms of cytotoxicity.

References

1. Newton JF, Hook JB. Isolated perfused rat kidney. In: Jakoby WB, ed. *Methods in Enzymology*, Vol. 77. New York: Academic Press, 1981:94–105.

2. Bekersky I. Use of the isolated perfused kidney as a tool in drug disposition studies. *Drug Metab Rev* 1983;14:931–960.

3. Cox PGF, Moons MM, Slegers JFG, Russel FGM, van Ginneken CAM. Isolated perfused rat kidney as a tool in the investigation of renal handling and effects of nonsteroidal anti-inflammatory drugs. *J Pharmacol Methods* 1990;24:89–103.

4. Ross BD, Epstein FH, Leaf A. Sodium reabsorption in the perfused rat kidney. *Am J Physiol* 1973;225:1,165–1,171.

5. Ross BD. The isolated perfused rat kidney. *Clin Sci Mol Med* 1978;55:513–521.

6. Alcorn D, Emslie KR, Ross BD, Ryan GB, Tange JD. Selective distal nephron damage during isolated kidney perfusion. *Kidney Int* 1981;19:638–647.

7. Schurek HJ, Kriz W. Morphologic and functional evidence for oxygen deficiency in the isolated perfused rat kidney. *Lab Invest* 1985;53:145–155.

8. Brezis M, Rosen S, Silva P, Epstein FH. Selective vulnerability of the medullary thick ascending limb to anoxia in the isolated perfused rat kidney. *J Clin Invest* 1984;73:182–190.

9. Lieberthal W, Stephens GW, Wolf EF, Rennke HG, Vasilevsky ML, Valeri CR, Levinsky NG. Effect of erythrocytes on the function and morphology of the isolated perfused rat kidney. *Renal Physiol* 1987;10:14–24.

10. Brezis M, Rosen S, Spokes K, Silva P, Epstein FH. Transport-dependent anoxic cell injury in the isolated perfused rat kidney. *Am J Pathol* 1984;116:327–341.

11. Tarloff JB, Goldstein RS, Hook JB. Xenobiotic biotransformation by the kidney: pharmacological and toxicological aspects. *Prog Drug Metab* 1990;12:1–39.

12. Goldstein RS, Schnellmann RG. Toxic responses of the kidney. In: Klaassen CD, ed. *Casarett and Doull's Toxicology: The Basic Science of Poisons*, 5th ed. New York: McGraw-Hill, 1996:417–442.

13. Ross BD, Tange J, Emslie K, Hart S, Smail M, Calder I. Paracetamol metabolism by the isolated perfused rat kidney. *Kidney Int* 1980;18:562–570.

14. Mitchell JR, Jollow DJ, Potter WZ, David DC, Gillette JR, Brodie BB. Acetaminophen-induced hepatic necrosis. I. Role of drug metabolism. *J Pharmacol Exp Therapeut* 1973;187:185–194.

15. Hinson JA, Pohl LR, Monks TJ, Gillette JR. Acetaminophen-induced hepatotoxicity. *Life Sci* 1981;29:107–116.

16. McMurtry RJ, Snodgrass WR, Mitchell JR. Renal necrosis, glutathione depletion, and covalent binding after acetaminophen. *Toxicol Appl Pharmacol* 1978;46:87–100.

17. Newton JF, Yoshimoto M, Bernstein J, Rush GF, Hook JB. Acetaminophen nephrotoxicity in the rat. I. Strain differences in nephrotoxicity and metabolism. *Toxicol Appl Pharmacol* 1983;69:291–306.

18. Newton JF, Bailie MB, Hook JB. Acetaminophen nephrotoxicity in the rat. Renal metabolic activation *in vitro*. *Toxicol Appl Pharmacol* 1983;70:433–444.

19. Newton JF, Braselton WE, Kuo CH, Kluwe WM, Gemborys MW, Mudge GH, Hook JB. Metabolism of acetaminophen by the isolated perfused kidney. *J Pharmacol Exp Therapeut* 1982;221:76–79.

20. Trumper L, Monasterolo LA, Elias MM. Probenecid protects against *in vivo* acetaminophen-induced nephrotoxicity in male Wistar rats. *J Pharmacol Exp Therapeut* 1998;284:606–610.

21. Gerson RJ, Casini A, Gilfor D, Serroni A, Farber JL. Oxygen-mediated cell injury in the killing of cultured hepatocytes by acetaminophen. *Biochem Biophys Res Commun* 1985;126:1,129–1,137.

22. Adamson GM, Harman AW. A role for the glutathione peroxidase/reductase enzyme system in the protection from paracetamol toxicity in isolated mouse hepatocytes. *Biochem Pharmacol* 1989;38:3,323–3,330.

23. Madias NE, Harrington JT. Platinum nephrotoxicity. *Am J Med* 1978;65:307–314.

24. Blachley JD, Hill JB. Renal and electrolyte disturbances associated with cisplatin. *Ann Intern Med* 1981;95:628–632.

25. Meijer S, Sleijfer DT, Mulder NH, Sluiter WJ, Marrink J, Koops HS, Brouwers TM, Oldhoff J, van der Hem GK, Mandema E. Some effects of combination chemotherapy with cis-platinum on renal function in patients with nonseminomatous testicular carcinoma. *Cancer* 1983;51:2,035–2,040.

26. Miura K, Goldstein RS, Pasino DA, Hook JB. Cisplatin nephrotoxicity: role of filtration and tubular transport of cisplatin in isolated perfused kidneys. *Toxicology* 1987;44:147–158.

27. Berndt WO, Davis ME. Renal methods for toxicology. In: Hayes AW, ed. *Principles and Methods of Toxicology*, 2nd ed. New York: Raven Press, 1989:629–648.

28. Goldstein RS, Hook JB, Bond JT. Renal tubular transport of saccharin. *Biochem Pharmacol* 1978;204:690–695.

29. Mann WA, Welzel GE, Goldstein RS, Sozio RS, Cyronak MJ, Kao J, Kinter LB. Characterization of the renal effects and renal elimination of sulotroban in the dog. *Biochem Pharmacol* 1991;259:1,231–1,240.

30. Cacini W, Keller MB, Grund VR. Accumulation of cimetidine by kidney cortex slices. *Biochem Pharmacol* 1982;221:342–346.

31. Cross RJ, Taggart VJ. Renal tubular transport: Accumulation of *p*-aminohippurate by rabbit kidney slices. *Biochem Pharmacol* 1950;161:181–190.

32. Goldstein RS, Pasino DA, Hewitt WR, Hook JB. Biochemical mechanisms of cephaloridine nephrotoxicity: time and concentration dependence of peroxidative injury. *Toxicol Appl Pharmacol* 1986;83:261–270.

33. Smith JH. The use of renal cortical slices from the Fischer 344 rat as an *in vitro* model to evaluate nephrotoxicity. *Fund Appl Toxicol* 1988;11:132–142.

34. Smith JH, Hewitt WR, Hook JB. Role of intrarenal biotransformation in chloroform-induced nephrotoxicity in rats. *Toxicol Appl Pharmacol* 1985;79:166–174.

35. Ruegg CE, Gandolfi AJ, Nagle RB, Krumdieck CL, Brendel KP. Preparation of positional renal slices for study of cell-specific toxicity. *J Pharmacol Methods* 1987;17:111–123.

36. Blackmore M, Richardson JC, Rhodes SA, Patterson L, Spencer AJ, Gray TJB. Rat renal cortical slices: maintenance of viability and use in *in vitro* nephrotoxicity testing. *Toxicol In Vitro* 1997;11:723–729.

37. Kluwe WM. Renal function tests as indicators of kidney injury in subacute toxicity studies. *Toxicol Appl Pharmacol* 1981;57:414–424.

38. Kyle GM, Luthra R, Bruckner JV, MacKenzie WF, Acosta D. Assessment of functional, morphological, and enzymatic tests for acute nephrotoxicity induced by mercuric chloride. *J Toxicol Environ Health* 1983;12:99–117.

39. Miyajima H, Hewitt WR, Cote MG, Plaa GL. Relationships between histological and functional indices of acute chemically induced nephrotoxicity. *Fund Appl Toxicol* 1983;3:543–551.

40. Tarloff JB, Goldstein RS, Silver AC, Hewitt WR, Hook JB. Intrinsic susceptibility of the kidney to acetaminophen toxicity in middle-aged rats. *Toxicol Lett* 1990;52:101–110.

41. Atkinson RM, Currie JP, Davis B, Pratt DAH, Sharpe HM, Tonich EG. Acute toxicity of cephaloridine, an antibiotic derived from cephalosporin C. *Toxicol Appl Pharmacol* 1966;8:398–406.

42. Wold JS. Antibiotic nephropathies. In: Hook JB, ed. *Toxicology of the Kidney*. New York: Raven Press, 1981:251–266.

43. Kuo CH, Hook JB. Depletion of renal glutathione content and nephrotoxicity of cephaloridine in rabbits, rats and mice. *Toxicol Appl Pharmacol* 1982;63:292–302.

44. Tune BM. Relationship between the transport and toxicity of cephalosporins in the kidney. *J Infect Dis* 1975;122:33–44.

45. Tune BM. Effect of organic acid transport inhibitors on renal cortical uptake and proximal tubular toxicity of cephaloridine. *J Pharmacol Exp Therapeut* 1972;181:250–256.

46. Wold JS, Turnipseed SA. The effect of renal cation transport inhibitors on the *in vivo* and *in vitro* accumulation and efflux of cephaloridine. *Life Sci* 1980;27:2,559–2,564.

47. Kuo CH, Maita K, Sleight SD, Hook JB. Lipid peroxidation: a possible mechanism of cephaloridine-induced nephrotoxicity. *Toxicol Appl Pharmacol* 1983;67:78–88.

48. Cojocel C, Laeschke KH, Inselmann G, Baumann K. Inhibition of cephaloridine-induced lipid peroxidation. *Toxicology* 1985;35:295–305.

49. Goldstein RS, Contardi LR, Pasino DA, Hook JB. Mechanisms mediating cephaloridine inhibition of renal gluconeogenesis. *Toxicol Appl Pharmacol* 1987;87:297–305.

50. Ruegg CE, Gandolfi AJ, Brendel K. Differential patterns of injury to the proximal tubule of renal cortical slices following *in vitro* exposure to mercuric chloride, potassium dichromate or hypoxic conditions. *Toxicol Appl Pharmacol* 1987;90:261–273.

51. Phelps JS, Gandolfi AJ, Brendel K, Dorr RT. Cisplatin nephrotoxicity: *in vitro* studies with precision-cut rabbit renal cortical slices. *Toxicol Appl Pharmacol* 1987;90:501–512.

52. Wolfgang GHI, Gandolfi AJ, Brendel K. Evaluation of organic nephrotoxins in rabbit renal cortical slices. *Toxicol In Vitro* 1989;31:341–350.

53. Wolfgang GHI, Gandolfi AJ, Nagle RB, Brendel K, Stevens JL. Assessment of S-(1,2-dichlorovinyl)-L-cysteine induced toxic events in rabbit renal cortical slices. Biochemical and histological evaluation of uptake, covalent binding and toxicity. *Chem-Biol Interact* 1990;75:153–170.

54. Ruegg CE, Mandel LJ. Bulk isolation of renal PCT and PST. II. Differential responses to anoxia and hypoxia. *Am J Physiol* 1990;259:F176–F185.

55. Porter KE, Dawson AG. Inhibition of respiration and gluconeogenesis by paracetamol in rat kidney preparations. *Biochem Pharmacol* 1979;28:3,057–3,062.

56. Cojocel C, Maita K, Pasino DA, Kuo C-H, Hook JB. Metabolic heterogeneity of the proximal and distal kidney tubules. *Life Sci* 1983;33:855–861.

57. Bastin J, Cambon N, Thompson M, Lowry OH, Burch HB. Change in energy reserves in different segments of the nephron during brief ischemia. *Kidney Int* 1987;31:1,239–1,247.

58. Uchida S, Endou H. Substrate specificity to maintain cellular ATP along the mouse nephron. *Am J Physiol* 1988;255:F977–F983.

59. Aleo MD, Rankin GO, Cross TJ, Schnellmann RG. Toxicity of *N*-(3,5-dichlorophenyl) succinimide and metabolites to rat renal proximal tubules and mitochondria. *Chem-Biol Interact* 1991;78:109–121.

60. Aleo MD, Wyatt RD, Schnellmann RG. The role of altered mitochondrial function in citrinin-induced toxicity to rat renal proximal tubule suspensions. *Toxicol Appl Pharmacol* 1991;109:455–463.

61. Shao R, Tarloff JB. Lack of correlation between *para*-aminophenol toxicity *in vivo* and *in vitro* in female Sprague-Dawley rats. *Fundam Appl Toxicol* 1996;31:268–278.

62. Schnellmann RG, Mandel LJ. Cellular toxicity of bromobenzene and bromobenzene metabolites to rabbit proximal tubules: the role and mechanism of 2-bromohydro-quinone. *J Pharmacol Exp Therapeut* 1986;237:456–461.

63. Schnellmann RG, Mandel LJ. Multiple effects of presumed glutathione depletors on rabbit renal proximal tubules. *Kidney Int* 1986;29:858–862.

64. Sina JF, Noble C, Bean CL, Bradley MO. Renal tubules *in vitro* as a model for nephrotoxicity. In: McQueen C, ed. *In Vitro Toxicology: Model Systems and Methods.* Caldwell, NJ: Telford Press, 1989:263–290.

65. Rush GF, Ponsler GD. Cephaloridine-induced biochemical changes and cytotoxicity in suspensions of rabbit isolated proximal tubules. *Toxicol Appl Pharmacol* 1991;109:314–326.

66. Balaban RS, Soltoff S, Storey JM, Mandel LJ. Improved renal cortical tubule suspension: spectrophotometric study of O_2 delivery. *Am J Physiol* 1980;238:F50–F59.

67. Vinay P, Gougoux A, Lemieux G. Isolation of a pure suspension of rat proximal tubules. *Am J Physiol* 1981;241:F403–F411.

68. Sina JF, Bean CL, Noble C, Bradley MO. Isolation and characterization of proximal tubule suspensions from rabbits for *in vitro* studies. *In Vitro Toxicol* 1986;1:5–12.

69. Sina JF, Bean CL, Bland JA, MacDonald JS, Noble C, Robertson RT, Bradley MO. An *in vitro* assay for cytotoxicity to proximal tubule suspensions from rabbit kidney. *In Vitro Toxicol* 1986;1:13–22.

70. Gesek FA, Wolff DW, Strandhoy JW. Improved separation method for rat proximal and distal renal tubules. *Am J Physiol* 1987;253:F358–F365.

71. Beach RE, Watts BA, Good DW, Benedict CR, DuBose TD. Effects of graded oxygen tension on adenosine release by renal medullary and thick ascending limb suspensions. *Kidney Int* 1991;39:836–842.

72. Gaudio KM, Thulin G, Ardito T, Kashgarian M, Siegel NJ. Metabolic alterations in proximal tubule suspensions obtained from ischemic kidneys. *Am J Physiol* 1989;257:F383–F389.

73. Venkatachalam MA, Bernard DB, Donohoe JF, Levinsky NG. Ischemic damage and repair in the rat proximal tubule: differences among the S1, S2, and S3 segments. *Kidney Int* 1978;14:31–49.

74. Shanley PF, Rosen MD, Brezis M, Silva P, Epstein FH, Rosen S. Topography of focal

proximal tubular necrosis after ischemia with reflow in the rat kidney. *Am J Pathol* 1986;122:462–468.

75. Newton JF, Yoshimoto M, Bernstein J, Rush GF, Hook JB. Acetaminophen nephrotoxicity in the rat. II. Strain differences in nephrotoxicity and metabolism of *p*-aminophenol, a metabolite of acetaminophen. *Toxicol Appl Pharmacol* 1983;69:307–318.

76. Tarloff JB, Goldstein RS, Morgan DG, Hook JB. Acetaminophen and *p*-aminophenol nephrotoxicity in aging male Sprague-Dawley and Fischer 344 rats. *Fundam Appl Toxicol* 1989;12:78–91.

77. Gartland KPR, Eason CT, Bonner FW, Nicholson JK. Effects of biliary cannulation and buthionine sulphoximine pretreatment on the nephrotoxicity of para-aminophenol in the Fisher 344 rat. *Arch Toxicol* 1990;64:14–25.

78. Klos C, Koob M, Kramer C, Dekant W. *p*-Aminophenol nephrotoxicity: biosynthesis of toxic glutathione conjugates. *Toxicol Appl Pharmacol* 1992;115:98–106.

79. Fowler LM, Moore RB, Foster JR, Lock EA. Nephrotoxicity of 4-aminophenol glutathione conjugates. *Human Exp Toxicol* 1991;10:451–459.

80. Fowler LM, Foster JR, Lock EA. Nephrotoxicity of 4-amino-3-*S*-glutathionylphenol and its modulation by metabolism or transport inhibitors. *Arch Toxicol* 1994;68:15–23.

81. Anthony ML, Beddell CR, Lindon JC, Nicholson JK. Studies on the effects of L(áS,5S)-á-amino-3-chloro-4,5-dihydro-5-isoxazoleacetic acid (AT-125) on 4-aminophenol-induced nephrotoxicity in the Fischer 344 rat. *Arch Toxicol* 1993;67:696–705.

82. Lau SS, Monks TJ, Gillette JR. Identification of 2-bromohydroquinone as a metabolite of bromobenzene and *o*-bromophenol: implications for bromobenzene-induced nephrotoxicity. *J Pharmacol Exp Therapeut* 1984;230:360–366.

83. Lau SS, Monks TJ, Greene KE, Gillette JR. The role of *ortho*-bromophenol in the nephrotoxicity of bromobenzene in rats. *Toxicol Appl Pharmacol* 1984;72:539–549.

84. Shao R, Ring SC, Tarloff JB. Coincubation of rat renal proximal tubules with hepatic subcellular fractions potentiates the effects of *para*-aminophenol. *Fundam Appl Toxicol* 1997;39:101–108.

85. Lock EA, Cross TJ, Schnellmann RG. Studies on the mechanism of 4-aminophenol-induced toxicity to renal proximal tubules. *Human Exp Toxicol* 1993;12:383–388.

86. Kinne R. New approaches to study renal metabolism: isolated single cells. *Min Elect Metab* 1983;9:270–275.

87. Ormstad K, Orrenius S, Jones DP. Preparation and characteristics of isolated kidney cells. In: Jakoby WB, ed. *Methods in Enzymology*, Vol. 77. New York: Academic Press, 1981:137–146.

88. Boogaard PJ, Nagelkerke JF, Mulder GJ. Renal proximal tubular cells in suspension or in primary culture as in vitro models to study nephrotoxicity. *Chem-Biol Interact* 1990;76:251–292.

89. Eveloff J, Haase W, Kinne R. Separation of renal medullary cells: isolation of cells from the thick ascending limb of Henle's loop. *J Cell Biol* 1980;87:672–681.

90. Lash LH, Tokarz JJ. Isolation of two distinct populations of cells from rat kidney cortex and their use in the study of chemical-induced toxicity. *Anal Biochem* 1989;182:271–279.

91. Boogaard PJ, Mulder GJ, Nagelkerke JF. Isolated proximal tubular cells from rat kidney as an *in vitro* model for studies on nephrotoxicity. I. An improved method for preparation of proximal tubular cells and their functional characterization by 4-methylglucose uptake. *Toxicol Appl Pharmacol* 1989;101:135–143.

92. Lash LH, Tokarz JJ. Oxidative stress in isolated rat renal proximal and distal tubular cells. *Am J Physiol* 1990;259:F338–F347.

93. Lash LH, Woods EB. Cytotoxicity of alkylating agents in isolated rat kidney proximal tubular and distal tubular cells. *Arch Biochem Biophys* 1991;286:46–56.

94. Tune BM, Fravert D. Mechanisms of cephalosporin nephrotoxicity: a comparison of cephaloridine and cephaloglycin. *Kidney Int* 1980;18:591–600.

95. Reed DJ. Regulation of reductive processes by glutathione. *Biochem Pharmacol* 1986;35:7–13.

96. Lash LH, Anders MW, Jones DP. Glutathione homeostasis and glutathione-S-conjugate toxicity in the kidney. *Rev Biochem Toxicol* 1988;9:29–67.

97. Streeter AJ, Nims RW, Sheffels PR, Heur YH, Yang CS, Mico BA, Gombar CT, Keeler LK. Metabolic denitrosation of *N*-nitrosodimethylamine *in vivo* in the rat. *Cancer Res* 1990;50:1,144–1,150.

98. Terracini B, Parker VH. A pathological study on the toxicity of S-dichlorovinyl-L-cysteine. *Food Cosmet Toxicol* 1965;3:67–74.

99. Gandolfi AJ, Nagle RB, Soltis JJ, Plescia FH. Nephrotoxicity of halogenated vinyl cysteine compounds. *Res Commun Chem Pathol Pharmacol* 1981;33:249–261.

100. Elfarra AA, Anders MW. Renal processing of glutathione conjugates. Role in nephrotoxicity. *Biochem Pharmacol* 1984;33:3,729–3,732.

101. Elfarra AA, Lash LH, Anders MW. Metabolic activation and detoxication of nephrotoxic cysteine and homocysteine-S-conjugates. *Proc Natl Acad Sci U S A* 1986;83:2,667–2,671.

102. Jones DP, Moldeus P, Stead AH, Ormstad K, Jornvall H, Orrenius S. Metabolism of glutathione and a glutathione conjugate by isolated kidney cells. *J Biol Chem* 1979;254:2,787–2,792.

103. Hassall CD, Gandolfi AJ, Duhamel RC, Brendel K. The formation and biotransformation of cysteine conjugates of halogenated ethylenes by rabbit renal tubules. *Chem-Biol Interact* 1984;49:283–297.

104. Elfarra AA, Jakobson I, Anders MW. Mechanism of S-(1,2-dichlorovinyl)glutathione-induced nephrotoxicity. *Biochem Pharmacol* 1986;35:283–288.

105. Parker VH. A biochemical study of the toxicity of S-dichlorovinyl-L-cysteine. *Food Cosmet Toxicol* 1965;3:75–87.

106. Stonard MD, Parker VH. The metabolism of S-(1,2-dichlorovinyl)-L-cysteine by rat liver mitochondria. *Biochem Pharmacol* 1971;20:2,429–2,437.

107. Lash LH, Anders MW. Cytotoxicity of S-(1,2-dichlorovinyl)glutathione and S-(1,2-dichlorovinyl)-L-cysteine in isolated rat kidney cells. *J Biol Chem* 1986;261:13,076–13,081.

108. Mandel LJ, Murphy E. Regulation of cytosolic free calcium in rabbit proximal renal tubules. *J Biol Chem* 1984;259:11,188–11,196.

109. Murphy E, Mandel LJ. Cytosolic free calcium levels in rabbit proximal kidney tubules. *Am J Physiol* 1982;242:C124–C128.

110. Gaush CR, Hard WL, Smith TF. Characterization of an established line of canine kidney cells (MDCK). *Proc Soc Exp Biol Med* 1966;122:931–935.

111. Cereijido M, Robbins ES, Dolan WJ, Rotunno CA, Sabatini DD. Polarized monolayers formed by epithelial cells on a permeable and translucent support. *J Cell Biol* 1978;77:853–880.

112. Rindler MJ, Chuman LM, Shaffer L, Saier MH. Retention of differentiated properties in an established dog kidney epithelial cell line (MDCK). *J Cell Biol* 1979;81:635–648.

113. Hull RN, Cherry WR, Weaver GW. The origin and characteristics of a pig kidney cell strain, LLC-PK$_1$. *In Vitro* 1976;12:670–677.

114. Handler JS, Perkins FM, Johnson JP. Studies of renal cell function using cell culture techniques. *Am J Physiol* 1980;238:F1–F9.

115. Malstrom K, Stange G, Murer H. Identification of proximal tubular transport functions in the established kidney cell line, OK. *Biochim Biophys Acta* 1987;902:269–277.

116. Van den Bosh L, DeSmedt H, Borghgraef R. Characteristics of Na$^+$-dependent hexose transport in OK, an established renal epithelial cell line. *Biochim Biophys Acta* 1989;979:91–98.

117. Boogaard PJ, Zoeteweij JP, van Berkel TJC, van't Noordende JM, Mulder GJ, Nagelkerke JF. Primary culture of proximal tubular cells from normal rat kidney as an *in vitro* model to study mechanisms of nephrotoxicity. Toxicity of nephrotoxicants at low concentrations during prolonged exposure. *Biochem Pharmacol* 1990;39:1,335–1,345.

118. Leffert H, Paul D. Serum dependent growth of primary cultured differentiated fetal rat hepatocytes in arginine-deficient medium. *J Cell Physiol* 1973;81:113–124.

119. Gilbert SF, Migeon BR. D-Valine as a selective agent for normal human and rodent epithelial cells in culture. *Cell* 1975;5:11–17.

120. Elliget KA, Trump BF. Primary cultures of normal rat kidney proximal tubule epithelial cells for studies of renal cell injury. *In Vitro Cell Dev Biol* 1991;27A:739–748.

121. Ford SM, Williams PD, Grassl S, Holohan PD. Transepithelial acidification by cultures of rabbit proximal tubules grown on filters. *Am J Physiol* 1990;259:C103–C109.

122. Rabito CA. Occluding junctions in a renal cell lime (LLC-PK$_1$) with characteristics of proximal tubular cells. *Am J Physiol* 1986;250:F734–F743.

123. Chabardes D, Imbert-Teboul M, Montegut M, Clique A, Morel F. Distribution of calcitonin-sensitive adenylate cyclase activity along the rabbit kidney tubule. *Proc Natl Acad Sci U S A* 1976;73:3,608–3,612.

124. Morel F, Chabardes D, Imbert-Teboul M. Vasopressin action sites along the nephron. *J Physiol Paris* 1981;77:615–620.

125. Helman SI, Grantham JJ, Burg MB. Effect of vasopressin on electrical resistance of renal cortical collecting tubules. *Am J Physiol* 1971;220:1,825–1,832.

126. Ausiello DA, Hall DH, Dayer J-M. Modulation of cyclic AMP-dependent protein kinase by vasopressin and calcitonin in cultured porcine renal LLC-PK₁ cells. *Biochem J* 1980;186:773–780.

127. Gstraunthaler G, Handler JS. Isolation, growth, and characterization of a gluconeogenic strain of renal cells. *Am J Physiol* 1987;252:C232–C238.

128. Koechlin N, Pisam M, Poujeol P, Tauc M, Rambourg A. Conversion of a rabbit proximal convoluted tubule (PCT) into a cell monolayer: ultrastructural study of cell dedifferentiation and redifferentiation. *Eur J Cell Biol* 1991;54:224–236.

129. Bridges JW, Wiebkin P, Fry JR. Rate-limiting factors in xenobiotic metabolism by cytochrome P-450, sulphotransferase, and glucuronyl transferase. In: Coon MJ, Conney AH, Estabrook RW, Gelboin HV, Gillette JR, O'Brien PJ, eds. *Microsomes, Drug Oxidations, and Chemical Carcinogenesis*, Vol. 2. New York: Academic Press, 1980:619–627.

130. Holme JA, Soderlund E, Dybing E. Drug metabolism activities of isolated rat hepatocytes in monolayer culture. *Acta Pharmacol Toxicol* 1983;52:348–356.

131. Niemann C, Gauthier JC, Richert L, Ivanov MA, Melcion C, Cordier A. Rat adult hepatocytes in primary pure and mixed monolayer culture. Comparison of the maintenance of mixed function oxidase and conjugation pathways of drug metabolism. *Biochem Pharmacol* 1991;42:373–379.

132. Bruggeman IM, Mertens JJWM, Temmink JHM, Lans MC, Vos RME, van Bladeren PJ. Use of monolayers of primary rat kidney cortex cells for nephrotoxicity studies. *Toxicol In Vitro* 1989;3:261–269.

133. Williams PD, Laska DA, Tay LK, Hottendorf GH. Comparative toxicities of cephalosporin antibiotics in a rabbit kidney cell line (LLC-RK1). *Antimicrob Agents Chemother* 1988;32:314–318.

134. Schuetz EG, Li D, Omiecinski CJ, Muller-Eberhard U, Kleinman HK, Elswick B, Guzelian PS. Regulation of gene expression in adult rat hepatocytes cultured on a basement membrane matrix. *J Cell Physiol* 1988;134:309–323.

135. Vamvakas S, Sharma VK, Sheu SS, Anders MW. Pertubations of intracellular calcium distribution in kidney cells by nephrotoxic haloalkenyl cysteine S-conjugates. *Mol Pharmacol* 1990;38:455–461.

136. Chen Q, Jones TW, Brown PC, Stevens JL. The mechanism of cysteine conjugate cytotoxicity in renal epithelial cells. Covalent binding leads to thiol depletion and lipid peroxidation. *J Biol Chem* 1990;265:21,603–21,611.

137. Farber JL, Leonard TB, Kyle ME, Nakae D, Serroni A, Rogers SA. Peroxidation-dependent and peroxidation-independent mechanisms by which acetaminophen kills cultured hepatocytes. *Arch Biochem Biophys* 1988;267:640–650.

138. Albano E, Carini R, Parola M, Bellomo G, Gori-Gatti L, Poli G, Dianzani MU. Effects of carbon tetrachloride on calcium homeostasis. A critical reconsideration. *Biochem Pharmacol* 1989;38:2,719–2,725.

139. Kaloyanides GJ, Pastoriza-Munoz E. Aminoglycoside nephrotoxicity. *Kidney Int* 1980;18:571–582.

140. Humes HD. Aminoglycoside nephrotoxicity. *Kidney Int* 1988;33:900–911.

141. Gilbert DN, Houghton DC, Bennett WM, Plamp CE, Reger K, Porter G. Reversibility of gentamicin nephrotoxicity in rats: recovery during continuous drug administration. *Proc Soc Exp Biol Med* 1979;160:99–103.

142. Saito H, Inui KI, Hori R. Mechanisms of gentamicin transport in kidney epithelial cell line (LLC-PK₁). *J Pharmacol Exp Therapeut* 1986;238:1,071–1,076.

143. Schwertz DW, Kreisberg JI, Venkatachalam MA. Gentamicin-induced alterations in pig kidney epithelial (LLC-PK₁) cells in culture. *J Pharmacol Exp Therapeut* 1986;236:254–262.

144. Ramsammy LS, Josepovitz C, Lane B, Kaloyanides GJ. Effect of gentamicin on phospholipid metabolism in cultured rabbit proximal tubular cells. *Am J Physiol* 1989;256:C204–C213.

11

Primary Hepatocyte Culture as an *In Vitro* Toxicologic System of the Liver

Shayne Cox Gad

Gad Consulting Services, Raleigh, North Carolina

The body's second largest organ, the liver, is the key organ in toxicology. A large variety of chemicals, including pharmaceuticals and environmental agents, are known to induce liver toxicity in the human population. Examples of liver toxicants include (1) the widely used over-the-counter analgesic, acetaminophen, which, when taken in large quantities, is responsible for a significant number of suicidal or accidental hepatotoxic events, sometimes resulting in death; (2) the key ingredient of alcoholic beverages, ethanol, prolonged exposure to which is known to lead to liver cirrhosis in both laboratory animals and humans; and (3) the naturally occurring mushroom hepatotoxins, the thermal stable cyclic octapeptides, α-, β-, and γ-amanitins, which are believed to be responsible for most of the clinically observed hepatotoxicity after ingestion of the poisonous mushroom *Amanita phalloides*.

Additionally, one must consider what firms of markers are to be used to evaluate the effect of interest. Initially, such markers have been exclusively either morphologic (does a change in microscopic structure occur?), observational (is the cell/preparation dead or alive or has some gross characteristic changed?), or functional (does the model still operate as it did before?). Recently, it has become clear that more sensitive models do not just generate a single endpoint type of data, but rather a multiple set of measures that in aggregate provide a much more powerful set of answers.

Several approaches to *in vitro* models exist for evaluating or studying hepato-toxicity.

The first and oldest approach is that of the isolated organ preparation. Perfused and superfused tissues and organs have been used in physiology and pharmacology since the late 19th century. A vast range of these tissues and organs is available, and a number of them have been widely used in toxicology. Almost any endpoint can be evaluated in most target organs, and these are closest to the *in vivo* situation and, therefore, generally the easiest from which to extrapolate or conceptualize. Those things that can be measured or evaluated in the intact organism can largely also be evaluated in an isolated tissue or organ preparation. However, the drawbacks or limitations of this approach are also compelling.

An intact animal generally produces on tissue preparation. Such a preparation is viable generally for a day or less before it degrades to the point of losing utility. As a result, such preparations are useful as screens only for agents having rapidly reversible (generally pharmacologic or biomechanic) mechanisms of action. They are superb for evaluating mechanisms of action at the organ level for agents acting rapidly, but not generally for cellular effects or for agents acting over a course of more than a day.

The second approach is to use tissue or organ culture. Such cultures are attractive, owing to maintaining the ability for multiple cell types to interact in at least a near-physiologic manner. They are generally not as complex as perfused organs, but they are stable over a longer period of time, somewhat increasing their utility as screens. They are truly a middle ground between perfused organs and cultured cells. Only for relatively simple organs (such as the skin and bone marrow) are good models performing in a manner representative of the *in vitro* organ available.

The third and most common approach is that of cultured cell models. These cells can be either primary or transformed (immortalized), but the former have significant advantages when used as predictive target organ models. Such cell culture systems can be used to identify and evaluate interactions at the cellular, subcellular, and molecular level on an organ- and species-specific basis [1]. The advantages of cell culture are that single organisms can generate multiple cultures for use, that these cultures are stable and useful for protracted periods of time, and that effects can be studied precisely at the cellular and molecular levels. The disadvantages are that isolated cells cannot mimic the interactive architecture of the intact organ, and they will respond over time in a manner that becomes decreasingly representative of what happens *in vivo*. An additional concern is that with the exceptions of hepatocyte cultures, the influence of systemic metabolism is not factored in unless extra steps are taken. Stammati et al. [2] and Tyson and Stacey [3] present some excellent reviews of the use of cell culture in toxicology. Any such cellular systems would be more likely to be accurate and sensitive predictors of adverse effects if their function and integrity were evaluated while they were operational. For example, cultured nerve cells should be excited while being exposed and evaluated.

A wide range of target–organ-specific models have already been developed and used. Their incorporation into a library-type approach requires that they be evaluated for reproducibility of response, ease of use, and predictive characteristics under the intended conditions of use. These evaluations are probably at least somewhat specific to any individual situation. Table 11.1 presents an overview of representative systems for a range of hepatic target organs. This table does not mention any of the new coculture systems in which hepatocytes are "joined up" in culture with a target cell type to produce a metabolically competent cellular system.

TABLE 11.1.   Representative *in vitro* test systems for hepatic toxicity[a]

System	Endpoint	Evaluation	References
Primary hepatocytes (S, M)	Multiple: · Biotransformation · Genotoxicity · Peroxisome proliferation · Biliary dysfunction · Membrane damage · Ion regulation · Energy regulation · Protein synthesis · Genomic studies	Can be used for species/species evaluations and comparisons	[2–6]
Hamster hepatocytes (S)	Functional: biochemical	Correlates with *in vivo* effects of acetaminophen	[7]
Rat liver slices (S)	Functional: alterations in ion content, leakage of damage markers, changes in biosynthetic capability Morphologic: histopathologic evaluation metabolic activity and iso-enzyme evaluation	Rank correlation with *in vivo* findings for a wide range of chemicals	[8–12]
Isolated perfused liver (M)	Functional: biochemical and metabolic	Correlation with *in vivo* findings for a wide range of chemicals	[13,14]

[a]Tyson and Stacey [3] estimated in 1989 that 800 unpublished studies of a toxicologic nature exist on cultured hepatocytes.

NOTE: Letters in parentheses indicate primary employment of system: S = screening system; M = mechanistic tool; NA = not available.

Besides being one of the organs sensitive to toxic agents, the liver plays an important role in metabolic transformation, leading to detoxification or activation (metabolism of the relatively inert parent compound to highly reactive metabolites) of blood-borne xenobiotics. The parenchymal cells in the liver perform this metabolic transformation, commonly called xenobiotic metabolism. The parenchymal cells, also commonly referred to as hepatocytes, which constitute over 80% of the liver by weight, have a high level of the inducible P450 mixed function oxygenases

(MFOs). The MFOs are membrane-bound enzymes, mainly located on the endoplasmic reticulum, especially the smooth endoplasmic reticulum. The MFOs are responsible for the phase I reactions, such as oxidations, reductions, and hydrolyses, and are found to be isozymes that can be distinguished based on antigenicity and substrate specificity. Different animal species are known to have different isozyme patterns. This difference in MFO activity is believed to play a significant role in the species differences in toxicity of some toxicants. Oxidation of xenobiotics by the MFO is followed by phase II metabolism, in which the metabolites are made more polar by conjugation to highly polar small molecules, such as glutathione, sulfate, and glucuronic acid. Phase II conjugation enzymes, including glutathione-S-transferases (GSTs), sulfatases, and uridine diphosphate (UDP)-glucuronosyltransferases, are abundant in the liver, found both in the cytosol and on the endoplasmic reticulum. The conjugated metabolites are usually excreted directly from the liver as bile, or from the kidney as components of the urine. The importance of metabolism in chemical toxicity in the human population can be illustrated by the finding that alcoholics are more sensitive to chemical toxicity. Normal doses of acetaminophen, for instance, can induce hepatotoxicity—and, in some cases, liver failure—in the alcoholic population. One explanation for this observation is that is alcohol consumption leads to the induction of P450 MFO activities, thereby increasing the rate of formation of the toxic metabolites from acetaminophen.

Because of the metabolic and toxicologic importance of the liver, *in vitro* experimental systems have been developed. These systems include the use of subcellular fractions, isolated hepatocytes, liver slices, and isolated perfused livers. Of these systems, the most versatile and well characterized is the cultured parenchymal cells or hepatocytes. In this chapter, the culturing of hepatocytes and their applications in toxicologic studies are reviewed.

11.1 ISOLATION OF PRIMARY HEPATOCYTES

The liver contains multiple cell types. Based on stereologic analysis, the parenchymal cells, commonly referred to as hepatocytes, are found to constitute 92.5% of the total volume of liver cells. The remainder of the cells are sinusoidal and perisinusoidal cells, including the fenestrated endothelial cells, the stationary macrophage Kupffer cells, and the vitamin-A-storing stellate cells or lipocytes [15].

Dispersion of liver cells is commonly performed using a two-step perfusion procedure, first by perfusing the liver with a calcium removal agent, and then by perfusing with a solution containing a digestive enzyme, such as collagenase, to achieve cell dissociation. In small, experimental animals, the procedure can be performed *in situ* [16–18]. After achieving anesthesia, the animal's abdomen is cut open and the portal vein is cannulated with a cannula connected to a perfusion solution. The inferior vena cava is severed, and perfusion is initiated. The initial perfusion solution usually consists of an isotonic calcium-free buffer, sometimes containing a chelating agent, such as ethylene guanine thioacetamide (EGTA), serving to clear the blood in the liver as well as to prevent blood coagulation. The removal of calcium ions by EGTA has been reported to enhance the viability of the isolated hepatocytes. The perfusion solution is then changed to that containing collagenase, which serves to dissociate the liver cells. Upon completion of collagenase digestion, the liver is removed, placed in an isotonic buffer solution, and gently agitated to release the dissociated cells.

Although the *in situ* procedure is appropriate for small animals such as rats and mice, the "biopsy" procedure is used for larger animals, with which a dissected portion of the liver rather than the whole organ is perfused. The biopsy procedure, first described by Reese and Byard [19], involved the perfusion of the dissected lobe via one of the major blood vessels exposed at the cut surface (Fig. 11.1). This procedure is useful with livers from small animals as well as with liver portions from large animals, including human. The advantage of the "biopsy" procedure is that surgical specimens, which sometimes are the only specimens available from humans, can be used for hepatocyte isolation. Furthermore, although flow rate is extremely critical to hepatocyte viability for the *in situ* procedure, it is less critical for the biopsy procedure, probably because of the numerous channels at the cut surface for the exudation of perfusate in the biopsy samples, resulting in a lower

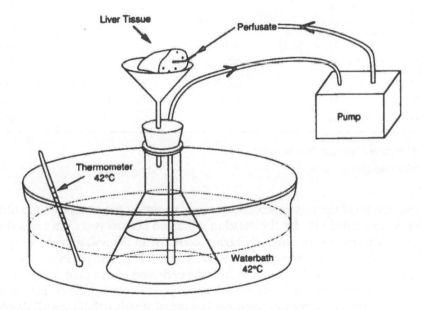

FIG. 11.1. Schematic drawing of the perfusion system for the isolation of hepatocytes. This procedure is commonly referred to as "biopsy perfusion," developed by Reese and Byard [19]. The procedure has been used successfully in our laboratory for the isolation of hepatocytes from multiple animal species, including humans.

TABLE 11.2. Isolation of hepatocytes from human liver portions: results of 19 consecutive isolations

Donor			Hepatocytes	
Sex	Age	Weight of liver portion (g)	Yield (x 106)	Viability(%)[a]
M	73	8.3	13.0	92.0
M	66	17.7	52.0	85.0
F	40	19.6	212.3	94.7
F	37	17.6	68.5	86.0
M	76	32.8	114.0	92.0
F	27	9.4	52.0	91.2
F	27	12.3	96.0	90.5
F	69	19.4	129.0	90.2
M	79	11.8	136.0	93.4
M	73	9.0	34.5	90.2
M	53	3.2	11.3	82.0
F	66	10.7	28.5	95.0
M	b	4.9	41.3	92.7
M	49	8.2	51.5	91.5
F	34	7.2	10.8	63.2
M	57	11.3	53.0	77.7
M	b	46.7	671.3	84.0
F	43	38.3	1010.3	83.0
M	58	19.2	195.0	91.0

[a]Determined by trypan blue exclusion.
[b]Not available.

pressure build up. In our laboratory, using a form of the procedure modified by Reese and Byard [19] for the isolation of human hepatocytes from surgical specimen [42], we have an extremely high success rate in the isolation of highly viable (over 80% viability) human hepatocytes (Table 11.2). The parameters we found critical to a successful isolation include the perfusion of a limited size (up to 50 g), the use of specimen with only one cut surface and intact capsule on the uncut surfaces, perfusion via one or sometimes two blood vessels rather than all visible vessels (to allow circulation), and the use of an appropriate concentration of collagenase. The ability to culture hepatocytes from multiple animal species, including humans, allows one to perform experiments on species comparison of xenobiotic metabolism and toxicity.

After cell dissociation, the most common procedure is to partially purify the parenchymal cells by several low-speed (50-g) centrifugation steps. The parenchymal cells will pellet, whereas the nonparenchymal cells will stay in the supernatant. Further purification of each cell population can be performed via density centrifugation. Using Percoll, Smedsrod and Pertoft [20] developed a procedure for the purification of parenchymal cells and nonparenchymal cells. The Kupffer and endothelial cell populations can be further separated from each other via differential attachment: The Kupffer cells, like most macrophages, attach quickly onto a plastic substratum and are thereby separated from the less adherent endothelial cells.

For most toxicologic studies, the parenchymal cells are used. Although the endothelial and Kupffer cells are known to play critical roles in liver toxicity, including parenchymal cell toxicity and regeneration, they have been generally ignored by toxicologists in their studies. In terms of xenobiotic metabolism, the nonparenchymal cells probably play a rather minor role when compared with the parenchymal cells. In our laboratory, we found that xenobiotic metabolism measured based on the activation of promutagens dimethyl nitrosamine and 3-methyl cholanthrene, and the measurement of 7-deoxyycoumarin-*O*-deethylase activity was nearly undetectable in the nonparenchymal cells isolated from rat livers [21]. Our study, therefore, confirms that parenchymal cells, rather than the nonparenchymal cells, are the cells primarily responsible for xenobiotic metabolism in the liver.

11.2 CULTURING OF ISOLATED HEPATOCYTES

Freshly isolated hepatocytes are routinely cultured as monolayer cells on collagen-coated plastic, although extremely short-term experiments (hours) can be performed with hepatocytes in suspension. In suspension, the freshly isolated hepatocytes rapidly lose viability, with nearly 100% cell death after a 24-hour period. As monolayer culture on plastic or collagen-coated plastic surfaces, the hepatocytes assume an epithelial cell morphology (Fig. 11.2). The cells are polygonal, with distinct nuclei and cytoplasmic inclusions. Binucleated cells are often seen, reflecting the occurrence of both mononucleated and binucleated cells in the liver *in vivo*. The cells retain viability, but gradually (over a period of several days) they will lose their liver functions, including albumin synthesis and P450 MFO activities. The phase II conjugating enzymes, however, do not appear to decrease with culturing time, at least for the initial several days, during which the phase I enzymes show a dramatic decrease [22].

Recently, hepatocytes have been found to retain more of their functional characteristics when cultured on a murine tumor cell Engelbreth–Holm–Swarm sarcoma (EHS)-derived reconstituted basement membrane matrix called Matrigel as compared with culturing on collagen [23,24]. Matrigel is a mixture of components extracted from the EHS tumor spontaneously forming a stable gel at 37°C. The components of Matrigel are similar to that found in basement membrane, including laminin (60%), type IV collagen (30%), heparin sulfate proteoglycan (3%), nidogen (5%), and entactin (1%) [25]. On Matrigel, the hepatocytes assume a rounded cell morphology (Fig. 11.2). Hepatocytes cultured on Matrigel were found to secrete higher levels of liver-specific proteins, including albumin, transferrin, haptoglobin, and hemopexin. The interaction of the basement membrane matrix components with membrane receptors, as well as the rounded cell shape, may be involved in the prolonged maintenance of differentiated functions.

Another approach to maintaining the differentiated properties is to coculture the hepatocytes with another cell type. Hepatocytes cocultured with rat liver epithelial cells are found to express high levels of liver functions, including acute-phase protein synthesis and both phase I and phase II drug metabolism [26,27], even after weeks of culturing. After coculturing with a variety of transformed epithelial-like cell lines, Donato et al. [22] reported that in general the hepatocytes retained a higher level of xenobiotic metabolism activities than did pure cultures after several days in culture.

Methods for the culturing of hepatocytes as cell aggregates (multicellular spher-

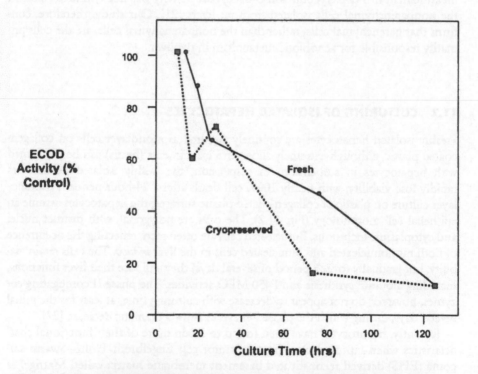

FIG. 11.2. Decline of 7-ethoxycoumarin-O-deethylase (ECOD) activity for freshly isolated and cryopreserved adult rat hepatocytes after culturing. The cells were cultured on collagen-coated plastic. ECOD activity is commonly measured to represent P450 mixed function oxygenase (MFO) activity. Although a steady drop in activity was observed, at 24-hour postplating, the cells still retained over 50% of the original activity.

oids) have been developed [28,29]. Such a culturing system allows hepatocytes to retain the cuboidal cell shape and three-dimensional cell–cell contact, similar to the liver *in vivo*. Hepatocytes cultured as spheroids may maintain differentiated functions better than monolayer cultures. Detailed characterization of hepatocytes cultured as spheroids and their applications in toxicology studies have yet to be reported. Another development is the culturing of hepatocytes as entrapped aggregates in a packed bed bioreactor [30]. The hepatocytes in the bioreactor resemble those of the liver *in vivo*, with the cuboidal cell shape, three-dimensional cell–cell contact, and nutrient perfusion. The hepatocyte bioreactor has the potential to be a more realistic model of the liver *in vivo* in drug metabolism and toxicology.

In spite of the above-mentioned advances in maintaining differentiation of hepatocytes in culture, in toxicology, hepatocytes are still mainly used as short-term suspension cultures and monolayer cultures on collagen-coated surfaces. However, it has now been shown possible to cryopreserve primary hepatocytes, thawing them later with a retention of metabolic function [31].

11.3 XENOBIOTIC METABOLISM

11.3.1 Phase I Oxidation: P450 Isozymes

A major problem with prolonged hepatocyte cultures is the decline of cytochrome P450 MFO activities with time in culture (Fig. 11.2). Moreover, different P450 isozymes appear to decline at different rates [32], and amidst the decline of most of the isozymes, induction of a few specific isozymes has been observed [33]. Attempts to maintain P450 content as well as MFO isozymes include the supplementation of medium with multiple agents, such as thyroid and pituitary hormones, sex steroids, glucorticoid hormones, and dimetylsulfoxide (DMSO) [34–38], the culturing of hepatocytes on reconstituted basement membrane [40], and coculturing hepatocytes with other cell types [26,22,39]. However, none of these approaches has been shown to maintain the multiple P450 isozymes. For instance, although coculturing of hepatocytes with liver epithelial cells is believed to be the most effective, differential decline of P450 isozymes is still observed [41]. Short-term cultures of freshly isolated hepatocytes, therefore, remain preferable over prolonged cultures for metabolism studies.

Although the P450 isozymes decline in culture, some are found to be inducible by the addition of specific inducers in the culture medium, usually after a 2- to 3-day incubation period. The inducibility of some of the isozymes appears to be a function of the culturing conditions. For instance, cytochrome P450IIE and IIB1/2 could only be induced when hepatocytes were cultured on reconstituted basement membrane. Phenobarbital, a known P450 inducer, was found to induce cytochrome P450IIB1/2 and P450IIIA1 [42] in rat hepatocytes cultured on Matrigel, a reconstituted basement membrane. Ethanol was found to induce cytochromes P450IIE, IIB1/2, and IIIA in cultured rat hepatocytes on Matrigel [43].

Similar isozymes are induced in cultured hepatocytes and *in vivo*, illustrating that isolated hepatocytes are a relevant experimental system for cytochrome P450 induction studies. It is interesting to note that P450 induction performed *in vivo* is found to lead to induced activities in the isolated hepatocytes for at least 24 hours in culture. After administration of several cytochrome P450 inducers (phenobarbitone, β-naphthoflavone, dexamethasone, and isoniazid), the induced cytochrome P450 enzyme activities found *in vivo* were found to be maintained in the isolated hepatocytes for 24 hours in culture on a positively charged plastic surface, namely, a Primaria culture plate [44]. This observation further confirms the appropriateness of using freshly isolated hepatocytes for xenobiotic metabolism studies.

Virtually all studies on xenobiotic metabolism are on the liver parenchymal cells (hepatocytes). The potential activity of nonparenchymal cells are mostly ignored. In our laboratory, we recently compared the potential xenobiotic metabolic activity of rat liver parenchymal and nonparenchymal cells and found that nearly all detectable activities were found in the parenchymal cells [21]. Our data, therefore, suggest that metabolite-induced damage to nonparenchymal cells, often found *in vivo*, are probably a result of the metabolites generated by the neighboring parenchymal cells.

11.3.2   Phase II Conjugation

Although the phase I metabolism has been studied in detail in cultured hepatocytes, relatively less information is available for phase II metabolism. In phase II, the highly reactive metabolites generated from phase I metabolism of xenobiotics are "inactivated" via glucuronidation and sulfation. Important phase II enzymes are (1) UDP-glucuronyltransferase, which catalyzes glucuronidation; (2) sulfotransferase, which catalyzes sulfation; and (3) the GST, which catalyzes glutathione conjugation. All three major conjugation pathways (glucuronidation, sulfation, and glutathione conjugation) have been observed in cultured hepatocytes.

Our study on acetaminophen conjugation illustrates the existence of glucuronidation and sulfation in cultured rat hepatocytes [45]. Rat hepatocytes from male and female animals were cultured either on collagen or Matrigel, and both the sulfate and the glucuronide conjugation of acetaminophen were studied. The *in vivo* sex differences in acetaminophen metabolism, with the male rat excreting more acetaminophen sulfate and less acetaminophen glucuronide than the female rat [46], was apparent in the 1- and 2-day cultures and became less apparent on day 4. We found the hepatocytes cultured on Matrigel to maintain sulfation better than do those cultured on collagen, a finding consistent with the general observation that Matrigel is a better substrate for the maintenance of hepatocyte differentiation.

The effect of culturing on GST has been studied. Both GST activity and mRNA levels are found to be affected by medium composition and culture time. But unlike cytochrome P450 isozymes, which tend to diminish with time in culture, GST tends to increase with prolonged culturing. Fetal calf serum, the addition of nicotinamide, or the inclusion of DMSO increases GST. A similar trend was observed for hepatocytes cocultured with liver epithelial cells [47]. A later, more detailed study showed a more complex picture. [48]: In conventional culture, after 6 days in culture, a decrease in GST subunits 1 and 2 occurred, maintenance of subunits 3 and 4, and an increase in subunit 7 when fetal calf serum was present. Omission of fetal calf serum led to an increase in subunits 3, 4, and 7. Coculturing with liver epithelial cells led to increases in all subunits. Exclusion of fetal calf serum in the cocultures led to a significant reduction in subunit 2.

These findings with phase II metabolism again confirm that freshly isolated hepatocytes are representative of the live *in vivo*.

11.3.3 Species Comparison Studies

The species difference in xenobiotic metabolism is a well-documented phenomenon and is believed to be one of the major factors accountable for the observed species difference in sensitivity toward chemical toxicants. Because the liver is the major organ for xenobiotic metabolism, hepatocytes cultured from multiple animal species are an attractive experimental system to evaluate this phenomenon, especially as a system to extrapolate data from laboratory animals to humans.

In our laboratory, the parallelogram approach is used to extrapolate data obtained from laboratory animals and cultured animal and human cells to humans *in vivo* (Fig. 11.3). By understanding the relationship between hepatocytes and *in vivo* cell results obtained from laboratory animals, one can logically predict human *in vivo* results. Successful application of this approach is dependent on the ability of the cultured hepatocytes to retain the species differences found *in vivo*.

Results with several model chemicals seem to indicate that isolated hepatocytes retain the species differences in xenobiotic metabolism. One such compound is amphetamine (AMP). AMP is metabolized via two major pathways: aromatic hydroxylation to parahydroxyamphetamine (pHA) and oxidative deamination (ben-

In Vivo Toxicity	X	X	X	Predict
In Vitro Toxicity	X	X	X	X
Mouse	Rat	Monkey	Human

FIG. 11.3. The "parallelogram approach" to the extrapolation of human toxicity. Human *in vivo* response (e.g., toxicity, metabolism) is to be derived from *in vitro* human hepatocyte data based on the *in vitro–in vivo* relationship observed in laboratory animals. The key to this approach is to have a thorough understanding of the data derived from laboratory animals, preferably using multiple animal species. A corollary to this approach is that one may use cultured hepatocytes to select an animal species closest to humans in xenobiotic metabolism.

zoic acid, hippuric acid). Species differences in AMP metabolism is known, with differences existing both in the rate of metabolism and in the ration of metabolites [50]. Rabbits metabolize AMP rapidly, with the majority of the metabolites being deamination products. Rats have a moderate rate of metabolism, and unlike most other species, they prefer the formation of aromatic hydroxylation products, such as pHA. Humans and squirrel monkeys metabolize AMP slowly into mainly deamination metabolites (acids). Green et al. [49] studied AMP metabolism in hepatocytes isolated from rats, rabbits, dogs, squirrel monkeys, and humans and found that isolated hepatocytes retained the species differences found *in vivo*. Similar results were also obtained in our laboratory with freshly prepared and cryopreserved rat and rhesus monkey hepatocytes. After a 4-hour incubation with 10-M AMP, freshly isolated hepatocytes (5 x 106, in a total volume of 10 ml) from rabbits, rats, and rhesus monkeys were found to metabolize 100%, 32%, and 13%, respectively, of the total AMP. The one human hepatocyte isolate studied showed only 12% metabolism after a 24-hour incubation period. High-performance liquid chromatography (HPLC) analysis of the metabolites showed that rat hepatocytes produced more pHA than did acids, whereas rabbits, rhesus monkeys, and humans produced more acids than did pHA (Table 11.3). Our data, therefore, are consistent with the *in vivo* findings of Dring et al. [50]. Other examples of the similarity between freshly isolated hepatocytes and whole animals in xenobiotic metabolism include studies on tolbutamide with rats, rabbits, dogs, and squirrel monkeys [51] and studies on diazepam with rats, rabbits, dogs, guinea pigs, and humans [52]. Monteith et al. [53] showed that acetylaminofluorine was metabolized by cultured human hepatocytes in a manner similar to that by humans *in vivo* [54]. These studies, therefore, support the usefulness of isolated hepatocytes in the evaluation of species differences in xenobiotic metabolism, which is especially important in human metabolism studies because experimentation with human *in vivo* is rarely possible.

TABLE 11.3. Metabolism of amphetamine (AMP) by hepatocytes isolated from multiple animal species into p-hydroxyamphetamine (pHA) and deamination-derived benzoic and hippuric acids

Animal (%)	4 hr			24 hr		
	Acids (%)	pHA (%)	AMP (%)	Acids (%)	pHA (%)	AMP (%)
Rat	10.4	32.5	48.8	23.5	52.9	15.7
Rabbit	54.2	10.0	0.0	Not analyzed		
Monkey	5.7	1.4	86.9	22.1	2.1	68.7
Human	Not analyzed			5.1	3.4	88.1

Discrepancy between hepatocytes and *in vivo* observation also exists. An example is that reported by Humpel et al. [55] with four radiolabeled drugs: lonazolac, bromerguride, lisuride, and terguride. Although the metabolite profile generated from hepatocytes for lonazolac and bromerguride were similar to *in vivo* findings, a significantly lower number of metabolites was found for lisuride and terguride in hepatocytes than found *in vivo*. Exhepatic metabolism, for instance, by other organs or by gut flora may be a reason for the differences. Hepatocytes isolated from rats, guinea pigs, beagle dogs, and cynomolgus monkeys were used in study. Although good correlation was found between hepatocytes and *in vivo* findings in metabolic stability, no quantitative correlation was observed. This finding illustrates the complexity of data extrapolation. In addition to the lack of nonhepatic tissues and other host factors for the hepatocytes in culture, several important differences exist between the *in vitro* system and the whole animal that most investigators do not consider: Although hepatocyte experiments are performed using a static pool of medium containing the drugs, *in vivo* experiments are performed with a bolus injection. The hepatocytes *in vitro* are, therefore, exposed to a relatively constant concentration of the substrate, whereas the liver *in vivo* is exposed to a continuum of concentrations, depending on the route of administration and the rate of clearance. Furthermore, one needs to account for the difference in cell mass between *in vitro* experiments (which usually employ millions of hepatocytes) and *in vivo* experiments with the intact liver (which has a substantially higher number of hepatocytes).

11.4 TOXICOLOGY

Hepatocytes are used extensively in hepatotoxicity and carcinogenicity studies. Toxicity assays can be divided into the following main categories: cytotoxicity, genotoxicity, and enzyme induction. As illustrated by the various examples cited below, the advantage of cultured hepatocytes is that one can combine cytotoxicity measured with various other biochemical measurements and cotreatment of metabolic inhibitors to allow mechanistic evaluation of toxicity, which cannot be easily performed *in vivo* or using cell-free systems, such as microsomal preparations.

11.4.1 Cytotoxicity

Measurement of cytotoxicity is probably the most widely used aspect of cultured hepatocytes. A variety of endpoints has been developed to quantify cytotoxicity.

Morphology. As mentioned earlier, conventional (i.e., on collagen-coated surface) monolayer cultures of hepatocytes exhibit an epithelial cell morphology with a polygonal shape and prominent nuclei (Fig. 11.2). With the use of light microscopy, cytotoxicity can sometimes be visualized using a light microscopy by the rounding of the cells, blebbing, and detachment. With the use of electron microscopy, cytotoxicity can be observed by alterations in ultrastructure—for instance, swelling of mitochondrial membranes, appearance of lysosomal inclusion bodies, and so on. The study of Gross et al. [39] illustrates the induction of mitochondrial swelling and lamella inclusion bodies by amiodarone, a hepatotoxic drug (see below). Morphologic observation is usually performed as a qualitative, but not quantitative, measurement of cytotoxicity. Sorensen [56] developed a quantitative morphometric analysis procedure with hepatocytes damaged by indomethacin and showed good correlation with cytotoxicity measured by enzyme release (see below). The procedures, however, have not yet been widely applied.

Trypan blue uptake.   Trypan blue is a high-molecular-weight chemical relatively impermeable to viable cells. Cells with damaged plasma membrane, however, will allow rapid permeation of the dye into the cytoplasm, thereby staining the cell nuclei blue. Trypan blue exclusion is commonly used to evaluate cell viability during hepatocyte isolation, and it has been applied to short-term suspension cultures to evaluate chemical cytotoxicity. Thompson et al. [57], for instance, evaluated the toxicity of eugenol (4-allyl-smethoxyphenol), a chemical known to be present in various food and medicinal preparations, in isolated rat hepatocytes. A good correlation between the loss of intracellular glutathione and cell death as measured by trypan blue uptake was obtained. The correlation between glutathione depletion and cytotoxicity induced by eugenol was further confirmed by (1) the prevention of cytotoxicity by *N*-acetylcysteine and (2) the augmentation of cytotoxicity by the glutathione depletion agent, diethylmaleate. This study illustrates the application of cultured hepatocytes, combining metabolism and cytotoxicity studies, toward elucidating the mode of action of a hepatotoxic agent. Trypan blue exclusion can be performed quantitatively, but it is tedious, requiring manual counting. This counting is usually performed for experiments with a limited number of samples. Trypan blue uptake is commonly expressed as the number of cells exhibiting trypan blue, divided by the total cell population counted. A significant increase in this ratio over untreated or solvent-treated controls would indicate cytotoxicity.

Release of cytoplasmic enzyme.   Assaying for plasma levels of liver-specific enzymes, alanine aminotransferase (ALT; also known as serum glutamate pyruvate transaminase, SGPT), and aspartate aminotransferase (AST; also known as serum glutamate oxaloacetate transaminase, SGOT), is an approach well accepted for the evaluation of liver damage *in vivo*, both in laboratory animals [58] and in humans [59]. Although both ALT and AST are present in the cytosol, the mitochondria of hepatocytes contain only AST [70]. Analysis of both ALT and AST, therefore, can allow one to distinguish plasma membrane damage from mitochondria damage.

This analysis of enzyme release has been applied toward cultured hepatocytes. Freshly isolated hepatocytes are allowed to attach, usually overnight. Medium is then changed to that containing the substances to be tested. After another culturing period, usually 24–96 hours, the amount of cytoplasmic enzyme released into the culture medium is determined. In addition to ALT and AST. The release of another cytoplasmic enzyme, lactate dehydrogenase (LDH), is also commonly used. Because of the universal presence of LDH in cells, increase of plasma LDH for *in vivo* experiments does not necessarily indicate liver damage. Under *in vitro* conditions, however, as only hepatocytes are used, LDH release into the culture medium usually reflects plasma membrane damage as well as ALT and AST. Data on enzyme leakage usually is presented as percent of total cytoplasmic enzyme content, using detergent treatment (e.g., triton X-100) to induce 100% lysis.

An example of the application of enzyme release is the evaluation of amiodarone, an iodinated benzofuran widely used for life-threatening cardiac dysrhythmias [61]. Amiodarone is known to induce an increased level of LDH, ALT, and AST in patients receiving chronic therapy. Cultured rat hepatocytes, when treated with amiodarone and the metabolite desethylamiodarone, were found to release LDH, AST, and ALT into the medium in a dose-dependent and time-dependent manner, thereby reproducing the clinical findings. This study, therefore, illustrates the application of *in vitro* cytotoxicity in the evaluation of *in vivo* hepatotoxicity.

Because of the availability of automatic clinical analyzers that can accurately measure LDH, ALT, and AST, a large number of samples can be assayed with ease. Enzyme release is, therefore, widely used to quantify cytotoxic responses of cultured hepatocytes to toxic agents. A point of caution in using enzyme release as an endpoint is that one needs to ensure that the chemicals tested do not inactivate the enzyme measured. -Mercapto acids, for instance, are known to inactivate LDH [62].

Other endpoints. Although the above-mentioned endpoints are the most commonly used, a variety of other endpoints can be used to monitor cytotoxicity. Such endpoints include (1) macromolecular incorporation (mainly measured by radiolabeled amino acid and ribonucleotide incorporation, because hepatocytes in culture have limited DNA synthesis activity) and (2) functional endpoints, including albumin synthesis, glycogen synthesis, and urea synthesis. Gomez-Lechon et al. [63] evaluated a variety of hepatotoxins with multiple cytotoxicity endpoints in cultured rat hepatocytes and found that the sensitivity of each endpoint varies with each hepatotoxin. Although RNA and albumin synthesis were the most sensitive endpoints for α-amanitin, LDH release was the most sensitive for *d*-galactosamine and acetaminophen, and ureogenesis was the most sensitive for thioacetamide.

11.4.2 Genotoxicity

Genotoxic chemicals are agents that would interact with SNA, leading to heritable changes [64]. The role of liver metabolism in the "activation" of nongenotoxic parent compounds (promutagens) to highly reactive electrophiles is well established. Enzymes involved include cytochrome P450 isozymes (see above), peroxidases, monoamine oxidase, and flavin-containing oxygenase [65]. Postmitochondrial supernatant (S9) and microsomes prepared from livers of Aroclor-1254–treated rats are commonly used in mutagenesis assays as an exogenous-activating system for promutagens [66,67]. Hepatocytes themselves are also used as a promutagen-activating system, as well as a target cell to measure genotoxicity [88].

Promutagen activation. This process is accomplished by coculturing primary hepatocytes with a reporter cell that has little or no P450 metabolism but with which mutation can be quantified. An example is the coculturing of Chinese hamster ovary (CHO) cells with hepatocytes, followed by mutagen treatment and quantification of mutants at the hypo-xanthine-guanine phosphoribosyl transferase (HGPRT) gene locus [68]. Using this procedure, we found that rat hepatocytes could be cryopreserved to retain promutagen-activating activity [69], and that the parenchymal cells in rat liver had substantial activity, whereas the nonparenchymal cells in rat liver had virtually no activity in the activation of two promutagens, 3-methylcholanthrene and dimethyl nitrosamine [21]. Although liver homogenate is routinely used in mutagenicity assays, evidence exists that activation of promutagens by intact hepatocytes provide data more relevant to *in vivo* carcinogenicity. For instance, after comparison of mutagenicity of benzo(*a*)pyrene and four major metabolites in the Ames test, Glat et al. [70] found that the relative mutagenicity of the compounds obtained using intact hepatocytes was closer to the *in vivo* ranking of carcinogenicity than were data obtained with liver homogenate.

Induction of unscheduled DNA synthesis. After exposure to a mutagen, a certain level of DNA repair activity is induced. Some of the DNA lesions are excised and replaced with new DNA, a process known as "unscheduled" DNA synthesis (UDS) (as opposed to the "scheduled" DNA synthesis during normal DNA replication). This activity can be measured in virtually all mammalian cells, including hepatocytes. The measurement of UDS in hepatocytes as an assay for mutagens was pioneered by Williams [71]. After allowing freshly isolated hepatocytes to attach on collagen-coated coverslips, the test substance is added simultaneously with radiolabeled thymidine. Autoradiography is used to quantify radiolabeled thymidine incorporation, with the results expressed as a number of grains in the emulsion over the nuclei.

Others. Several other approaches have been used to evaluate the interaction of toxicants with hepatocyte DNA. Using the induction of single-strand breaks as an endpoint, Liu and Castonguay [72] showed that (+)-catechin, a plant flavonoid, could inhibit the metabolism and genotoxicity of a tobacco-specific carcinogen. Cole et al. [73] measured the binding of aflatoxin B_1 and 2-acetyl-aminofluorine to the DNA of hepatocytes cultured from male and female rats and humans. The known sex and species differences in carcinogenicity were found to correlate well with the DNA binding levels.

11.4.3 Enzyme Induction

The liver contains a host of enzymes that can be up regulated. From a toxicologist's point of view, the important inducible enzymes are the peroxisomal enzymes and P450 MFO isozymes. As described below, a correlation often exists between the induction of these enzymes and hepatocarcinogenicity. Whether enzyme induction and carcinogenicity are merely coexisting events or mechanistically linked is yet to be elucidated. We have recently hypothesized that some enzyme inducers may also induce the expression of other cellular genes, including cellular oncogenes. Such enzyme inducers, therefore, have the potential to promote the expression of preexisting mutations in these genes, leading to cancer development [69]. Our hypothesis led us to the development of a multiple endpoint hepatocyte assay in which cytotoxicity and enzyme induction (peroxisome induction and P450 induction) were used as toxicity endpoints [74].

Because of the interanimal variability, the difficulty of quantitation of organ dose, and the activities of extrahepatic factors, it is often difficult to study enzyme induction *in vivo*, especially when mechanistic or quantitative (e.g., dose-response relationship) information is to be obtained. A cell culture system, such as cultured hepatocytes maintained in a chemically defined medium, is, therefore, ideal for enzyme induction studies. It is important to emphasize here that cultured hepatocytes remain the only *in vitro* liver system with which enzyme induction can be studied. The relatively long time period (days) required for induction precludes the use of systems, such as liver slices or isolated perfused livers, the viability of which can only be maintained for a relatively short time period (hours).

Peroxisomal induction. Peroxisomes are organelles found ubiquitously in animal and plant cells. The organelle has a diameter of approximately 0.1–1.5 m, with a limiting membrane and a fine granular matrix. In normal hepatocytes, the peroxisome-to-mitochondria ration is 1:5 or 1:6. The organelles are believed to be involved in fatty acid oxidation, thermogenesis, and respiration. Peroxisome proliferation (increase in peroxisome number) has been known to be induced by high-fat diets, hormonal alterations, such as hyperthyroidism, and, as described below, certain chemicals, such as some hypolipidemic agents and plasticizers. The biologic effects of peroxisome proliferators *in vivo* in rodents include increases in smooth endoplasmic reticulum and P450 MDO activity, hepatomegaly, hypolipidemia, and hepatocarcinogenesis. Peroxisome proliferators have been suggested to be a novel class of hepatocarcinogens [75]. These agents include the hypolipidemic agent clofibrate and the plasticizer di-(2-ethylhexyl)phthalate. They are unique in that they do not have genotoxicity as measured by the conventional mutagenicity assays. Peroxisome induction can be measured morphologically by counting the number of peroxisomes per cell or biochemically by monitoring cyanide-insensitive palmitoyl CoA oxidation. Induction of peroxisomal β-oxidation in cultured hepatocytes by agents known to induce peroxisomes in rodents *in vivo* has been shown [76].

Whether peroxisome proliferators are human carcinogens is still debatable. Liver biopsies from patients clinically exposed to peroxisomal inducers clofibrate [77], gemfibrozil [78], and fenofibrate [79] did not show evidence of increased peroxisome proliferation relative to untreated controls. This species difference in response was also observed in cultured hepatocytes. Elcomb and Mitchell [80] showed that mono(2-ethylhexyl)phthalate and two of its active metabolites significantly induced palmitoyl CoA oxidation in rat hepatocytes but not in marmoset or human hepatocytes. Foxworthy et al. [81] showed that the monkey hepatocytes were significantly less responsive than were rat hepatocytes to the peroxisome induction effects of ciprofibrate, bezafibrate, and LY 171883, thereby supporting the rodent–primate difference observed *in vivo*. In our laboratory, we have evaluated four potent rodent peroxisome proliferators [clofibrate, diethylhexylphthalate (DEHP), lactofen, and Wy-14,653] and found consistent activation of palmitoyl CoA oxidase activity in rat hepatocytes but not in three separate isolations of human hepatocytes (Table 11.4). However, because the carcinogenic mechanism of the peroxisome proliferators is not yet defined, it is difficult to project from these observations to human carcinogenicity.

11.4.4 P450 Induction

As the major organ for xenobiotic metabolism, the cytochrome P450 MFO system in the liver is well characterized. As indicated earlier, cultured hepatocytes are used extensively to study xenobiotic metabolism. The recent findings on the induction of the P450 MFOs are reviewed here.

P450 MFOs exist as isozymes coded for by the P450 gene superfamily. It is

TABLE 11.4. Effect of peroxisome proliferators on palmitoyl CoA oxidase activity in cultured human hepatocytes

Treatment	Palmitoyl CoA oxidase activity (nmol/min/mg protein)[a]		
	I	II	III
DMSO control	0.07	0.00;0.23	0.20;0.23
Clofibrate (1mM)	0.06	0.00;0.01	0.26;0.24
DEHP (1 mM)	0.03	0.08;0.13	0.26;0.27
Lactofen (0.1 mM)	—	0.00;0.02	0.20;0.27
Wy-14,646 (0.1 mM)	0.05	—	—

[a]Activity of individual plates shown. Because of the limitation in the number of human hepatocytes available, treatments were performed as duplicates for II and III and as single plates for I. I: 17-year-old girl; II: 41-year-old man; III: 35-year-old woman. All donors were Caucasian.

often confusing to compare results of different publications because of the multiple names assigned to the isozymes. Recently, a standardized nomenclature for the existing isozymes was proposed based on sequence homology [82]. In this review, the precious "nonstandard" nomenclature is converted to this "standardized" nomenclature to allow an easier comparison of results.

Findings from the laboratory of Guzelian et al. [32] are probably the most influential for our understanding of P450 induction in cultured hepatocytes. Earlier results from this laboratory showed that P450 IA1 in cultured hepatocytes was induced by 3-methylcholanthrene, β-naphthoflavone, 3,4,3',4'-tetrachlorobiphenyl, and Aroclor 1254 [97]. Incubation of cultured rat hepatocytes with the inducers led to a 5–15-fold increase in P450 IA1 protein (measured by immunoprecipitation) accompanied by increases in the synthesis of mRNA and protein for this isozyme. In the same study, the authors showed that safrole, Aroclor 1254, and 3,4,5,2'4'5'-hexachlorobiphenyl prevented the loss of P450 IA2 from the cultured hepatocytes, whereas 3-methylcholanthrene, b-naphthoflavone, and 3,4,3',4'-tetrachlorobiphenyl had no effect. The authors concluded that modifications of both synthesis and degradation are the major mechanisms for the regulation of P450 isozyme levels, and that different isozymes might be regulated by different mechanisms. The same laboratory also showed that P450 IIIA1 was inducible by glucocorticoids in cultured hepatocytes [83]. Dexamethasone was found to be more effective than were the other glucocorticoids tested, which include betamethasone, α-methylprednisolone, 16α-carbonitrile, triamcinolone, corticosterone, and hydrocortisone. The induction was shown to be caused by an increase in the rate of synthesis of the isozyme, and it was reversible after the withdrawal of the inducers from the cultured medium.

Although phenobarbital (PB) is a powerful P450 inducer *in vivo*, its ability to induce cultured hepatocytes is often referred to as "difficult" to demonstrate. An earlier report by Michalopoulous et al. [84] showed the PB induced both P450 content and smooth endoplasmic reticulum in rat hepatocytes cultured on floating collagen gel. Whereas induction by 3-methylchloanthrene required only 2 days, a 5-day induction period was needed for PB. This report was followed by that of Stenberg and Gustaffsson [85], showing that PB-induced androstenedione hydroxylation was similar to that found *in vivo*. The authors concluded that the inclusion of rat serum in the culture medium was important to study P450 induction. The recent study of Wortelboer et al. [86] showed that PB only induced a marginal induction of P450 activities in cultured rat hepatocytes measured using a variety of endpoints, such as total P450 content, dealkylation of 7-ethoxyresorufin and 7-pentoxyresorufin, and testosterone hydroxylation. The induction found in the cultured hepatocytes was significantly lower than that found *in vivo*. This finding is consistent with the earlier finding of Schuetz et al. [23], which showed that hepatocytes cultured on collagen were only minimally induced by PB, whereas cells cultured on Matrigel had significant induction. Via the measurement of mRNA [40] and the measurement of cytochrome P450 activities [43], PB was found to induce mainly the isozymes P450 IIB1/IIB2.

By studying hepatocytes cultured on Matrigel, Schuetz et al. [40] further showed that growth hormone may directly affect the expression of some inducible P450 isozymes. Coincubation of hepatocytes with PB and human growth hormone (somatotropin) was shown to completely block the induction of P450 IIB1/IIB2 gene expression. The study was the first to show inhibitory control growth hormone on P450 induction. This observation demonstrates further the complex mechanisms involved in the regulation of P450 isozymes and the usefulness of the cultured hepatocyte system in the elucidation of the mechanisms.

11.5 HUMAN HEPATOCYTES

As stated earlier, one of the powerful applications of hepatocyte culture systems is the extrapolation to humans *in vivo* using data obtained from hepatocytes cultured from laboratory animals and humans and from laboratory animals *in vivo*, an approach we termed the *parallelogram approach* (Fig. 11.3). The key findings on human hepatocytes with toxicologic implications are summarized here.

The establishment of reproducible procedures to isolate and culture human hepatocytes, combined with the characterization of the properties of cultured human hepatocytes, is critical to this application. Many laboratories, including ours, have reported the successful isolation and culturing of human hepatocytes. Guguen-Guillouzo et al. [27] showed the isolation of highly viable hepatocytes from three human livers obtained from kidney donors. By cannulating a portal branch of the left lobe and ligating the arteries, successful perfusion of a small (10% of total) area was achieved, leading to hepatocytes with high viability. In our laboratory, we have experience with the isolation of hepatocytes from whole-human livers and surgical fragments. We found that perfusion of small liver portions, preferably 50 g or less, was the most efficient. We have an extremely high success rate (over 90%), yielding hepatocytes with high viability, as judged by dye exclusion and attachment [28]. The isolated human hepatocytes are responsive to the induction of DNA synthesis by epidermal growth factor [87] and can be transiently transfected with exogenous DNA [85].

Grant et al. [89] studied the stability of P450 isozymes in human hepatocytes cultured as monolayer cells in tissue culture plastic. The results showed that the decline in isozyme activity was apparently slower than that reported for rat hepatocytes. *O*-dealkylation activities of ethoxyresorufin, pentoxyresorufin, and benzyloxyresorufin at 72 hours after culturing were found to be approximately 64%, 162%, and 100%, respectively, of that found in freshly isolated cells. However, nicotinamide adenine dinucleotide phosphate (NADPH)-cytochrome c reductase and NADH-cytochrome b5 reductase decreased to 32% and 22% of the level found in fresh cells, respectively. During this time period, glutathione levels remained unaltered. This study represents one of the first to study the phase I oxidative and phase II conjugative metabolisms of human hepatocytes in culture.

Several studies illustrated the use of human hepatocytes in the evaluation of species differences in toxicity. Begue et al. [90] compared human and rat hepatocytes, showing that the naturally occurring flavonoid (=)-cyanidanol-3 could protect cultured rat hepatocytes against the cytotoxicity of aflatoxin-Bs, but it did not have protective effects toward human hepatocytes. The results suggest species differences in metabolism of aflotoxinB1 or (=)-cyanidanol-3. As mentioned earlier, LeBot et al. [91] showed differences between rat and human hepatocytes toward the cytotoxicity of anthracycline antibiotics. Using hepatocytes as the exogenous-activating system for promutagens, and by the direct measurement of DNA adduct formation, Hsu et al. [92] showed that aflotoxinB1 was more potent as a mutagen in rat hepatocytes than it was in mouse hepatocytes, with the potency of human hepatocytes falling somewhere in between. The difference in potency appears to agree with the known rat–mouse difference in carcinogenicity. This finding was later confirmed by ultra-structural studies of human, rat, and mouse hepatocytes, showing a similar order of potency ranking for the induction of nucleolar segregation [93]. These studies suggest that human hepatocytes in culture may be a relevant model for the evaluation of the sensitivity of human liver cells to carcinogens.

Donato et al. [94] showed that 3-methylcholanthrene, PB, and ethanol could induce cytochrome P450 MFO in cultured human hepatocytes. The level of induction (1.5- to 2-fold) was lower than that reported for cultured rat hepatocytes. Dexamethasone had no induction effect on human hepatocytes. This study is consistent with findings in our laboratory that cultured human hepatocytes are generally less responsive to enzyme inducers than are rat hepatocytes.

The most difficult problem with the use of human hepatocytes is the low availability of human liver tissue for research. Access to surgical wastes via collaboration with surgeons and having a laboratory in the vicinity of the operation room are important factors for a fruitful program. One approach to increase the utility of human hepatocytes is to develop a reproducible procedure for cryopreservation, so the cells can be stored and transported. We have developed a procedure with which human and rodent hepatocytes could be cryopreserved, with the recovered cells retaining both viability and xenobiotic metabolism [69,95]. The cryopreserved cells were later used to study radiopharmaceutical metabolism and found to yield metabolites similar to fresh cells, but at a lower rate [96]. Because of the potential complications of cryopreservation-related cellular damages, freshly isolated cells are still preferred over cryopreserved cells in toxicity studies in our laboratory.

11.6 INDUSTRIAL APPLICATIONS

Toxicologic properties play an important role in industrial product development. Not only do the products need to have the desired efficacy, they must also have the acceptable level of toxicity. It is not an uncommon event to have a potential product with extremely desirable efficacy rejected after the end of the battery of toxicologic studies. Product candidates with an acceptable preclinical toxicologic profile may also be rejected because of human toxicity discovered during clinical trials. In fact, hepatotoxicity is one of the toxicity endpoints easily measured in human subjects, and this has been the demise of a number of potential products.

The challenge is, therefore, how does one predict animal toxicity before expensive and time-consuming toxicology studies, and how does one ensure nontoxicity in humans? One approach is to perform short-term toxicology screening assays on related chemical structures with desired efficacy. For compounds with a structure that may have hepatotoxic effects, toxicity assays with cultured hepatocytes may be performed. In our laboratory, we have developed a concept called "preference index (P.I.)," which is simply a ranking of preference based on both efficacy and hepatocyte cytotoxicity for these compounds. When a series of structurally related compounds is to be compared, we first develop an efficacy ranking using a quantitative measure called "relative efficacy (R.E.)," where

R.E. = efficacy (chemical X)/efficacy (reference chemical).

This ranking is followed by the calculation of "relative cytotoxicity (R.C.)," where

R.C. = cytotoxicity (chemical X)/cytotoxicity (reference chemical).

The P.I. is then calculated as

P.I. = R.E./R.C.

The P.I. is, therefore, simply a ratio of the desired properties to the undesired properties. One can envision adding other factors to the equation, applying weights to each factor, and leading to the development of an equation for P.I. encompassing all known properties of the potential product. Compounds with high P.I. values should have the highest potential to be developed into a commercialized product.

For pharmaceutical development, metabolic stability and fate are also critical for product development. A chemical with an extremely short half-life may not be a practical drug, because continuous infusion may be required to sustain the therapeutic level. Using cultured hepatocytes, one can compare simultaneously the disappearance rate of a series of structural analogues, allowing the selection of, for instance, a structure with the highest stability. Although this approach ignores metabolism by other organs, one has to ensure that the liver is the major organ for metabolism of the chemicals studied. This approach may be studied via a comparison of *in vivo* and *in vitro* (cultured hepatocytes) metabolite profiles.

Extrapolation of human toxicity is an extremely valuable application. When one encounters a product candidate with obvious species differences in toxicity after testing in multiple species of laboratory animals, one has the question of which animal species is more similar to the human species. Using the parallelogram approach described earlier, data can be developed with cultured hepatocytes from the laboratory animals tested as well as from human hepatocytes. Understanding the correlation between the *in vivo* and *in vitro* studies with laboratory animals and the respective hepatocytes may allow one to evaluate which animal species is the closest to humans (Fig. 11.3).

11.7 CONCLUSIONS

It can be seen from the information presented in this chapter that cultured hepatocytes represent an exciting *in vivo* toxicologic system. Hepatocytes can now be isolated and cultured from virtually any mammalian species, including humans. Short-term hepatocyte cultures are known to retain most of the liver functions, including species differences in xenobiotic metabolism. Culturing conditions are constantly being refined to allow prolonged maintenance of the differentiated properties, with the most significant finding being the use of a reconstituted basement membrane fraction as attachment substratum. Various toxicity endpoints have been developed to allow research on hepatotoxicity, mutagenicity, and carcinogenicity. Enzyme induction in cultured hepatocytes, including the induction of peroxisomal enzymes and various isozymes, have been demonstrated. To be able to study metabolism, enzyme induction and toxicity in one system is definitely the strength of cultured hepatocytes as compared with other *in vitro* systems. With cultured hepatocytes, one can study (1) toxicity mechanisms, (2) correlation of toxicity and metabolism, and (3) the mechanism of enzyme induction under defined experimental conditions. If one were to use liver homogenates, besides the pitfalls in membrane disruption leading to potential artifacts, one could study neither toxicity nor enzyme induction. Problems with liver slices include (1) cell damage during preparation, (2) nonphysiological distribution of oxygen, nutrients, and test chemical to cells as different regions of the slices, and (3) the inability to study enzyme induction because of the short lifespan of the system. However, it is necessary to point out that the nonhepatocyte approaches are also important. With cell homogenate or microsome fractions, one can easily modify cofactor components and substrate concentrations, allowing one to evaluate specific biochemical parameters (e.g., K_m and V_{max}) for individual metabolic pathways. With liver slices, one can evaluate hepatotoxic effects on the nonparenchymal cell populations that are essentially absent in hepatocyte cultures. A successful investigator will be the one choosing complementary experiment systems that can produce the most comprehensive information.

Besides improving the culturing conditions, some aspects of the application of hepatocyte culturing systems in toxicology require further development. Based on our increasing knowledge in liver biology, mechanistic-based toxicity endpoints should be developed and validated with relevant *in vivo* systems. This development should include the studying of toxicity toward nonparenchymal cells, which is known to occur *in vivo* but virtually ignored by *in vitro* toxicologists. Sata need to be obtained with more model chemicals, especially with human hepatocytes, to validate the parallelogram approach for the prediction of human health risk. The approach of using cultured hepatocytes in the selection of an animal species that has the highest resemblance to humans in xenobiotic metabolism for *in vivo* toxicology testing should be seriously evaluated. A combination of cultured hepatocytes and other differentiated cells (e.g., intestinal cells) should be evaluated for the development of *in vitro* systems in response to multiorgan metabolism or toxicity. Hepatocyte culture as an experimental system already has a significant impact on toxicology. A thorough understanding of the system, knowing both the strengths and limitations, will allow its utilization to the fullest in our continual quest toward the prediction and understanding of chemical toxicity in humans.

References

1. Acosta D, Sorenson EMB, Anuforo DC, et al. An *in vitro* approach to the study of target organ toxicity of drugs and chemicals. *In Vitro Cellular Dev Biol* 1985;21:495–504.

2. Stammati AP, Silano V, Zucco F. Toxicology investigations with cell culture systems. *Toxicology* 1981;20:91–153.

3. Tyson CA, Stacey NH. *In vitro* screens from CNS, liver and kidney for systemic toxicity. In: Mehlman M, ed. *Benchmarks: Alternative Methods in Toxicology*. Princeton, NJ: Princeton Scientific, 1989:111–136.

4. Amacher DE, Schomaker SJ, Fasulo LM, Kenny CV. Assessment of 7-ethoxycoumarin 0-deethylase activities in primary rat, dog, and monkey hepatocytes under standardized culture conditions. *In Vitro Toxicol* 1997;10:373–382.

5. Shultz VD, Campbell W, Karr S, Hixon DC, Thompson NL. *TA1* oncofetal rat liver cDNA and putative amino acid permease: temporal correlation with c-myc during acute CCl_4 liver injury and variation of RNA levels in response to amino acids in hepatocyte cultures. *Toxicol Appl Pharmacol* 1999;154:84–96.

6. Terada M, Niguheshi F, Nurata K, Tomita T. Nepanipyrim, a new fungicide inhibits intracellular transport of very low density lipoprotein in rat hepatocytes. *Toxicol Appl Pharmacol* 1999;154:1–11.

7. Harman AW, Fisher LJ. Hamster hepatocytes in culture as a model for acetaminophen toxicity: studies with inhibitors of drug metabolism. *Toxicol Appl Phamacol* 1983;71:330–341.

8. Gandolfi AJ, Brendel K, Tisher R, Azri S, Hanan G, Waters SJ, Hanzlick RP, Thomas CM. Utilization of precision-cut liver slices to profile and rank-order potential hepatotoxins. Presented at PMA-Drusafe East Spring Meeting, May 2, 1989.

9. Adams PE. *In vitro* methods to study hepatic drug metabolism. *Emphasis* 1995;6(2).

10. Fisher RL, Hasal SJ, Sanuik JT, Gandolfi J, Brendel K. Determination of optimal incubation media and suitable slice diameter in precision-cut tissue slice culture, Part 2. *Toxicol Methods* 1995;5:115–130.

11. Brendel K, Fisher RL, Krandieck CL, Gandolfi AJ. Precision-cut rat liver slices in dynamic organ culture for structure-toxicity studies. In: Lapon CA, Frazier JM, eds. *In Vitro Biological Systems*. San Diego: Academic Press, 1993:222–230.

12. Drahushuk AT, McGarrigle BP, Slezak BP, Stegeman JJ, Olson JR. Time- and concentration-dependent induction of CYP1A1 and CYP1A2 in precision-cut rat liver slices incubated in dynamic organ culture in the presence of 2,3,7,8-tetrachlorodibenzo-*p*-dioxin. *Toxicol Appl Pharmacol* 1999;155:127–138.

13. Mehendale HM. Application of isolated organ techniques in toxicology. In: Hayes AW, ed. *Principles and Methods of Toxicology*. New York: Raven Press, 1989:699–740.

14. Wyman J, Stokes JS, Goehring M, Buring M, Moore T. Data collection interface for isolated perfused rat liver: recording oxygen consumption, perfusion pressure and pH. *Toxicol Methods* 1995;5:1–14.

15. Blouin A, Bolender RP, Weibel ER. Distribution of organelles and membranes between hepatocytes and nonhepatocytes in the rat liver parenchyma. *J Cell Biol* 1977;72:441–445.

16. Berry MN, Friend DS. High yield preparation of isolated rat liver parenchymal cells. *J Cell Biol* 1969;43,506–43,520.

17. Seglen PO. Preparation of liver cells. *Exp Cell Res* 1973;82:391–398.

18. Seglen PO. Preparation of isolated liver cells. In: Prescott DM, ed. *Methods in Cell Biology*. New York: Academic Press, 1976:30–78.

19. Reese JA, Byard JL. Isolation and culture of adult hepatocytes from liver biopsies. *In Vitro* 1981;17:935–940.

20. Smedsrod B, Pertoft H. Preparation of pure hepatocytes and reticuloendothelial cells in high yield from a rat liver by means of Percall centrifugation and selective adherence. *J Leokocyte Biol* 1985;38:213–230.

21. Teepe AG, Beck DJ, Li AP. A comparison of rat liver parenchymal and nonparenchymal in the activation of promutagens. *Environ Mol Mutagen* 1992;42:173–184.

22. Donato MT, Castell JV, Gomez-Lechon MJ. Co-cultures of hepatocytes with epithelial-like cell lines: expression of drug biotransformation activities by hepatocytes. *Cell Biol Toxicol* 1991;7:1–14.

23. Schuetz EG, Li D, Omiecinski CJ, Muller-Eberhard U, Kleinman HK, Elswick B, Guzelian PS. Regulation of gene expression in adult rat hepatocytes cultured on a basement membrane matrix. *J Cell Physiol* 1988;134:309–323.

24. Sinclair PR, Schuetz EG, Bement W, Haugen SA, Sinclair JF, May BK, Li D, Guzelian PS. Role of heme in phenobarbital induction of cytochromes P450 and 5-aminolevulinate synthase in cultured rat hepatocytes maintained on an extracellular matrix. *Arch Biochem Biophys* 1990;282:386–392.

25. Kleinman HK, McGarvey ML, Hassell JR, Star VL, Cannon FB, Laurie GW, Martin GR. Basement membrane complexes with biological activity. *Biochemistry* 1985;25:312–318.

26. Begue JM, Le Bigot JF, Guguen-Guillouzo C, Kiechel JR, Guillouzo A. Cultured human adult hepatocytes: a new model for drug metabolism studies. *Biochem Pharmacol* 1983;32:1,643–1,646.

27. Guguen-Guillouzo C, Clement B, Baffet G, Beaumont C, Morel-Chany E, Glaise D,

Guillouzo A. Maintenance and reversibility of active albumin secretion by adult rat hepatocytes co-cultured with another liver epithelial cell type. *Exp Cell Res* 1983;143:47–54.

28. Li AP, Colburn SM, Beck DJ. A simplified method for the culturing of primary adult rat and human hepatocytes as multicellular spheroids. *In Vitro Cell Dev Biol* 1992;28A:673–677.

29. Tong JZ, De Lagausie P, Furlan V, Cresteil T, Bernard O, Alvarez F. Long-term culture of adult rat hepatocyte spheroids. *Exp Cell Res* 1992;200:326–332.

30. Li AP, Barker G, Beck DJ, Colburn S, Monsell R, Pellegrin C. Culturing of primary hepatocytes as entrapped aggregates in a packed bed bioreactor: a potential bioartificial liver. *In Vitro Cell Dev Biol* 1993;29A:249–254.

31. Ruegg CE, Silber PM, Shughal RA, Ismail J, Lu C, Bode DC, Li AP. Cytochrome P-450 induction and conjugated metabolism in primary human hepatocytes after cryopreservation. *In Vitro Toxicol* 1997;10:207–216.

32. Guzelian PS, Bissel DM, Meyer VA. Drug metabolism in adult rat hepatocytes in primary monolayer culture. *Gastroenterology* 1977;72:1,232–1,239.

33. Emi Y, Chijiiwa C, Omura T. A different cytochrome P-450 form is induced in primary cultures of rat hepatocytes. *Proc Natl Acad Sci U S A* 1990;87:9,746–9,750.

34. Decad GM, Hsieh DPH, Byard JL. Maintenance of cytochrome P-450 and metabolism of aflatoxin B1 in primary hepatocyte cultures. *Biochem Biophys Res Commun* 1977;78:279–287.

35. Evarts RP, Marsden E, Thorgeirsson SS. Regulation of heme metabolism and cytochrome P-450 levels in primary culture of rat hepatocytes in a defined medium. *Biochem Pharmacol* 1984;33:565–569.

36. Grant MH, Melvin MA, Shaw P, Melvin WT, Burke MD. Studies on the maintenance of chromes P-450 and b5, monooxygenases and cytochrome reductases in primary cultures of rat hepatocytes. *FEBS Lett* 1985;190:99–103.

37. Muakkassah-Kelly SF, Bieri F, Waechter F, Bentley P, Staubli W. Long-term maintenance of hepatocytes in primary culture in the presence of DMSO: further characterization and effect of nafenopin, a peroxisome proliferator. *Exp Cell Res* 1987;171:37–51.

38. Pain AJ. The maintenance of cytochrome P-450 in rat hepatocyte culture: some applications of liver cell cultures to the study of drug metabolism, toxicity and the induction of the P-450 system. *Chem-Biol Interact* 1990;74:1–31.

39. Gross SA, Bandyopadhyay S, Klaunig JE, Somani P. Amiodarone and desethylamiodarone toxicity in isolated hepatocytes in culture. *Proc Soc Exp Biol Med* 1991;190:163–169.

40. Schuetz EG, Schuetz JD, May B, Guzelian P. Regulation of cytochrome P-450b/e and P-450p gene expression by growth hormone in adult rat hepatocytes cultured on a reconstituted basement membrane. *J Biol Chem* 1990;265:1,188–1,192.

41. Niemann C, Gauthier J, Richert L, Ivanov M, Melcion C, Cordier A. Rat adult hepatocytes in primary pure and mixed monolayer culture. Comparison of the maintenance of mixed function oxidase and conjugation pathways of drug metabolism. *Biochem Pharmacol* 1991;42:373–379.

42. Kocarek T, Schuetz EG, Guzelian PS. Differentiated induction of cytochrome p450b/e and P450p mRNAs by dose of penobarbital in primary cultures of adult rat hepatocytes. *Mol Pharmacol* 1990;38:440–444.

43. Sinclair JF, McCaffrey J, Sinclair PR, Bement WJ, Lambrecht LK, Wood SG, Smith EL, Schenkman JB, Guzelian PS, Park SS, Gelboin HV. Ethanol increases cytochromes P450 IIE, IIB1/2, and IIIA in cultured rat hepatocytes. *Arch Biochem Biophys* 1991;284:360–365.

44. Hammond AH, Fry JR. The *in vivo* induction of rat hepatic cytochrome p-450–dependent enzyme activities and their maintenance in culture. *Biochem Pharmacol* 1990;40:637–642.

45. Kane RE, Tector J, Brems J, Li A, Kaminski D. Sulfation and glucuronidation of acetaminophen by cultured hepatocytes reproducing *in vivo* sex-differences in conjugation on Matrigel and type I collagen. *In Vitro Cell Dev Biol* 1991;27A:953–960.

46. Green MD, Fisher LJ. Age- and sex-related differences in acetaminophen metabolism in the rat. *Life Sci* 1981;29:2,421–2,428.

47. Morel R, Vanderberghe Y, Pemble S, Taylor JB, Ratanasavanh D, Rogiers V, Ketterer B, Guillouzo A. Regulation of glutathione s-transferase subunits 3 and 4 in cultured rat hepatocytes. *FEBS Lett* 1989;258:99–102.

48. Vandenberghe Y, Foriers A, Rogiers V, Vercruysse A. Changes in expression and "*de novo*" synthesis of glutathione s-transferase subunits in cultured adult rat hepatocytes. *Biochem Pharmacol* 1990;39:685–690.

49. Green CE, LeValley Se, Tyson Ca. Comparison of amphetamine metabolism using isolate hepatocytes from five species including human. *J Pharmacol Exp Ther* 1986;237:931–936.

50. Dring LG, Smith RL, Williams RT. The metabolic fate of amphetamine in man and other species. *Biochem J* 1970;116:425–435.

51. Gee SJ, Green CE, Tyson CA. Comparative metabolism of tolbutamide by isolated hepatocytes from rat, rabbit, dog, and squirrel monkey. *Drug Metab Dispos* 1984;12:174–178.

52. Chenery RJ, Ayrton A, Oldham HG, Standring P, Norman SJ, Seddon T, Kirbt R. Diazepam metabolism in cultured hepatocytes from rat, rabbit, dog, guinea pig and man. *Drug Metab Dispos* 1987;15:312–317.

53. Monteith DK, Michalopoulos G, Strom SC. Metabolism of acetylaminofluorene in primary cultures of human hepatocytes: dose-response over a four-log range. *Carcinogenesis* 1988;9:1,835–1,841.

54. Weisburger JH, Grantham PH, Vanhorn E, Steigbigel NH, Rall DP, Weisburger EK. Activation and detoxification of n-2-fluorenylacetamide in man. *Cancer Res* 1964;24:475–479.

55. Humpel M, Sostarek D, Gieschen H, Labitzky C. Studies on the biotransformation of lonazolac, bromerguride, lisuride and terguride in laboratory animals and their hepatocytes. *Xenobiotica* 1989;19:361–377.

56. Sorensen EMB. Validation of a morphometric analysis procedure using indomethacin-induced alterations in cultured hepatocytes. *Toxicol Lett* 1989;45:101–110.

57. Thompson DC, Constantin-Teodosiu D, Moldeus P. Metabolism and cytotoxicity of eugenol in isolated rat hepatocytes. *Chem-Biol Interact* 1991;77:137–147.

58. Charbonneau M, Brodeur J, duSouich P, Plaa GL. Correlation between acetone potentiated CCl4-induced liver injury and blood concentrations after inhalation or oral administration. *Toxicol Appl Pharmacol* 1986;84:286–294.

59. Zimmerman HJ. Function and integrity of the liver. In: Henry JH, ed. *Clinical Diagnosis and Management by Laboratory Methods*. Philadelphia: Saunders, 1984:217–250.

60. Kachman JF, Moss DW. Enzymes. In: Tietz, N, ed. *Fundamentals of Clinical Chemistry*. Philadelphia: Saunders, 1976:565–698.

61. Somani P, Baudyopadhyay S, Klaunig JE, Gross SA. Amiodarone- and desethylamiodarone-induced myelinoid bodies and toxicity in cultured rat hepatocytes. *Hepatology* 1990;11:81–92.

62. Chaffee RRJ, Barlett WL. Inhibition of lactate dehydrogenase by α-mercaptoacids. *Biochim Biophys Acta* 1960;39:370–372.

63. Gomez-Lechon MJ, Monloya A, Lopez P, Donato T, Larrauri A, Castell JV. The potential use of cultured hepatocytes in predicting the hepatotoxicity of xenobiotics. *Xenobiotica* 1988;18:725–735.

64. Li AP, Heflich RH. *Genetic Toxicology*. Boca Raton, FL:CRC Press, 1991.

65. Heflich RH. Chemical mutagens. In: Li AP, Heflich RH, eds. *Genetic Toxicology*. Boca Raton, FL: CRC Press, 1990:143–202.

66. Ames BN, McCann J, Yamasaki E. Methods for detecting carcinogens and mutagens with the salmonella/mammalian microsome mutagenicity test. *Mutat Res* 1975;31:347–364.

67. Li AP. Hypothesis: modification of oncogene expression as a major mechanism of action of "non-genotoxic" carcinogens. *Environ Mol Mutagen* 1989;14:113–114.

68. Li AP, Gupta RS, Heflich RH, Wassom JS. A review and analysis of the Chinese hamster ovary/hyposanthine guanine transferase assay to determine the mutagenicity of chemical agents. A report of Phase III of the U.S. Environmental Protection Agency Gene-Tox Program. *Mutat Res* 1988;196:17–36.

69. Loretz LJ, Wilson AGE, Li AP. Promutagen activation by freshly isolated and cryopreserved rat hepatocytes. *Environ Mol Mutagen* 1988;12:335–341.

70. Glat HR, Billings R, Platt KL, Oesch F. Improvement of the correlation of bacterial mutagenicity with carcinogenicity of benzo(a)pyrene and four of its major metabolites by activation with intact liver cells instead of cell homogenate. *Cancer Res* 1981;41:270–277.

71. Williams GM. Detection of chemical carcinogens by unscheduled DNA synthesis in rat liver primary cultures. *Cancer Res* 1981;37:1,845–1,851.

72. Liu L, Castonguay A. Inhibition of the metabolism and genotoxicity of 4-(methylnitrosamino)-1-(3-pyridyl)-1-butanone (NNK) in rat hepatocytes by (+)-catechin. *Carcinogenesis* 1991;12:1,203–1,208.

73. Cole KE, Jones TW, Lipsky MM, Trump BF, Hsu IC. *In vitro* binding of aflatoxin B1 and 2-acetylaminofluorene to rat, mouse and human hepatocyte DNA: the relationship of DNA binding to carcinogenicity. *Carcinogenesis* 1988;9:711–716.

74. Li AP, Merrill JC. Development of a cultured hepatocyte system to study nongenotoxic mechanism of hepatocarcinogenesis. *Toxicologist* 1989;9:63.

75. Teddy JK, Azarnoff DL, Hignite CE. Hypolipidaemic hepatic peroxisome proliferators form a novel class of chemical carcinogens. *Nature* 1980;283:397–398.

76. Lake BG, Evans JG, Gray TJ, Korosi SA, North CJ. Comparative studies on nafenopin-induced peroxisome proliferation in the rat, Syrian hamster, guinea pig, and marmoset. *Toxicol Appl Pharmacol* 1989;99:148–160.

77. Hanefeld M, Kemmer C, Leonhardt W, Kunze KD, Jaross W, Haller H. Effects of p-dichlorophenoxy-isobutyric acid (CPIB) on the human liver. *Atherosclerosis* 1980;36:159–172.

78. de la Iglesia FA, Penn SM, Lucas J, McGuire EJ. Quantitative stereology of peroxisomes in hepatocytes from hyperlipoproteinemic patients receiving gemfibrozil. *Micron* 1981;12:97–98.

79. Blumcke S, Schwartzkopff W, Lobeck H, Edmondson NA, Prentice DE, Blane GF. Influence of fenofibrate on cellular and subcellular liver structure in hyperlipidemic patients. *Atherosclerosis* 1983;46:105–116.

80. Elcomb CR, Mitchell AM. Peroxisome proliferation due to di(2-ethylhexly)philhalate (DEHP): Species differences and possible mechanisms. *Environ Health Perspect* 1986;70:211–219.

81. Foxworthy PS, White SL, Hoover DM, Eacho PI. Effect of ciprofibrate, bezafibrate, and LY170883 on peroxisome beta-oxidation in cultured rat, dog, and rhesus monkey hepatocytes. *Toxicol Appl Pharmacol* 1990;104:386–394.

82. Nebert DW, Adesnik M, Coon MJ, Estabrook RW, Gonzalez FJ, Guengerich FP, Gunsalus IC, Johnson EF, Kemper B, Levin W, Phillips IR, Sato R, Waterman MR. The P450 gene superfamily: recommended nomenclature. *DNA* 1987;6:1–11.

83. Schuetz EG, Wrighton SA, Barwick JL, Guzelian PS. Induction of cytochrome p-450 by glucocorticoids in rat liver. 1. Evidence that glucocorticoids and pregnenolone 16-carbonitrile regulate de novo synthesis of a common form of cytochrome P-450 in cultures of adult rat hepatocytes and in the liver *in vivo. J Biol Chem* 1984;259:1,999–2,006.

84. Michalopoulos G, Sattler CA, Sattler G, Pitot H. Cytochrome P-450 induction by phenobarbital and 3-methylcholantherene in primary cultures of hepatocytes. *Science* 1976;193:907–909.

85. Stenberg A, Gustaffsson J. Induction of cytochrome P-450-dependent hydroxylases in primary monolayer cultures of rat hepatocytes. *Biochem Biophys Acta* 1978;540:402–407.

86. Wortelboer HM, de Kruiff CA, van Iessel AA, Falke HE, Noordhoek J, Blaauboer BJ. Comparison of cytochrome P450 isozyme profiles in rat liver and hepatocyte cultures. The effects of model inducers on apoproteins and biotransformation activities. *Biochem Pharmacol* 1991;42:381–390.

87. Li AP, Myers CA, Roque MA, Kaminski DL. Epidermal growth factor, DNA synthesis and human hepatocytes. *In Vitro Cell Dev Biol* 1991;27A:831–833.

88. Li AP, Myers CA, Kaminski DL. Gene transfer in primary cultures of human hepatocytes. *In Vitro Cell Dev Biol* 1992;28:678–684

89. Grant MH, Burke MD, Hawksworth GM, Duthie SJ, Engeset J, Petrie JC. Human adult hepatocytes in primary monolayer culture. Maintenance of mixed function oxidase and conjugation pathways of drug metabolism. *Biochem Pharmacol* 1987;36:2,311–2,616.

90. Begue JM, Baffet G, Campion JP, Guillouzo A. Differential response of primary cultures of human an rat hepatocytes to aflatoxin B1-induced cytotoxicity and protection by the hepatoprotective agent (+)-cyanidanol-3. *Biol Cell* 1988;63:327–333.

91. LeBot MA, Begue JM, Kernaleguen D, Robert J, Ratanasarah D, Airau J, Riche C, Guillouzo A. Different cytotoxicity and metabolism of doxorubicin, daunorubicin, epirubicin, esorubicin, and idarubicin in cultured human and rat hepatocytes. *Biochem Pharmacol* 1988;37:3,877–3,887.

92. Hsu IC, Harris CC, Lipsky MM, Snyder S, Trump BF. Cell and species differences in metabolic activation of chemical carcinogens. *Mutat Res* 1987;177:1–7.

93. Cole KE, Jones TW, Lipsky MM, Trump B, Hsu IC. Comparative effects of three carcinogens on human, rat and mouse hepatocytes. *Carcinogenesis* 1989;10:139–143.

94. Donato MT, Gomez-Lechon MJ, Castell JV. Effects of xenobiotics on monooxygenase activities in cultured human hepatocytes. *Biochem Pharmacol* 1990;39:1,321–1,326.

95. Loretz LJ, Li AP, Flye MW, Wilson AGE. Optimization of cryopreservation procedures for rat and human hepatocytes. *Xenobiotica* 1989;19:489–498.

96. Moerlein SM, Weisman RA, Beck D, Li AP, Welch MJ. Metabolism *in vitro* of radioiodinated *N*-isopropyl-para-iodoamphetamine by isolated hepatocytes. *Nucl Med Biol* 1992.

97. Edward AR, Wrighton SA, Pasco DS, Fagan JB, Li D, Guzelian PS. Synthesis and degradation of 3-methylcholanthrene-inducible cytochromes P-450 and their mRNAs in primary monolayer cultures of adult rat hepatocytes. *Arch Biochem Biophys* 1985;241:495–508.

12

Application of *In Vitro* Model Systems to Study Cardiovascular Toxicity

Kenneth Ramos[1] and Daniel Acosta[2]

[1]Department of Physiology and Pharmacology, College of Veterinary Medicine,
Texas A&M University, College Station, Texas

[2]College of Pharmacy, University of Cincinnati Medical Center, Cincinnati, Ohio

The cardiovascular system consists of the heart and blood vessels forming a circuit for the transport of oxygen and nutrients to all tissues throughout the body and the removal of waste products of metabolism. A highly regulated sequence of excitation/contraction coupling cycles from the atria to the ventricles allows the heart to pump blood through the vascular network. Cardiac output is initially received by the aorta, which in turn delivers blood to arterioles, capillaries, and post-capillary venules. Oxygen and nutrient exchange at the tissue level is regulated by discrete changes of microvascular resistance in response to intrinsic metabolic demands. Blood returns to the heart through capacitance vessels of the venous compartment. In serving these circulatory functions, cardiovascular cells are repeatedly exposed to blood-borne toxicants and their metabolic byproducts.

Although cardiovascular function is carried out in a coordinated fashion, the cardiovascular system is characterized by a large degree of structural and functional heterogeneity. From a morphologic standpoint, cardiac muscle consists of nodal tissue, Purkinje tissue, and ordinary muscle. Nodal tissue exhibits a high degree of automaticity, that is, the capacity to depolarize spontaneously, whereas

Purkinje tissue is specialized for the conduction of electrical impulses. Ordinary muscle cells exhibit variable degrees of electromechanic and pharmacomechanic coupling in response to contractile stimuli. The walls of the heart consist of three distinct layers. The epicardium is the external layer originating from visceral connective tissue. The middle layer is referred to as the myocardium and consists exclusively of muscle cells. The innermost layer of the heart is the endocardium, which is formed by a thin sheet of endothelial cells extending from the coronary vessels to line the chambers and valves of the heart.

The blood vessel wall of large- and medium-sized arteries consists of three distinct layers. The innermost layer, referred to as the *tunica intima*, represents a layer of endothelial cells resting on a thin basal lamina. In large vessels, such as the human aorta, a distinct subendothelial layer can also be identified. The medial layer consists of several sheets of smooth muscle cells dispersed in a matrix of collagen and elastin. The outermost layer consists of fibroblastic cells providing structural support to the vessel wall and contributing to the regulation of smooth muscle function. With the exception of capillaries, vessels of small diameter share many of the features described above. However, in these vessels, the media is less elastic and often limited to a few layers of smooth muscle cells. Capillaries are endothelial tubes resting on a thin basal lamina to which pericytes readily attach.

Cardiovascular toxicity can result from excessive accumulation of toxic chemicals within the tissue, cardiovascular-specific bioactivation of protoxicants, or chemical interference with specialized cellular functions. Because cardiotoxic insult interferes with the ability of the heart to pump blood through the vasculature, blood flow to major organs is often compromised. In contrast to the immediacy of cardiotoxic responses, vascular toxicities are often characterized by slow onsets and long latency periods. Angiotoxicity may cause alterations of arterial pressure, blood flow control, and vascular growth and remodeling. The assessment of cardiovascular toxicity *in vivo* is often complicated by the presence of complex humoral, neuronal, and endocrine influences.

A considerable increase in research activity directed at the elucidation of oxidative mechanisms in the pathogenesis of cardiovascular injury occurred during the past 5 years. This explosion was fueled by the recognition that cardiovascular disorders are characterized by loss of redox homeostasis, as often seen after many forms of toxicant injury. It is now established that oxidative modification of low-density lipoproteins (ox-LDL) plays a critical role in atherosclerosis and related disorders. Although precise mechanisms have not been elucidated, several groups have suggested that ox-LDL interferes with nitric oxide (NO) function, stimulation of cytokines and chemotactic factors or redox-regulated transcription factors, such as AP-1 and NF-kb. A central role for oxidant mechanisms in cardiovascular disease is consistent with studies showing evidence for beneficial effects by antioxidants in reducing cardiovascular events in patients with coronary heart disease [1] or ischemia/reperfusion injury [2].

The vascular production of reactive oxygen metabolites increases substantially in disease states, such as atherosclerosis and hypertension, and dietary lowering of cholesterol improves endothelium-dependent vascular relaxation and normalizes endothelial superoxide production [3]. Recent evidence suggests that the major source of vascular superoxide anion and hydrogen peroxide is a membrane-bound, reduced nicotinamide adenine dinucleotide (NADH)-dependent oxidase. In the context of this chapter, it is worth noting that *in vitro* technologies have shed light on the long-suspected link between cardiovascular/cerebrovascular disorders and emotional stress. Studies by Yu et al. [4] have shown that adrenaline toxicity to vascular endothelial cells during periods of emotional stress may involve deamination by monoamine oxidase A to form methylamine, a product further deaminated by semicarbazide-sensitive amine oxidase (SSAO) to formaldehyde, hydrogen peroxide, and ammonia. These findings are intriguing in light of the established role of SSAO in cardiovascular toxicity [5]. From our perspective, these findings underscore the usefulness of *in vitro* technologies in advancing our understanding of critical biologic mechanisms and chemical toxicity. This chapter presents an update of *in vitro* models commonly used in cardiovascular toxicity testing.

12.1 *IN VITRO* MODEL SYSTEMS IN TOXICITY TESTING

12.1.1 Perfused Organ Preparations

Perfused heart preparations, including the modified Langerdorff technique and the working heart preparation, have been used to evaluate various aspects of myocardial function *in vitro* after exposure to xenobiotics [6–8]. The Langerdorff heart preparation constitutes a beating heart that does not perform, whereas the working heart preparation is perfused through the left atrium to generate a left-sided working preparation. In the working heart preparation, perfusion fluid passes from the left atrium to the ventricle, where it is ejected via the aorta into a chamber against hydrostatic pressure to mimic physiologic resistance to flow [9]. Potassium-arrested hearts can also be used to examine flow-dependent effects in the absence of myocardial function [10]. Blood vessel segments from vascular beds can be isolated and processed for perfusion *in vitro*. Aortic preparations are most often preferred over other macrovascular preparations. Aortic segments can be readily excised and perfused and superfused with appropriate buffered solutions [11]. Smaller vessels can also be isolated and processed for perfusion *in vitro* [12]. Although the application of microvascular preparations in toxicologic studies is not widespread because of complexities associated with handling and processing vessels with diameters <100 μm, several laboratories now routinely evaluate the responses of microvessels to oxidants and other reactive species. For instance, work in Hein and Kuo's laboratory [13] using perfused preparations of coronary arterioles has recently shown that ox-LDL inhibits endothelium-dependent dilation in a manner comparable with the impairment observed in atherosclerotic vessels. Interestingly, their work suggested that initiation of superoxide anion production and subsequent deficiency of L-arginine may be responsible for the detrimental effect of ox-LDL.

Perfused organ systems are probably more representative of the *in vivo* situation than are other *in vitro* preparations. Perfused preparations are particularly advantageous because they retain the level of structural organization found *in vivo* without the influence of extraneous variables. Using these preparations, toxin-induced changes in physiologic/pharmacologic sensitivity and changes in excitability or contractility can be readily evaluated. In the case of vascular preparations, studies can be conducted in the presence or absence of endothelial cells to assess the interactions of luminal and medial cells. In this fashion, the biologic actions of NO, a soluble gas synthesized by the endothelium, were first discovered, and to date, much has been learned about the biology of this versatile molecule using perfused preparations. As noted previously, a growing list of conditions, including hypercholestrolemia, are associated with diminished release of NO into the arterial wall either because of impaired synthesis or excessive oxidative degradation. Diminished NO contributes to myocardial ischemia in patients with coronary artery disease and may facilitate vascular inflammation leading to oxidation of lipoproteins and foam cell formation [14]. It must be noted, however, that whether NO is useful or harmful depends on its chemical fate and the rate and location of its production within the tissue [15]. The balance between NO and superoxide is believed to be a critical determinant in many human diseases, including atherosclerosis, neurodegenerative disease, ischemia-reperfusion, and cancer [16].

The most significant limitation of perfused preparations in toxicity testing is that only a small number of replicate preparations can be processed at one time. The time required for isolation and placement of the tissue under physiologic conditions is critical to the preservation of tissue viability. Caution must also be exercised to ensure that initial fiber length (preload) and the force that must be overcome for muscle shortening (afterload) approximate those encountered *in vivo*. Perfused organ systems can only be used for short periods of time because of rapid loss of viability.

Parameters commonly used to evaluate xenobiotic-induced cardiotoxicity include time to peak tension, maximal rate of tension development, and tension development. The oxygen concentration of the perfusate entering the aorta and leaving the pulmonary artery can also be monitored as an index of myocardial oxygen consumption. Coronary flow can be measured by collecting effluent fractions as a function of time, and a latex balloon connected to a pressure transducer can be inserted in the left atria to monitor ventricular pressure. Pin electrodes connected to the right atrium, apex, and pulmonary vessels can be used to obtain electrocardiographic recordings. In the case of vascular preparations, the cannula employed for vessel perfusion can be equipped with side arms for pressure recording. As with cardiac preparations, measurements of contractility and stress development can also be used to evaluate the vascular effects of drugs and chemicals.

12.1.2 Isolated Muscle Preparations

Strips of atrial, ventricular, or papillary muscles [17], as well as segments from various vascular beds [18], can be placed in a bath and superfused with oxygenated physiologic solutions for measurements of tension development. Isolated preparations operate under constant conditions of carbon dioxide exchange, ionic gradients, and diffusions of byproducts of cellular metabolism. In the case of vascular preparations, spiral strips are preferred over longitudinal strips to avoid alterations in the geometry of muscle fibers. Alternatively, simple ring preparations can be prepared from most vessels. After equilibration in a physiologic solution, isolated muscle preparations are subjected to multiple stress/relation cycles to define optimal length—that is, the length at which maximal contractility occurs in response to a contractile agonist. The length of the equilibration period is dictated by intrinsic mechanic properties of the tissue. Preparations devoid of plasma membrane restrictions in permeability can be obtained by stripping off the plasmalemma using detergent or mechanical disruption [19–21]. Contractions can be elicited using standard solutions of free ionized calcium within the physiologic range, as described by Fabiato and Fabiato [22].

As with other *in vitro* models, exogenous influences of neuronal and humoral origin are excluded. The preload and afterload placed on the tissue *in vitro* can be controlled accurately. Experiments can be conducted to evaluate (1) isometric force development, which precludes muscle shortening; (2) isotonic force development, in which the afterload is predetermined and the muscle is able to shorten; and (3) quick-release contractions in which the afterload is varied during contraction [17]. Oxygenation of the tissue is caused by diffusion, and thus, the thickness of the strips and the concentration of oxygen in the bath must be carefully monitored. Concentration-response relationships can be constructed for selected contractile agonists in the absence or presence of toxin [23,24]. These relationships are obtained by cumulative increases in the concentration of each agonist without intervening washout until attainment of maximal developed force. However, before subsequent agonist addition, preparations should be allowed to reach steady state. In the case of vascular preparations, the effects of relaxing agents can also be evaluated. The vessel segment is precontracted to about 70–80% of the maximal contraction and then challenged with the relaxing agent. Isolated preparations can be controlled with precision, but their stability is limited to brief periods of time.

Significant progress was made during the past 5 years in advancing the application of isolated vascular preparations to study angiotoxic insult. For instance, studies by Haklar et al. [25] examined the effects of a high-cholesterol diet on the production of reactive oxygen species (ROS) in rabbit aortic rings, and the protective effects of vitamin E and probucol in preventing peroxidative changes. Their results indicated that cholesterol feeding increased ROS production, as reflected by changes in lucigenin and luminol chemiluminescence. Both vitamin E and probucol were effective as scavengers of free radicals, but the effect of vitamin E was more pronounced. In other studies [26], aortic strips were employed to asssess the ability of superoxide dismutase (SOD) mimetics to protect NO from destruction by oxidant stress in rabbit aorta. Oxidant stress was generated in isolated aortic rings by inactivation of endogenous Cu/Zn SOD with diethyldithiocarbamate, either alone or in combination with xanthine oxidase to generate superoxide. MnTMPyP, Mn (III) tetrakis [1-methyl-4-pyridyl]porphyrin, was the only SOD mimetic to restore NO-dependent relaxation in conditions of both extracellular and intracellular oxidant stress.

Ma et al. [27] conducted studies using combined *in vivo/in vitro* regimens to evaluate changes of the glutathione (GSH) system against peroxynitrite (ONOO-) in hypercholeterolemia, and the effects of carvedilol, a beta-blocker with free radical-scavenging activity, in this system. New Zealand white rabbits were fed either a normal diet, a high-cholesterol diet, or a high-cholesterol diet supplemented with either carvedilol or propranolol for 8 weeks. Hypercholesterolemia decreased tissue GSH content, attenuated the vasorelaxation response to ONOO-, reduced NO regeneration from ONOO-, and potentiated ONOO(-)-induced vascular tissue injury. Carvedilol afforded tissue protection from oxidant injury. These results suggest that hypercholesterolemia impairs GSH-mediated detoxification against ONOO- and renders vascular tissue more susceptible to oxidative injury. These studies exemplify the utility of *in vitro* approaches in defining mechanisms of toxic action within the context of additional adverse environment factors. The vasculotoxic effects of quinones were recently examined in aortic rings at rest or depolarized with 80-mM K^+ in the presence of nifedipine. 2,5-di-t-butyl-1, 4-benzoquinone evoked a slow tonic contraction that could be antagonized by Ni^{2+}. The myotonic effect was dependent on activation of Ca^{2+} influx via a Ni^{2+}-sensitive pathway, whereas its myolytic activity involved antagonism of Ca^{2+} entry via L-type channels or depletion of intracellular Ca^{2+} stores [28].

The contractile response of normal and atherosclerotic arteries to the aqueous component of cigarette smoke extract (CSE) was evaluated by Sugiyama et al. [29]. Thoracic aortas were isolated from control rabbits or 1.5% cholesterol-fed rabbits, all of which had visible, advanced atheromatous surface changes on the aortas. Rings were suspended in organ chambers and tested with CSE (0.01–3.0 'l/ml) after precontraction with phenylephrine. The contractile response to CSE was significantly greater in atherosclerotic aortas than it was in control aortas. The magnitude of the precontractions by phenylephrine was not different between control and atherosclerotic aortas. Exogenous addition of SOD significantly attenuated CSE-induced contractions in both control and atherosclerotic aortas and pretreatment of aortic rings with diethyldithiocarbamate to deplete endogenous vascular CuZn-SOD-activity potentiated CSE-induced contraction in control, but not atherosclerotic aortas. The vascular SOD activity was significantly lower in atherosclerotic aortas than it was in control aortas, indicating that atherosclerotic arteries may be supersensitive to the constrictor effect of superoxide anion derived from CSE. Reduced endogenous vascular SOD activity may partly contribute to increased susceptibility to oxidative stress in atherosclerotic arteries. In related studies [30], H_2O_2 (20–200 'm) was shown to elicit concentration-dependent contractions in rat aortic rings. Pretreatment with N-nitro-L-arginine, or removal of the endothelium, augmented the contractile response. Extracellular Ca^{2+} entry through voltage-dependent Ca^{2+} channels and intracellular Ca^{2+} from caffeine- and noradrenaline-sensitive stores appeared to participate in H_2O_2-induced contractions. More recently, aortic ring preparations were used by Conklin and Boor [31] to evaluate the vasculotoxic actions of allylamine, a toxic amine mimicking atherosclerosis by induction of oxidative stress by acrolein and H_2O_2 and formation of cellular lesions. Aortic rings from allylamine-treated rats were more sensitive to the contractile effects of high potassium or norepinephrine than were anatomically matched control rings. No difference in response to acetylcholine- or sodium nitroprusside-induced relaxation was detected between control and allylamine-treated rat aortic rings. These findings demonstrate that allylamine injury alters contractile functions at the level of vascular smooth muscle.

Environmental tobacco smoke (ETS) is associated with increased coronary artery disease, and some of the mechanisms involved in endothelial dysfunction have been elucidated using aortic rings [32]. ETS has been shown to reduce endothelium-dependent acetylcholine-induced relaxation, a response antagonized by L-arginine. From these studies, it was concluded that chronic dietary supplementation with an NO donor, such as L-arginine, offsets the endothelial dysfunction associated with ETS in normocholesterolemic rabbits, possibly through substrate loading of the NO pathway. These findings may be a major significance in advancing our understanding of pathogenetic mechanisms associated with ETS. In related studies, Satake et al. [33] showed that the cytochrome P-450 (CYP) system may be involved in the endothelium-dependent relaxation of vascular smooth muscle by isoproterenol. The role of free radical mechanisms in the regulation of vascular function is consistent with studies by Ito et al. [34], who showed that treatment with diethyldithiocarbamate (DDC) decreases total SOD activity and increases superoxide anion levels in aortic rings. Compromised antioxidant capacity was in turn associated with increases in the steady-state mRNA levels of glyceraldehyde-3-phosphate dehydrogenase (GAPDH). Although decreased biologic activity of endothelium-derived NO was indicated by lower basal cyclic guanosine monophosphate (cGMP) levels in aortic rings treated with DDC compared with those in control rings, neither endothelium denudation nor NG-nitro-L-arginine methyl ester had any effects on the steady-state mRNA levels of GAPDH. These results suggest that oxidative stress causes induction of GAPDH gene expression and that ROS may be involved in this response.

Isolated preparations have recently been employed to examine the angiotoxic effects of ethanol [35], acetaldehyde [36], palytoxin [37], and cadmium [38]. In the latter study, cadmium was administered intraperitoneally at a dose of 1 mg/kg/day for 15 days and shown to cause a significant increase in mean arterial blood pressure. Endothelin-1 and noradrenaline produced concentration-dependent contractions of aortic rings that attained a lower maximal contraction in cadmium-injected rats compared with control rats. On the other hand, responses of aortic rings to different concentrations of potassium chloride did not show a significant difference between groups. Decreased responsiveness of the aortae of cadmium-hypertensive rats to endothelin-1 and noradrenaline was hypothesized to be caused by an interaction of cadmium with receptors or intracellular signal transduction pathways of these agents, or adaptive changes in vascular tissues after hypertension development.

Regional differences in physiologic and pharmacologic responsiveness must be considered in developing strategies that examine the vasculotoxic response. For instance, aortic rings exhibit higher sensitivity to norepinephrine than do mesenteric artery ring counterparts, whereas the sensitivity of the mesenteric artery to serotonin was considerably greater than that of aortic ring segments [39]. However, no differences in sensitivity to KCl or CaCl$_2$ between aorta and mesenteric artery were found. The phasic contraction of the aorta to norepinephrine was greater than the phasic contraction of the mesenteric artery, whereas the magnitude of phasic contraction of the aorta to serotonin was lower than that of the mesenteric artery. Thus, differences between the two vessels are likely dependent on the agonist's ability to mobilize calcium from intracellular stores.

12.1.3 Organ Culture

Organ culture systems are powerful tools to study cardiovascular cell biology. Ingwall et al. [40] described the use of whole fetal hearts in culture to study processes associated with myocardial cell injury. A variation of the method using the right atria has been reported by Speralakis and Shigenoubu [41] and by Tanaka et al. [42]. A similar approach was reported by Gotlieb and Boden [43], who described the preparation of organ cultures of aortic tissue. Organ culture preparations offer long-term stability relative to other *in vitro* preparations. In the case of blood vessels, this technology allows study of cell–cell and cell–substratum interactions, but it also allows study of structural functional relationships of the vessel wall matrix [44]. Long-term organ cultures of porcine aorta have led to the recognition that neointimal formation is caused primarily by cell proliferation of preexisting intimal smooth muscle cells. Interestingly, neointimal formation in these cultures is more pronounced in the presence of an endothelium that is turning over. In endothelial wound repair studies, the endothelium of the organ culture shows important differences when compared with tissue culture studies in monolayer culture showing that cell–cell interactions are important in the regulation of vascular functions. Nakazawa et al. [45] recently studied the basic characteristics of the rat embryonic circulation and the hemodynamic effects of several drugs, including α- and β-agonists, digitalis, and atrial natriuretic peptide, using a modified organ culture system.

Organ culture of rat aortic rings for as little as 24 hours results in significant loss of contractile responsiveness to different agonists. Wang et al. [46] showed that active tension development can be preserved when aortic segments are cultured in retinol-supplemented medium. The protection afforded by retinol is lost when rings are denuded of endothelium before culture, suggesting the endothelial cell layer as a mediator of this effect. In complementary experiments, retinol was found to have no direct effect on proliferation of cells of the A7r5 smooth muscle cell line, but it was found to augment growth inhibition of A7r5 cells grown in co-culture with bovine aortic endothelial cells.

12.1.4 Tissue Slices

Slice preparations of cardiac tissue have been developed and characterized as models to evaluate the toxicity of xenobiotics [47]. This system has proven useful in attempting to delineate interspecies comparisons of biotransformation and toxicity. The lack of sample reproducibility and limited viability of these preparations has been improved using automated slicers reproducibly generating thin tissue slices. The development of dynamic organ cultures of tissue slices, as recently described for liver, may prove useful for the study cardiotoxic events [48].

12.1.5 Single-Cell Suspensions

Various combinations of physical, enzymatic, and chelating agents have been successfully employed to isolate cells of cardiovascular origin. Suspensions of embryonic or neonatal cells derived from ventricular, atrial, or whole-heart tissue can be easily prepared by enzymatic or mechanical dissociation. Isolation of cardiac myocytes typically requires the use of calcium-free solutions to weaken the connective tissue matrix. The isolation of adult myocytes exhibiting tolerance to physiologic calcium concentrations requires the use of more sophisticated techniques. Jacobson [49] was one of the first to describe the use of a dissociation apparatus consisting of finger-like projections (bristles) providing abrasive action for the isolation of adult heart cells. Bkailey et al. [50] suggested replacement of calcium with strontium or barium to prevent loss of physiologic regulation. Adult hearts can also be dissociated by a modified, recirculating Langerdorff perfusion [51] yielding a high proportion of cells remaining rod shaped and quiescent in medium containing physiologic calcium levels. Welder et al. [52] have reported a method for the isolation of adult rat cardiac myocytes using a custom-made glass apparatus for simultaneous perfusion of multiple hearts.

The anatomic distribution of cells within the mammalian vessel wall of large- and medium-sized vessels facilitates the isolation of relatively pure suspensions of fibroblastic, endothelial, or smooth muscle cells. In contrast to cardiac preparations, vascular cells from embryonic, neonatal, and adult vessels can be efficiently isolated in calcium- and magnesium-containing solutions. This discrepancy may be based on differences in metabolic demand allowing vascular cells to remain viable under hypoxic conditions for longer periods. Endothelial cells are typically isolated by collagenase perfusion of intact cylinders, whereas collagenase/elastase mixtures or trypsin are employed to isolate medial smooth muscle cells or fibroblasts. Enzymes, temperature, osmolarity/pH, and time of incubation are important determinants of cell viability immediately after isolation. Most investigators would agree that purified collagenase alone is not adequate for the isolation of cardiovascular cells, which suggests that other constituents present in the crude enzyme preparation are essential to the isolation process. Caution must be exercised when purchasing commercially available enzyme preparations because different lots and suppliers often exhibit different activities. Other proteolytic enzymes, such as trypsin or papain, have been successfully used to isolate neonatal cardiac myocytes and adult vascular myocytes. Although trypsin is useful for the isolation of myocytes, caution must be exercised to avoid deleterious effects, such as increased cellular aggregation, removal of surface enzymes, reduction in receptor binding sites, and chromosomal damage. Carrying out the enzymatic digestion at an optimal low temperature can minimize cell damage during the isolation procedure. In most experimental systems, the use of trypsin in the isolation of endothelial cells is avoided to minimize cellular injury.

Myocardial cell suspensions represent a heterogeneous population of muscle and nonmuscle cells. Neonatal myocytes are remarkably resistant to injury and exhibit variable degrees of beating shortly after isolation. In contrast, spontaneous beating of adult cardiac myocytes is thought to be caused by uncontrolled leakage of calcium through a permeable plasma membrane. This phenomenon is thought to represent a calcium-paradox phenomenon similar to that originally described by Zimmerman and Hulsmann [53]. Under these conditions, release of intracellular protein, accumulation of calcium and sodium within the cell, and loss of intracellular potassium characterize cellular injury. These cells are said to be calcium-intolerant and often can be identified by their rounded appearance. However, assessment of cell shape alone may be misleading because rod-shaped cardiomyocytes may display abnormal resting membrane potentials [50]. Viable adult myocytes should exhibit normal sarcomere length and remain quiescent at low external calcium concentrations and contract in response to increasing calcium concentrations. The fact that adult cardiac myocytes are mechanically at rest when properly isolated suggests that fundamental differences in regulation exist between adult and neonatal cells.

Isolated cells can be microinjected with fluorescent dyes for the assessment of multiple cellular functions after exposure to toxic chemicals. Muscle cells in suspension can also be voltage-clamped to evaluate spontaneous and chemically induced electrophysiologic changes. In general, toxic responses may be of slower onset or differing magnitude relative to those observed in vivo. The viability of cells in suspension decreases rapidly as a function of time. For instance, the density of saturable β-adrenergic binding sites and coupling efficiency are significantly reduced in cardiac cell suspension [54]. More recently, isolated rat myocytes have been used to investigate mechanisms of cardiac preconditioning [55].

12.1.6 Cell Culture Systems

Because cells in suspension exhibit short-term stability, cell culture systems are preferred to evaluate chemical toxicity after prolonged exposures or after multiple challenges *in vitro*. Primary cultures can be established with relative ease from cell suspensions of cardiac and vascular tissue. Cultures must be characterized at the morphologic, ultrastructural, biochemical, or functional level. Vascular endothelial and smooth muscle cultures can also be established by the explant method, in which pieces of tissue are placed in a culture vessel to allow for cellular migration and proliferation *in vitro*. Although popular, the application of explant techniques may present limitations in attempting to define mechanisms of toxicity because this method affords cells with enhanced migratory potential a selective growth advantage *in vitro* [56]. Neonatal and embryonic cells of cardiac origin proliferate readily under appropriate conditions *in vitro* [57]. Although cardiac myocytes from adult animals do not divide in culture, recent studies have suggested that the ability of cardiac myocytes for cell division is repressed, but not completely lost [58]. This finding is consistent with studies showing that some atrial myocytes synthesize large amounts of DNA and undergo complete cell division after an infarct to the ventricle [59]. Myocardial cell division can also be stimulated by insertion of the large tumor antigen from the SV40 virus into the myocyte genome [60]. Wang et al. [61] successfully established a human fetal cardiac myocyte cell line by cotransfection of human cardiac myocytes with the SV40 large T antigen. These cells preserve many of the morphologic and functional features of human fetal cardiac myocytes in primary culture.

Vascular endothelial and smooth muscle cells derived from large- and medium-sized vessels of embryonic, neonatal, or adult animals proliferate readily under appropriate conditions *in vitro* [62]. Thus, cultures can be propagated to prepare cell strains retaining variable degrees of differentiation as a function of growth conditions *in vitro*. An early report on the isolation of endothelial cells from the heart microvasculature was presented by Simionescu and Simionescu [63]. Since then, several reports have appeared examining the effects of chemical injury on the microvasculature. Obeso et al. [64] described the development of a stable murine cell line from a mouse hemangioendothelioma as a model to study the responses of microvascular endothelial cells. These endothelioma cells synthesize angiotensin-converting enzyme, express surface receptors for acetylated LDL, produce thrombospodin, and show intracellular staining with an antibody to von Willebrand antigen. However, the application of such a model in toxicity testing may be limited because these cells express properties of a neoplastic phenotype. More recently, Bastaki et al. [65] completed studies to establish three mouse endothelial cell lines from aorta (MAECs), brain capillaries (MBECs), and heart capillaries (MHECs) to study angiogenesis *in vitro* and to compare their findings with established *in vivo* murine models. These lines were characterized for endothelial phenotypic markers, *in vivo* tumorigenic activity, basic fibroblast growth factor (bFGF) responsiveness. These cells were shown to express angiotensin-converting enzyme, acetylated LDL receptor, constitutive endothelial NO synthase, and vascular cell adhesion molecule-1 and bind Griffonia simplicifolia-I lectin. When injected subcutaneously in nude mice, MAECs induced the appearance of slow-growing vascular lesions reminiscent of epithelioid hemangioendothelioma, whereas MBEC xenografts grew rapidly, showing Kaposi's sarcoma-like morphologic features. No lesions were induced by injection of MHECs. Interestingly, all lines expressed both low-affinity heparan sulfate bFGT-binding sites and high-affinity tyrosine kinase receptors on their surfaces. These endothelial lines are thus suitable models to study mouse endothelium during angiogenesis and angioproliferative diseases. In other studies, an *in vitro* model of vascular injury by menadione-induced oxidative stress in bovine heart microvascular endothelial cells was developed [66]. Their results showed that menadione toxicity was mediated by poly[adenosine diphosphate (ADP)-ribose] polymerase activation by hydrogen peroxide.

In the preparation of primary cultures of heart cells, cardiac myocytes can be separated from nonmuscle cells by a differential pour-off technique based on the rate of attachment of cells in suspension to the substratum [67]. Most of the non-muscle cells attach to the dishes within 3 hours, whereas muscle cells remain in suspension for 16–19 hours. Because the percentage of fibroblasts in culture increases logarithmically, if no attempt is made to separate individual cell types, fibroblasts will eventually dominate the culture. This is a particular problem for cultures of prenatal or postnatal myoblasts because the cells resemble fibroblasts in their physical appearance. A monoclonal antibody to cell surface adhesion factors has been used to enrich preparations of cultured myocardial cells [68]. Addition of mitotic inhibitors and maintenance of cultures in glutamine-free or serum-free media have been successfully used to enhance culture purity [69]. By taking advantage of the anatomic distribution of cells within the vascular wall of large- and medium-sized arteries, relatively pure suspensions of vascular endothelial cells and smooth muscle cells can be prepared. Under most experimental conditions, vascular smooth muscle cells do not exhibit spontaneous contractility, but contract in response to pharmacologic stimulation [70]. Endothelial cells subjected to fluid mechanic forces associated with blood flow become elongated and orient themselves in the direct of shear stress [71]. Growth of endothelial cells on extracellular matrix (ECM) components, on layers of smooth muscle cells, and under flow may mimic some aspects of the vascular wall not found when grown on plastic.

The maintenance of most cells *in vitro* requires either serum or plasma for attachment, proliferation, and survival. Early on, Sato [72] demonstrated that the serum or plasma requirement for cell growth of several cell lines can be satisfied by addition of specific hormones and growth factors to synthetic media. The growth environment of cells in culture is an important determinant of cellular behavior *in vitro*. Although the mixture of collagens, noncollagen proteins, and carbohydrate-rich molecules comprising the ECM was originally described as a static support for cells *in vivo*, it is now recognized that matrix molecules modulate cellular behavior and toxicologic responsiveness. In this context, we have shown that the matrix regulates the phenotypic expression of aortic smooth muscle cells in culture [73]. Recent extensions of this work established a role for ECM interactions in the induction of proliferative vascular smooth muscle cells (vSMC) phenotypes after oxidative chemical injury [74]. Other important considerations related to the use of cultured cell systems in toxicology studies include recognition that the presence of serum modulates the antioxidant capabilities of cells *in vitro* [75] and that cardiovascular cells in culture undergo variable degrees of dedifferentiation, including loss of defined contractile features and cell-specific functions [76–78].

Of particular interest from a toxicologic perspective is that CYP activities are present in cells of the cardiovascular system and participate in metabolism-dependent toxicities. For instance, vertebrate cardiac endothelial cells of the marine scup express CYP1A1 [79]. This activity appears to be present in animals from contaminated environments, suggesting that endothelial may be important in the toxicologic effects of xenobiotics affecting the vasculature of the heart and other organ systems. Although CYPI family members are often associated with metabolism of exogenous substrates, more recent studies have linked P450s to eicosanoid formation [80,81]. The pattern of constitutive and inducible expression of CYPIA1 and CYPIB1 genes in cultured human vascular endothelial cells (ECs) and smooth muscle cells (SMCs) was recently defined [82]. The results of these studies showed that human vascular cells exhibit inducible CYPIA1 and CYPIB1 mRNA and associated enzymatic activities. In view of increasing recognition of the role of eicosanoid formation and macromolecular oxidations in atherogenesis, these results raise provocative questions regarding the connection between aromatic hydrocarbons and vascular disorders.

The occurrence of gross morphologic changes is routinely used to screen chemicals of unknown toxic potential and to access potentially significant toxic interactions. Flow cytometry and computerized evaluation of cell images has greatly expanded the usefulness of microscopic analysis in toxicity testing [83]. In the case of cardiac myocytes, toxicity can also be evaluated based on the arrhythmogenic potential of chemicals [84,85]. Arbitrary or computer-assisted grading systems can be implemented to evaluate these responses [86]. Because toxicity is often caused by interactions with, or disruption of antioxidant defense systems, measurements of the status of cells may be particularly useful in the elucidation of mechanisms of toxicity [87]. Ionic homeostasis can also be used as an index of disturbances in the structural and functional integrity of the plasma membrane [88,89]. Of particular interest is the application of coculture systems of muscle and nonmuscle cells to the assessment of cardiovascular toxicity. Cocultures of vascular endothelial and smooth muscle cells have been successfully used to study cell–cell interactions *in vitro* [90,91]. Cardiac myocytes have also been cocultured in the presence of neurons to attempt to replicate some of the relationships observed *in vivo*. These systems reconstruct the complexities of the cellular environment *in vivo*, but retain the advantages inherent to cell culture. The complexity of coculture systems is exemplified by studies showing that endothelial cells modulate the extent of binding, internalization, and degradation of LDL by arterial smooth muscle cells [92], and produce growth factors for both smooth muscle cells and fibroblasts [93]. Using a coculture technique, Vender [94] investigated the role of bovine pulmonary arterial endothelial-cell–derived factors in mediating abnormal smooth muscle cell growth under conditions of reduced oxygen tension. In this system, reduced oxygen tension of cultured endothelial cells stimulated the release of soluble factors that stimulated the proliferation of smooth muscle cells relative to normoxic conditions. These findings emphasize that if cell-to-cell interactions are critical in the progression the toxic response, the use of single cell culture systems is inherently limited.

Particularly relevant to the study of vascular toxicity are the paracrine influences exerted by inflammatory cells recruited to sites of injury, such as macrophages. To study such interactions, we have employed a system in which cells are seeded individually on separate surfaces that can be brought together to prepare cocultures. Smooth muscle cells are seeded on conventional tissue culture plates, whereas macrophages are seeded on tissue culture inserts made of a semipermeable membrane of varying pore sizes. Because this coculture model obviates direct cell–cell contact, individual cell populations can be separated at any time, while allowing soluble factors secreted from one cell to interact with the other. Using this technology, the modulatory effects of macrophages on smooth muscle cell proliferation and differentiation and the impact of dietary manipulation on these interactions have been studied [95–97].

Cardiovascular cells are continually exposed to a pressure gradient induced by the pulsatility of blood flow. The loss of this critical influence may contribute to the spontaneous changes in cellular phenotype when cells are placed in the static culture environment. In the case of vascular smooth muscle cells, the mechanism by which cells modulate from a quiescent-differentiated phenotype to a synthetic state often involves inactivation of contractile genes and loss of cell-specific contractile proteins. In spite of the long-held notion that the microenvironment of cultured cells is a powerful regulator of phenotypic expression, conventional cell culture technologies have often overlooked the impact of physical forces and ECM on the behavior of vascular cells. To address some of these limitations, we have recently used instrumentation replicating the pulsatility of blood flow *in vivo* by stretching cells seeded on a flexible matrix. This approach in cells is maintained in a static versus a mechanically active culture [98,99]. It is also worth noting that growth of vascular smooth muscle cells on different substrates and supplementation with vitamins, minerals, or factors can influence the extent to which differentiated properties are expressed *in vitro* [78].

As with isolated muscle preparations, free radical mechanisms have been extensively studied during the past 5 years using cell culture systems. Studies by Loegering et al. [100] showed the antioxidant, dietyldiothiocarbamate, can attenuate lipopolysaccharide (LPS)-stimulated induction of NO synthase in vascular smooth muscle cells and may thereby ameliorate impairment of vascular reactivity in aortic rings by ROS. Using cultured vascular smooth muscle cells, it has been established that the atherogenic response induced by benzo(a)pyrene (BaP) is associated with up-regulation of mitogen-activated *c-Ha-ras* and loss of coordinated cell-cycle progression in vascular smooth muscle cells. *c-Ha-ras* activation by BaP is mediated via electrophile-responsive cis-acting elements within the promoter region participating in redox regulation of gene transcription [101,102].

12.2 CONCLUDING REMARKS

The successful application of *in vitro* model systems to evaluate cardiovascular toxicity has evolved from a coordinated research effort over many years. Experiments using any one of the *in vitro* model systems described here, either alone or in combination, have been useful in advancing our understanding of biologic mechanisms of chemical toxicity. If *in vitro* models are to be used as predictors of the human response, preparations derived from species responding with fidelity to the toxic challenge should be employed. Clearly, the appeal of *in vitro* systems in toxicology will continue to grow as modern molecular technologies continue to be adapted to the repertoire of *in vitro* methodologies.

References

1. Napoli C. Low density lipoprotein oxidation and atherogenesis: from experimental models to clinical studies. *Giornale Italiano di Cardiologia* 1997;27:1,302-1,314.

2. Willy C, Thiery J, Menger M, Messmer K, Arfors KE, Lehr HA. Impact of vitamin E supplement in standard laboratory animal diet on microvascular manifestation of ischemia/reperfusion injury. *Free Rad Biol Med* 1995;19:919–926.

3. Harrison DG. Endothelial function and oxidant stress [review]. *Clin Cardiol* 1997;20:1, 111–117.

4. Yu PH, Lai CT, Zuo DM. Formation of formaldehyde from adrenalin *in vivo*; a potential risk factor for stress-related angiopathy. *Neurochem Res* 1997;22:615–620.

5. Ramos KS, Grossman SL, Cox LR. Allylamine-induced vascular toxicity *in vitro*: prevention by semicarbazide-sensitive amine oxidase inhibitors. *Toxicol Appl Pharmacol* 1988;95:61–71.

6. Pilcher GD, Langley AE. The effects of perfluoro-n-decanoic acid in the rat heart. *Toxicol Appl Pharmacol* 1986;85:389–397.

7. Hale PW, Poklis A. Cardiotoxicity of thioridazine and two steroisomeric forms of thioridazine 5-sufloxide in the isolated perfused rat heart. *Toxicol Appl Pharmacol* 1986;86:44–55.

8. Khatter JC, Agbanyo M, Navaratnam S, Nero B, Hoeschen RJ. Digitalis cardiotoxicity: cellular calcium overload a possible mechanism. *Basic Res Cardiol* 1989;84:553–563.

9. Neely JE, Liebermeister H, Battersby EJ, Morgan HE. Effect of pressure development on oxygen consumption in isolated rat hearts. *Am J Physiol* 1967;212:804–814.

10. McFaul SJ, McGrath JJ. Studies on the mechanism of carbon monoxide-induced vasodilation in the isolated perfused rat heart. *Toxicol Appl Pharmacol* 1987;87:464–473.

11. Crass MF, Hulsey SM, Bulkley TJ. Use of a new pulsatile perfused rat aorta preparation to study the characteristics of the vasodilator effect of parathyroid hormone. *J Pharmacol Exp Ther* 1988;245:723–734.

12. Granger HJ, Schelling ME, Lewis RE, Zaweija DC, Meininger CJ. Physiology and pathobiology of the microcirculation. *Am J Otolaryngol* 1988;9:264–277.

13. Hein TW, Kuo L. LDLs impair vasomotor function of the coronary microcirculation: role of superoxide anions. *Circ Res* 1998;83:404–414.

14. Cannon RO. Role of nitric oxide in cardiovascular disease: focus on the endothelium. *Clin Chem* 1998;44:1,809–1,819.

15. Mayer B, Hemmens B. Biosynthesis and action of nitric oxide in mammalian cells. *Trends Biochem Sci* 1997;22:477–481.

16. Darley-Usmar V, Wiseman H, Halliwell B. Nitric oxide and oxygen radicals: a question of balance. *FEBS Lett* 1995;369:131–135.

17. Foex P. Experimental models of myocardial ischemia. *Br J Anaesth* 1988;61:44–55.

18. Hester RK, Ramos K. Vessel cylinders. In: Tyson C, Frazier J, eds. *Methods in Toxicology*. San Diego, CA: Academic Press, 1991.

19. Ramos K. Sarcolemmal dependence of isosorbide dinitrate-induced relaxation of vascular smooth muscle. *Res Commun Chem Pathol Pharmacol* 1986;52:195–205.

20. Chatterjee M, Murphy RA. Calcium-dependent stress maintenance without myosin phosphorylation in "skinned" smooth muscle cells. *Science* 1983;221:464–466.

21. Fabiato A, Fabiato F. Activation of skinned cardiac cells. Subcellular effects of cardioactive drugs. *Eur J Cardiol* 1973;1:143–155.

22. Fabiato A, Fabiato F. Calculator programs for computing the composition of the solutions containing multiple metals and ligands used for experiments with skinned muscle cells. *J Physiol* 1979;75:363–505.

23. Gibbs CL, Woolley G, Kotsanas G, Gibson WR. Cardiac energetics in daunorubicin-induced cardiomyopathy. *J Mol Cell Cardiol* 1984;16:953–962.

24. Togna G, Dolci N, Caprino L. Inhibition of aortic vessel adenosine diphosphate degradation by cadmium and mercury. *Arch Toxicol* 1984;7:378–381.

25. Haklar G, Sirikci O, Ozer NK, Yalcin AS. Measurement of reactive oxygen species by chemiluminescence in diet-induced atherosclerosis: protective roles of vitamin E and probucol on different radical species. *Int J Clin Lab Res* 1998;28:122–126.

26. MacKenzie A, Martin W. Loss of endothelium-derived nitric oxide in rabbit aorta by oxidant stress: restoration by superoxide dismutase mimetics. *Br J Pharmacol* 1998;124:719–728.

27. Ma XL, Lopez BL, Liu GL, Christopher TA, Gao F, Guo Y, Feuerstein GZ, Ruffolo RR Jr, Barone FC, Yue TL. Hypercholesterolemia impairs a detoxification mechanism

against peroxynitrite and renders the vascular tissue more susceptible to oxidative injury. *Circ Res* 1997;80:894–901.

28. Fusi F, Gorelli B, Valoti M, Marazova K, Sgaragli GP. Effects of 2,5-di-t-butyl-1,4-benzo-hydroquinone (BHQ) on rat aorta smooth muscle. *Eur J Pharmacol Circ* 1998;346:237–243.

29. Sugiyama K, Ohgushi M, Matsumura T, Ota Y, Doi H, Ogata N, Oka H, Yasue H. Supersensitivity of atherosclerotic artery to constrictor effect of cigarette smoke extract. *Cardiovasc Res* 1998;38:508–515.

30. Sotnikova R. Investigation of the mechanisms underlying H_2O_2-evoked contraction in the isolated rat aorta. *Gen Pharmacol* 1998;31:115–119.

31. Conklin DJ, Boor PJ. Allylamine cardiovascular toxicity: evidence for aberrant vasoreactivity in rats. *Toxicol Appl Pharmacol* 1998;148:245–251.

32. Hutchison SJ, Reitz MS, Sudhir K, Sievers RE, Zhu BQ, Sun YP, Chou TM, Deedwania PC, Chatterjee K, Glantz SA, Parmley WW. Chronic dietary L-arginine prevents endothelial dysfunction secondary to environmental tobacco smoke in normocholesterolemic rabbits. *Hypertension* 1997;29:1,186–1,191.

33. Satake N, Shibata M, Shibata S. Endothelium- and cytochrome P-540-dependent relaxation induced by isoproterenol in rat aortic rings. *Eur J Pharmacol* 1997;319:37–41.

34. Ito Y, Pagano PJ, Tornheim K, Brecher P, Cohen RA. Oxidative stress increases glyceraldehyde-3-phosphate dehydrogenase mRNA levels in isolated rabbit aorta. *Am J Physiol* 1996;270:H81–H87.

35. Rhee HM, Song BJ, Cushman S, Shoaf SE. Vascular reactivity in alcoholic rat aortas: *in vitro* interactions between catecholamines and alcohol. *Neurotoxicology* 1995;16:179–185.

36. Brown RA, Savage AO. Effects of acute acetaldehyde, chronic ethanol, and pargyline treatment on agonist responses of the rat aorta. *Toxicol Appl Pharmacol* 1996;136:170–178.

37. Taylor TJ, Smith NC, Langford MJ, Parker GW, Jr. Effect of palytoxin on endothelium-dependent and -independent relaxation in rat aortic rings. *J Appl Toxicol* 1995;15:5–12.

38. Ozdem SS, Ogutman C. Responsiveness of aortic rings of cadmium-hypertensive rats to endothelin-1. *Pharmacology* 1997;54:328–332.

39. Adegunloye BI, Sofola OA. Differential responses of rat aorta and mesenteric artery to norepinephrine and serotonin *in vitro*. *Pharmacology* 1997;55:25–31.

40. Ingwall JS, DeLuca M, Sybers HD, Wildenthal K. Fetal mouse hearts: a model for studying ischemia. *Proc Natl Acad Sci U S A* 1975;72:2,809–2,813.

41. Speralakis N, Shigenoubu R. Organ cultured chick embryonic heart cells of various ages. Part I. Electrophysiology. *J Mol Cell Cardiol* 1974;6:449–471.

42. Tanaka H, Kasuya Y, Saito H, Shigenobu K. Organ culture of rat heart: maintained high sensitivity of fetal atria before innervation to norepinephrine. *Can J Physiol Pharmacol* 1987;66:901–906.

43. Gotlieb AI, Boden P. Porcine aortic organ culture: a model to study the cellular response to vascular injury. *In Vitro* 1984;20:535–542.

44. Koo EWY, Gottlieb AI. The use of organ cultures to study vessel wall pathobiology. *Scand Micro* 1992;6:827–835.

45. Nakazawa M, Morishima M, Tomita H, Tomita SM, Kajio F. Hemodynamics and ventricular function in the day-12 rat embryo: basic characteristics and the responses to cardiovascular drugs. *Pediatr Res* 1995;37:117–123.

46. Wang S, Wright G, Geng W, Wright GL. Retinol influences contractile function and exerts an anti-proliferative effect on vascular smooth muscle cells through an endothelium-dependent mechanism. *Pflugers Arch—Eur J Physiol* 1997;434:669–677.

47. Gandolfi AJ, Brendel K, Fisher RL, Michaud JP. Use of tissue slices in chemical mixture toxicology and interspecies investigations. *Toxicology* 1995;105:285–290.

48. Parrish AR, Gandolfi AJ, Brendel K. Precision-cut tissue slices: application in pharmacology and toxicology. *Life Sci* 1995;57:1,887–1,901.

49. Jacobson SL. Culture of spontaneously contracting myocardial cells from adult rats. *Cell Struct Funct* 1977;2:1–9.

50. Bkailey G, Speralakis N, Doane J. A new method for preparation of isolated single adult myocytes. *Am J Physiol* 1984;247:H1,018–H1,026.

51. Piper HM, Probst I, Schwartz P, Hutter FJ, Spieckermann PG. Culturing of calcium stable adult cardiac myocytes. *J Mol Cell Cardiol* 1982;14:397–412.

52. Welder A, Grant R, Bradlaw J, Acosta D. A primary culture system of adult rat heart cells for the study of toxicological agents. *In Vitro Cell Dev Biol* 1991;27:921–926.

53. Zimmerman ANE, Hulsmann WG. Paradoxical influences of calcium ions on the permeability of the cell membranes of the isolated heart. *Nature* 1966;211:646–647.

54. Welder AW, Machu T, Leslie SW, Wilcox RE, Bradlaw JD, Acosta D. Beta adrenergic receptor characteristics of postnatal rat myocardial cell preparations. *In Vitro Cell Dev Biol* 1988;24:771–777.

55. Cave AC, Adrian S, Apstein CS, Silverman HS. A model of anoxic preconditioning in the isolated rat cardiac myocyte. Importance of adenosine and adrenalin. *Basic Res Cardiol* 1996;91:210–218.

56. Alipui C, Ramos K, Tenner T. Alteration of aortic smooth muscle cell proliferation in diabetes mellitus. *Cardiovasc Res* 1993;27:1,229–1,232.

57. Kasten FR. Rat myocardial cells *in vitro*: mitosis and differentiated properties. *In Vitro* 1972;8:128–149.

58. Barnes DM. Joint Soviet-U.S. Attack on heart muscle dogma. *Science* 1988;242:193–195.

59. Rumyantsev PP. Interrelations of the proliferation and differentiation processes during cardiac myogenesis and regeneration. *Int Rev Cytol* 1977;51:187–193.

60. Claycomb WC, Lanson NA, Jr. Proto-oncogene expression in proliferating and differentiating cardiac and skeletal muscle. *Biochem J* 1987;247:701–706.

61. Wang Y-C, Neckelmann N, Mayne A, Herskoqit A, Srinivasan A, Sell KW, Ahmed Ansair A. Establishment of a human fetal cardiac myocyte cell line. *In Vitro Cell Dev Biol* 1991;27:63–74.

62. Ramos K, Cox LR. Primary cultures of rat aortic endothelial and smooth muscle cells:

an *in vitro* model to study xenobiotic-induced vascular cytotoxicity. *In Vitro Cell Dev Biol* 1987;23:288–296.

63. Simionescu M, Simionescu N. Isolation and characterization of endothelial cells from the heart microvasculature. *Microvasc Res* 1978;16:426–452.

64. Obeso J, Weber J, Auerbach R. A hemangioendothelionia-derived cell line: its use as a model for the study of endothelial cell biology. *Lab Invest* 1990;63:259–269.

65. Bastaki M, Nelli, EE, Dell'Era P, Rusnati M, Molinari-Tosatti MP, Parolini S, Auerbach R, Ruco LP, Possati L, Presta M. Basic fibroblast growth factor-induced angiogenic phenotype in mouse endothelium. A study of aortic and microvascular endothelial cell lines. *Arterios Thromb Vasc Biol* 1997;17:454–464.

66. Kossenjans W, Rymaszewski Z, Barankiewicz J, Bobst A, Ashraf M. Menadione-induced oxidative stress in bovine heart microvascular endothelial cells. *Microcirculation* 1996;3:39–47.

67. Ramos K, Acosta D. The heart. Primary cultures of newborn myocardial cells as a model system to evaluate the cardiotoxicity of drugs and chemicals. In: McQueen C, ed. *In Vitro Models in Toxicology*. Caldwell, NJ: Telford Press, 1989.

68. McDonagh JC, Cebrat EK, Nathan RD. Highly enriched preparations of cultured myocardial cells for biochemical and physiological analyses. *J Mol Cell Cardiol* 1987;19:785–793.

69. Wenzel DG, Cosma GN. A model system for measuring comparative toxicities of cardiotoxic drugs with cultured rat heart myocytes, endothelial cells and fibroblasts. I. Emetine, chloroquine and metronidazole. *Toxicology* 1984;33:103–115.

70. Cox LR, Ramos K. Allylamine-induced phenotypic modulation of aortic smooth muscle cells. *J Exp Pathol* 1989;87:43–48.

71. Ives CL, Eskin SG, McIntire LV. Mechanical effects on endothelial cell morphology: *in vitro* assessment. *In Vitro Cell Dev Biol* 1986;22:500–507.

72. Sato GH. The growth of cells in serum-free hormone supplemented media. *Methods Enzymol* 1979;58:94–109.

73. Ramos K, Weber TJ, Liau G. Vascular smooth muscle cell proliferation: influence of substratum and growth conditions *in vitro*. *Biochem J* 1993;289:57–63.

74. Parrish AR, Ramos KS. Differential processing of osteopontin characterizes the proliferative vascular smooth muscle cell phenotype induced by allylamine. *J Cell Biochem* 1997;65:267–275.

75. Bishop CT, Mirza Z, Crapo JD, Freeman BA. Free radical damage to cultured porcine aortic endothelial cells and lung fibroblasts: modulation by culture conditions. *In Vitro Cell Dev Biol* 1985;21:21–25.

76. Owens GK, Thompson MM. Developmental changes in isoactin expression in rat aortic smooth muscle cells *in vivo*. *J Biol Chem* 1986;261:13,373–13,380.

77. Simpson P, Savion S. Differentiation of rat myocytes in single cell cultures with and without proliferating non-myocardial cells. *Circ Res* 1982;50:101–116.

78. Ramos K. Cellular and molecular basis of xenobiotic-induced cardiovascular toxicity:

application of cell culture systems. In: Acosta D, ed. *Focus on Cellular Molecular Toxicology and In Vitro Toxicology.* Boca Raton, FL: CRC Press, 1990:139–155.

79. Stegeman JE, Miller MR, Hinton DE. Cytochrome P4501A1 induction and localization in endothelium of vertebrate (teleost) heart. *Mol Pharmacol* 1989;36:723–729.

80. Asakura T, Shichi H. 12(R)-hydroxyeicosatetraenoic acid synthesis by 3-methylcholan-threne- and clofibrate-inducible cytochrome P450 in porcine ciliary epithelium. *Biochem Biophys Res Commun* 1992;187:455–459.

81. Huang S, Gibson GG. Differential induction of cytochromes P450 and cytochrome P450-dependent arachidonic acid metabolism by 3,4,5,3',4'-pentachlorobiphenyl in the rat and guinea pig. *Toxicol Appl Pharmacol* 1991;108:86–95.

82. Zhao W, Parrish AR, Ramos KS. Constitutive and inducible expression of cytochrome P450IA1 and IB1 in human vascular endothelial and smooth muscle cells. *In Vitro Cell Dev Biol* 1998;34:671–673.

83. Luckhoff A. Measuring cytosolic free calcium concentration in endothelial cells with indo-1: the pitfalls of using the ratio of two fluorescence intensities recorded at different wavelengths. *Cell Calcium* 1986;7:233–248.

84. Wenzel DG, Innis JD. Arrhythmogenic and antiarrhythmic effects of lipolytic factors on cultured heart cells. *Res Commun Chem Pathol Pharmacol* 1983;41:383–396.

85. Aszalos A, Bradlaw JA, Reynaldo EF, Yang GC, El-Hage AN. Studies on the action of nystatin on cultured rat myocardial cells and cell membranes. *Biochem Pharmacol* 1984;33:3,779–3,786.

86. Yarom R, Hasin S, Raz S, Shimoni Y, Fixler R, Yagen B. T-2 toxin effect on cultured my-ocardial cells. *Toxicol Lett* 1986;31:1–8.

87. Ramos K, Combs AB, Acosta D. Cytotoxicity of isoproterenol to cultured heart cells: ef-fects of antioxidants on modifying membrane damage. *Toxicol Appl Pharmacol* 1983;70:317–323.

88. Ramos K, Combs AB, Acosta D. Role of calcium in isoproterenol cytotoxicity to cul-tured myocardial cells. *Biochem Pharmacol* 1984;33:1,898–1,992.

89. McCall D, Ryan K. The effect of ethanol and acetaldehyde on Na pump function in cul-tured rat heart cells. *J Mol Cell Cardiol* 1987;19:453–463.

90. Horrigan S, Campbell JH, Campbell GR. Effect of endothelium on VLDL metabolism by cultured smooth muscle cells of differing phenotype. *Atherosclerosis* 1988;71:57–69.

91. Weinberg CB, Bell E. A blood vessel model constructed from collagen and cultured vas-cular cells. *Science* 1986;231:397–399.

92. Davies PF, Truskey GA, Warren HB, O'Connor SE, Eisenhaure BA. Metabolic coopera-tion between vascular endothelial cells and smooth muscle cells in co-culture: changes in low-density lipoprotein metabolism. *J Cell Biol* 1995;101:871–879.

93. Saunders KB, D'Amore PA. An *in vitro* model for cell-cell interactions. *In Vitro Cell Dev Biol* 1992;28:521–528.

94. Vender RL. Role of endothelial cells in the proliferative response of cultured pulmonary

vascular smooth muscle cells to reduced oxygen tension. *In Vitro Cell Dev Biol* 1992;28:403–409.

95. Fan Y-Y, Chapkin RS, Ramos KS. A macrophage-smooth muscle cell co-culture model: applications in the study of atherogenesis. *In Vitro Cell Dev Biol* 1995;31:492–493.

96. Fan Y-Y, Ramos KS, Chapkin RS. Dietary gamma-linolenic acid modulates macrophage-vascular smooth muscle cell interactions. Evidence for a macrophage-derived soluble factor that downregulates DNA synthesis in smooth muscle cells. *Arterios Thromb Vasc Biol* 1995;15:1,397–1,403.

97. Fan Y-Y, Ramos KS, Chapkin RS. Dietary lipid source alters macrophage/vascular smooth muscle cell interactions *in vitro. J Nutr* 1996;126:2,083–2,088.

98. Lundberg MS, Sadhu DN, Grumman VE, Chilian WM, Ramos KS. Actin isoform and alpha(1_B)-adrenoceptor gene expression in aortic and coronary smooth muscle cells is influenced by cyclical stretch. *In Vitro Cell Dev Biol* 1995;31:595–600.

99. Lundberg MS, Chilian WM, Ramos KS. Differential response of rat aortic and coronary smooth muscle cell DNA synthesis in response to mechanical stretch *in vitro. In Vitro Cell Dev Biol* 1996;32:13–14.

100. Loegering DJ, Richard CA, Davison CB, Wirth GA. Diethyldithiocarbamate ameliorates the effect of lipoolysaccharide on both increase nitrite production by vascular smooth muscle cells and decreased contractile response of aortic rings. *Life Sci* 1995;57:169–176.

101. Bral CM, Ramos KS. Identification of novel benzo(a)pyrene-inducible cis acting elements within c-Ha-ras transcriptional regulatory sequences. *Mol Pharmacol* 1997;52:974–982.

102. Kerzee JK, Ramos KS. Redox regulation of c-Ha-ras gene expression in vascular smooth muscle cells. Submitted.

13

Gastrointestinal Toxicology: *In Vitro* Test Systems

Shayne Cox Gad

Gad Consulting Services, Raleigh, North Carolina

The gastrointestinal (GI) tract is a frequently overlooked potential target organ system for toxic effects. Specific GI toxicity is widely thought to be relatively rare and limited in form. Yet, for many potentially toxic agents, the length of the GI tract is not only the region of first contact, but also the route of entry into the body. The structure and function of the tract are also much more complex than was generally thought.

Two reasons why the GI tract has been thought of as a relatively insensitive target organ are as follows: (1) it cannot be easily observed directly *in vivo*, and (2) alterations in its functions are rarely expressed in a manner implicating the tract solely and directly. However, methodology does exist for studying GI function *in vivo*, and the tract is the primary target organ for some toxicants. Indeed, the tract expresses toxicity in a rich variety of manners.

Interest in various forms of GI toxicology has increased significantly since the mid-1980s. The reader is referred to reviews on the subject by Walsh [1,2] and Schiller [3] to gain a better understanding of the background, history, and methodology of the field.

At the same time, interest in having effective *in vitro* models for identifying and studying GI toxicity has developed from both the organ-system–specific problems with the *in vivo* model cited above and with a set of perceived general advantages for *in vitro* models. (These issues are discussed in detail elsewhere in this volume.)

13.1 TYPES OF GASTROINTESTINAL CELLS AND TOXICITY

The GI tract is not a monocytic structure of one or a few cell types, but a connected series of organs each composed of a complex of multifunctional cell types [4]. This organ system is made even more complex in its function by its resident symbiotic population of bacteria. The three large organs comprising the tract––the stomach and large and small intestines—serve not just as a highway for the passage of nutrients and water into (and waste out of) the body, but they also have significant metabolic homeostatic and immunologic functions. The GI tract has a range of expressions of toxicity. These expressions include irritation, cytotoxicity, malabsorption (of electrolytes, fluids, and nutrients), altered motility, altered secretion, and neoplasia. Each of these expressions can actually be subdivided further and, as such, can be expressed in a range of the cell types present in the tract.

Each of the expressions or types of toxicity in the GI tract deserves some specific consideration. As will be clear, some of the divisions among them are not cut and dried, with a range of interactions existing among each of the major toxicity types. However, the major types of toxicity can be considered as follows.

Irritation. A range of agents irritate the GI tract, causing erythema, disrupting membrane integrity, and serving to alter both absorption and GI motility. Some necrosis (cell death) may be present in more advanced stages leading to ulceration.

Cytotoxicity. GI cell populations may be killed selectively (one or two cell types affected) or generally (either a broad range of all types of cells dying). A range of cells displaying stages of dysfunction and structural breakdown will be observed, becoming increasingly more severe with some dying or dead. As some of the cell types "turn over" regularly, timing of "sampling" (observation) will influence the ability to detect and characterize responses to acute insults.

Malabsorption. Materials are absorbed by the GI tract by a range of mechanisms of both passive and active nature. Electrolytes (sodium, potassium and chloride ions, largely), fluids (mostly water), nutrients, and pharmaceutical agents have the tract as their chief route of entry into the body [7]. Kotsonis [22] has edited a volume on nutritional toxicology broadly addressing the adverse actions of xenobiotics on absorption and transport of a wide range of nutrients.

Altered secretory activity. A primary function of the GI tract is secretion of various entities into its contents (the major example being secretion of hydrogen ions by the parietal cells into the stomach). Agents irritating the mucosa can produce gastritis, which leads to reduced secretion. Other agents can stimulate gastric secretion, which can lead to acid-induced erosions and ulcers in the stomach and duodenum.

Altered motility. GI motility *in vivo* is influenced by the smooth muscle cells, entrinsic nervous system inputs, extrinsic nervous system input, hormone, and the constituents and volume of luminal content. Proper motility is essential for both absorptive functions and in influencing the potential of toxicants to do harm.

Neoplasia. This action is the transformation of constituent cell types into proliferative, nonfunctional forms. Some animal model GI neoplasias (such as rodent forestomach cancer) do not appear to have true human analogues. Some of the high turnover-rate cell populations in the GI tract are particularly prone to neoplasia.

Symbiotic population alteration Proper functioning of the GI tract, particularly in terms of absorption and metabolism, is dependent on a stable population (in terms of both organism types and numbers) of microorganisms exisiting in portions of the tract. The adverse effects of some antibiotics on tract function are well recognized because they alter the balance of the resident microbial population. These same populations play significant roles in the toxicity and carcinogenicity of a number of agents [5].

13.2 *IN VITRO* MODELS

The models available for use in toxicity testing can be classified in various ways. Table 13.1 presents one classification of models in general based on complexity, with intact organisms being the most complex and computer simulation the least. As the table broadly points out, each of these levels has advantages and disadvantages.

Isolated perfused organs and cultured cells provide valuable tools for evaluation of gastrointestinal toxicity. Isolated perfused organ preparations offer several advantages over experimentation with intact animals.

Perfused experiments lend themselves to a definitive evaluation of the role of a particular organ or tissue in the disposition of endogenous or exogenous chemicals. Although experimentation with whole-animal preparations may provide clues implicating a possible role of a particular organ in regulating the levels of a test toxin or an endogenous substance in response to a toxin, decisive conclusions may not be feasible.

Unlike *in vitro* homogenate preparations, organ perfusion studies allow the experimenter to retain the structural and functional integrity of the subject organ during experiments. Unlike in whole animals, perfusion experiments allow the experimenter to retain control over several experimental parameters, e.g., perfusion pressure and blood flow; in the intact animal, these measurements are likely to vary over the course of an experiment, especially in response to administration of the experimental toxic chemical. The concentrations of endogenous or exogenous

**TABLE 13.1.   Levels of models for toxicity
testing and research**

Level/Model	Advantages	Disadvantages
In vivo (intact higher organism)	Full range of organismic responses similar to target species	Costs Ethical/animal welfare concerns Species-to-species variability
Lower organisms (earthworms, fish)	Range of integrated organismic responses	Frequently lack responses of higher organism
Isolated organs	Intact yet isolated tissue and vascular system Controlled environmental and exposure conditions	Donor organism still required Time consuming and expensive No intact organismic responses Limited length of viability
Cultured cells	No intact animals directly involved Ability to carefully manipulate system Low costs Wide range of variables can be studied	Instability of system Limited enzymatic capabilities and viability of system No or limited integrated multicell or organismic responses
Chemical/ Biochemical systems	No donor organism problems Low cost Long-term stability of prep Wide range of variables can be studied Specificity of response	No *de facto* correlation to *in vivo* system Limited to investigation of single, defined mechanism
Computer simulations	No animal welfare concerns Speed and low per evaluation cost	Problematic predictive value beyond narrow range of structures Expensive to establish

substances and other factors can be under experimental control in isolated perfused organ studies. The isolated perfused organ lends itself to a broader range of concentrations of the drug of interest used in the study. That is, concentrations of drug at which the intact animal would not be expected to survive can be tested in isolated perfused organs. Determination of accurate and complete mass balance of the toxic chemical in question is possible throughout the perfusion experiment, because the compound must either be in the perfusate or in the tissue or be excreted via excretory fluids, such as bile and urine. Binding of the test drug to the glassware, tubing, and other components of the perfusion apparatus may occur, but this possibility can be explored in blank experiments, in which the perfusion experiment is conducted without the organ, from which appropriate correction factors are derived. Removal of such confounding factors is often technically feasible. Additionally, in perfusion studies, large blood or perfusate samples are available; thus, complete qualitative and qualitative analyses of minor and major biotransformation products of the test compound are feasible, because the volume of perfusate used in these experiments can be controlled. Also, tests with smaller quantities of test chemicals are feasible, which is important as limitations of either the availability of small quantities as the test materials or the cost of isotropically labeled, newly synthesized compounds can be formidable.

In perfused organ studies, it is also possible to maintain appropriate membrane barriers, not only between vascular and parenchymal sides, but also between individual cells; hence, the natural constraints of intact organs are retained throughout the experiment. It is clear that one may not be able to predict the qualitative and quantitative aspects of biotransformation of a test drug by intact organ based only on the results of *in vitro* experiments with homogenate preparation [23]. Factors governing the generation and availability of cofactors and transport of substrate to the site of biotransformation influence the final results in the intact perfused organ [24,25]. Such factors can remain operative in perfusion studies, unlike with other *in vitro* techniques, thereby enabling realistic extrapolation of the results to *in vivo* situations. Finally, experimental results have to be interpreted and extrapolated to the *in vivo* situation, in which intact organs interact continuously. Such interpretation and extrapolation are made easier by use of the intact perfused organs in toxicologic investigations.

Furthermore, the cell-to-cell interactions are preserved in an intact perfused organ, which might be either missing or at least compromised in isolated cells or other *in vitro* incubations. It is known that gap junctioning plays an important role in the regulation of cellular and tissue homeostatic mechanisms. These mechanisms would be preserved in intact perfused organs. The collagenase trypsin or other proteases used in procedures to isolate cells can alter the plasma membrane, thereby altering the permeability and even receptor characteristics of isolated cells. For example, in freshly prepared hepatocytes, glutathione levels are only half of normal values. Some essential and critical differences between the tissue slice experiments and perfusion studies are also of interest in this regard. Whereas the perfusion of intact organs allows entry of the chemical through the endothelium, which would be representative of what happens in the intact animal, tissue slice incubations permit entry of chemicals directly into the parenchymal cells through direct contact. Studies using tissue slice incubations may not represent the *in vivo* situation, because some chemicals might not be taken up through the endothelial barrier altogether or be taken up to a small extent. Hepatocytes or liver slices incubated with the calcium channel blockers do not have any influence on cellular calcium, whereas the perfused liver is responsive to these same calcium blockers. The latter findings clearly would be more representative of the *in vivo* situation than would be the former.

13.3 SPECIFIC ENDPOINT MODELS

13.3.1 Irritation

Irritation is usually evaluated by gross or microscopic evaluation of the surface of the tract after the selected test species has received or been exposed to the material of interest for a predetermined period of time. Such examinations are usually performed after terminal sacrifice of the test animals. One can also monitor the rate of loss of cells from the GI surface into the luminal fluid. Because irritation is more a tissue response than a cell response, it is harder to model *in vitro*.

13.3.2 Cytotoxicity

In vivo, the same methods used to evaluate irritation can be and are used to evaluate cytotoxicity. These methods can also be considered under the alternate label of "assessment of structural integrity" of the tract or portion of interest of the tract. Another gross measure of such integrity *in vivo* is fecal blood loss, which tells if significant damage has been done.

In vitro, assessment of cytotoxicity to either single-cell–type populations or mixed cultures is one of the most basic methodologies for all target tissues.

13.3.3 Malabsorption

The absorptive function is what is actually assessed. The simplest approach is to measure how much of a labeled material, which is administered, shows in the systemic fluids. Accurate measurement requires consideration of appropriate pharmacokinetic considerations [19]. The issues are (1) the determination of the GI absorption, metabolism, and excretion of toxic substances, and (2) factors affecting this process [26–28], and understanding the effect of toxic substances on the absorption of normal dietary constituents or orally administered therapeutic agents.

Numerous approaches are available for studying GI absorption, both *in vivo* and *in vitro* [29,30]. The technique chosen for a specific study will depend on the aspect of the absorption process that is of primary interest. Broadly speaking, the methodology can be categorized according to the procedure for administering the test substance to the absorptive surface of the gut and the technique for quantitating the extent or rate of absorption.

Numerous advances in the understanding of cellular mechanisms of electrolyte flux and nutrient absorption in the intestine have been made using *in vitro*, as opposed to *in vivo*, techniques. Kimmich [31] documented the impact of the development of *in vitro* techniques on the understanding of intestinal transport mechanisms of sugar. With *in vitro* procedures, physiologic variables, e.g., intestinal motility and mesenteric blood flow, can be eliminated or controlled. In addition, the experimenter has the option of control over such factors as the composition of the solution bath in the mucosal and the serosal sides of the intestine and the electrochemical potential difference between the mucosal and the serosal surfaces. Additionally, one has the ability to carefully control the stirring rate in the mucosal solution. The stirring rate influences the thickness of the unstirred water layer, which can impose significant resistance to the mucosal uptake of substrates, such as long-chain fatty acids, bile acids, cholesterol, and monosaccharides [32]. The layer of mucus adhering to the mucosal cell surface also can impede the diffusion of the nutrients, e.g., disaccharides, small peptides [33], and cholesterol [34]. The pronounced quantitative differences in the uptake rates of nutrients into various *in vitro* preparations of the small intestine [35] probably results, at least in part, from the differences in the resistances conferred by the unstirred water layer and mucus coat. *In vitro* techniques include those analogous to the *in vivo* methods already discussed. For example, investigators have studied the absorption of substances from isolated gut segments with perfusions of the lumen [36] and with perfusions of the vasculature [37].

The everted sac technique method [1] has been useful in the characterization of energy-dependent, carrier-mediated transport processes. In this procedure, small lengths of the intestine are everted, filled with fluid, and tied at both ends. Absorption is quantified by monitoring the appearance of a test substance inside the sac in the fluid bathing the serosal surface of the intestine. Unlike the *in vivo* condition, therefore, absorption of a test substance in this model is considered equivalent to its passage not only through the mucosa, but also through the submucosa, the external muscle layers, and serosal tissue of the gut wall. Problems with the everted segment technique include inadequate oxygen diffusion into the tissue and distension and hydration of the gut segment, requiring the preparation of the tissue and the experimental incubations to be short in duration. Everted sacs of the duodenum from rats exhibit structural abnormalities after a 5-minute incubation at $37^{\circ\circ}C$ [38].

A further refinement has been the development of methods for isolating gut mucosal cells [39]. Methods for recovering mucosal cells, e.g., scraping the inner surface of the gut with a glass slide or vibrating a gut segment everted on a glass spiral, have been improved to reduce contamination from cells of the lamina propria as well as to isolate crypt cells [40].

In vitro methods available include the inverted intestinal sac [6] and the use of rings cut from the whole wall of the intestine [17], which provides a better oxygenated tissue model, but is generally limited to measurement of accumulation of materials of interest in the ring tissue. The Ussing chamber is also available [8,9,18], which allows measurement of electrolyte flux across the intestinal membranes.

13.3.4 Altered Secretory Activity

Invasive and noninvasive *in vivo* techniques exist. The invasive technique requires surgical implantation of a sampling tube [10]. The noninvasive method requires quantitation of how much azure A is released from an azure A–resin complex in the stomach [11]. Available *in vitro* methods have been useful in understanding mechanisms [12]. One approach is the study of the epithelial tissue in a Ussing chamber. A segment of the stomach wall is removed from the animal, and the muscle layer is stripped off. The remaining tissue is mounted to separate the two solutions. Acid secretion and electrolyte flux can be monitored.

Nutrient digestion and absorption and alteration of these by xenobiotics [41] has been studied by the isolation of brush-border membranes, recovered as vesicles after differential centrifugation of tissue from intestinal segments [42,43] or biopsies [44]. The preparation permits analysis of membrane transport kinetics uncontaminated by cytosolic metabolism. Also, the production of membrane vesicles is reported to remove adherent mucus [33], so that this preparation is devoid of this diffusion barrier. Basolateral membranes of mucosal cells also have been isolated by the use of differential and discontinuous sucrose-gradient centrifugation [45]. Isolation of these membrane fractions has allowed the biochemical, structural, and genetic characterization of membrane transport proteins, such as the NA/glucose transporter in the brush-border membrane [46–49], and their regulation by signal transduction mechanisms. Methods based on the use of lipid-soluble fluorophores and florescent spectroscopy have permitted assessment of alterations in the fluidity of brush-border membranes induced by the membrane- perturbing agents [50].

The Ussing chamber [18] also is useful in the elucidation of mechanisms controlling electrolyte absorption and secretion, entailing *in vitro* short-term exposure of a segment of intestine to defined mucosal and serosal solutions. The electrical potential difference across the intestine is measured with a voltmeter and short-circuit current with an external microamp source. The flux of electrolytes is determined by the addition of an isotope to the solution bathing one surface of the intestinal segment and monitoring its accumulation in the tissue or solution bathing the other surface. Similarly, the unidirectional flux chamber designed by Schultz et al. [51] exposes a defined area of luminal surface of the intestinal wall to solution containing the test permeant. Brief exposure times are used to permit measurement of influx through the mucosa. This methodology permits assessment of the integrity of brush-border transport mechanisms and permits assessment of neuroregulation and its modification by exogenous substances [52,53].

An *in vitro* preparation with the advantage of more prolonged viability is the culture of mucosal biopsies [54]. This technique permits the *in vitro* maintenance of mucosal explants for 24–48 hours, depending on the species and region of the intestine undergoing biopsy. A major advantage of this approach is that the normal anatomic arrangement of the villus and its mucosal cell proliferation and differentiation can be studied, as in colonic carcinogenesis and chemotherapy. Toxicant accumulation and efflux can be characterized as well and related to biologic effects. This approach was used to provide evidence of the role of *p*-glycoprotein encoded by the *mrdl* (multidrug resistance) gene in colonic transport of the alkaloid vincristine [67]. Development of mucosal cell lines in culture (CaCo.$_2$, T$_{84}$, and HT$_{29}$) has further enhanced *in vitro* techniques for studying electrolyte transport, its regulation, and perturbation by toxicants. Dharmsathaphorn et al. [55] described the structural and functional characteristics of the human colonic carcinoma cell line, T$_{84}$, which forms a confluent monolayer in culture. These cells, when grown in a serum-supplemented medium, exhibit properties characteristic of the normal epithelium, including the presence of tight junctions, apical microvilli, and vectorial electrolyte transport. Transport by the monolayer, e.g., of sodium, mannitol, and inulin, can be monitored in a modified, low-turbulence Ussing chamber [56]. A collagen-coated filter substrate is preferable to plastic. A subclone of the HT_{29} colon adenocarcinoma cell line (HT-29-18N2) has been developed with a phenotype like the intestinal goblet cell and provides a methodologic approach for study of alteration in mucus secretion by xenobiotics [57]. The Ussing chamber and isolated parietal cells from gastric glands [8,9,12] are useful for both mechanistic and some limited test material screening-specific questions.

For improved definition of the physiology of individual cell types in the stomach, techniques have been developed to isolate and culture gastric mucus cells [58], the gastric gland, and its individual cell types. The gastric gland preparation is obtained by treatment of the intact mucosa with pronase or collagenase. This material is primarily composed of parietal and peptic cells in greater concentration than in the intact tissue.

Techniques also have been developed for isolation of parietal cells and chief cells from gastric glands. A calcium chelator, such as EDTA, increases dispersion of single cells from gastric glands [12]. A Percoll step-gradient purification or centrifugal elutriation [59] can be effectively used to separate parietal cells from other cell types in the tissue, based on the difference in their mass. With these single-cell preparations, no anatomical barrier exists between mucosal and serosal surfaces. Alternative approaches must therefore be used for measuring hydrogen ion secretion. Two techniques include quantitating oxygen consumption of the cells and measuring their accumulation of aminopyrine. Several types of studies have shown a close correlation between oxygen consumption and the energy-dependent process of acid secretion. Measurement of oxygen consumption with a Clark-type polarographic electrode or Gilson respirometer is one index of secretory activity. Another approach is based on monitoring the uptake of the weak base aminopyrine into the intracellular vesicular space of the parietal cell [9]. Aminopyrine, which has a pKa of 5, readily traverses plasma membranes in its unionized form as it exists at neutral pH. In the parietal cell under conditions in which acid secretion is occurring, the compound becomes ionized in acid secretory vesicles and remains trapped in this acidic fluid. The cellular content of the compound relative to that in the medium can be used as an index of acid formation by the parietal cell; the conventional approach is to use the ^{14}C isotope of aminopyrine and measure radioactivity in cells separated from the medium by filtration.

13.3.5 Altered Motility

The continued motility of the GI tract serves to move contents steadily through its length and out of the body. *In vivo*, one can simply measure the transit of one of a number of nonabsorbable intraluminal markers. Isolated superfused ileum can be used to evaluate specific test material responses [13], whereas isolated rabbit jejunum preparations [14] or cultured myocytes [15] can be used to study transit times and mechanistic questions, respectively.

A key variable in the GI disposition of a xenobiotic is the motility of the organ, which affects transit of the luminal contents. Numerous studies with pharmacologic agents, for example, have demonstrated that changes in gastric emptying can alter the overall absorption half-life of the compound. Generally, an increased delivery rate to the small intestine is associated with more rapid absorption because of the greater surface area and higher blood flow of this tissue compared with the stomach.

The motility patterns of the GI tract, i.e., the frequency and time course of contractions, differ throughout the organ. However, the major determinants are similar. One key component is the smooth muscle cells of the circular, longitudinal, and, in the stomach, oblique layers. These cells act as a syncytium and display a baseline electrophysiologic pattern with associated baseline contractile activity differing in various regions. A complex system of neurons in the gut wall, including the myenteric plexus, patterns after use of various procedures for rapid killing of the animal. A technique should be used to prevent movement of the marker in the gut before its assay. One approach, for example, is to put ligatures at 3- to 5-cm lengths of the small intestine. The technique introduced by Derblom et al. [60] permits detection of radioactivity along the continuous length of the GI tract. The organ is placed on a device that is moved at a constant rate under a scintillation detector, and radioactive counts are recorded continuously. Integration of the radioactive counts in a particular region, either instrumentally or by planimeter, then, permits reporting of marker content as a percentage of the total amount detected. One commonly used approach is to describe the small intestinal radioactivity for ten equal segments. Without equipment for continuous recording, the gut can be cut into segments individually counted in a conventional well-type gamma counter [61]. Care must be taken, however, that the marker is not lost in the process of cutting the segments.

Several procedures have been used for summarizing data on marker distribution to allow comparisons among treatments. In studies of gastric emptying, the percentage of the marker remaining in the stomach is calculated, and the mean for controls is compared with those in the experimental group by appropriate statistical methods. In studies of intestinal transit, one approach is to identify the most distal site that 50% of the marker has reached. This site, then, can be expressed as a percentage of the total distance of the small intestine, a value used in statistical analyses. Another approach is to compute the geometric center of marker distribution [62]. For each segment assayed, the marker content is determined, as a percentage of the total recovered, and is multiplied by the segment number, 1 being the most proximal and n being the nth and most distal. These values are summed and divided by n to generate the geometric center.

Several additional points should be noted about the experimental design of transit studies. First of all, the time at which the animal is killed should be chosen so that in control animals half the marker has been emptied from the stomach or has traversed to the midpoint of the small intestine after gastric or intraduodenal administration, respectively. This approach improves the likelihood of demonstrating inhibition or acceleration of transit by test substances. Second, in an investigation of the effect of a substance on GI transit, studies should be carried out at a series of times after various doses of the test substance. Because each animal provides only one measure of marker distribution, each experimental condition requires use of replicate animals.

Measurement of contractility of gastrointestinal smooth muscle *in vitro*. The contractile and electrophysiologic properties of smooth muscle can be assessed with relative simplicity using *in vitro* preparations of this organ. Such *in vitro* techniques are of value in screening potentially toxic substances for effects on GI smooth muscle and for elucidation of mechanisms of effects on propulsion observed with *in vivo* methodology. Use of *in vitro* techniques to study gastrointestinal motility has certain advantages over *in vivo* procedures. They are technically simpler to execute: Isolate the tissue from extrinsic neural and hormonal influences, and the tissue can be directly exposed to the test substance. These advantages are at the expense of loss of prediction of *in vivo* effects of a test substance.

The choice of a particular *in vitro* technique depends on the specific aim of the experimentation. Those techniques most commonly used differ in several ways. First, the species from which the gut segment is taken markedly affects the basal contractile activity. The rabbit jejunum, for example, maintains rhythmic contractions *in vitro* and is therefore especially useful for analysis of substances with inhibitory effects on GI smooth muscle [63]. The guinea pig ileum, in contrast, exhibits little spontaneous activity *in vitro*. This preparation is useful in the bioassay of agents causing contraction of smooth muscle. To test for depressant effects, the investigator must first induce contraction of the tissue with electrical stimulation, potassium depolarization, or pharmacologic agonists. Second, differences exist in the responses of GI smooth muscle depending on the site of the gut under investigation. This process limits the ability of the investigator to generalize an experiment carried out with a muscle preparation from a single region of the gut and reinforces the importance of strictly controlling the tissue region studied in a series of experiments.

In vitro study of GI smooth muscle can be evaluated with isolation of myocytes and measurement of contractile and electrophysiologic events of single cells [64,65]. This approach permits assessment of the direct effects of chemicals on these cells in isolation from their intrinsic innervation via the myenteric and submucosal plexi existing *in situ*. The isolated myocyte preparation has facilitated elucidation of membrane receptors, ionic channels, and excitation–contraction coupling mechanisms [67]. This approach provides an additional tool in toxicologic studies for precisely defining the mechanism of xenobiotic-induced, altered GI propulsion.

The guinea pig ileum preparation. The guinea pig ileum preparation has been extensively used in the study of the effects of chemicals on smooth muscle contractility. The guinea pig is euthanized without the use of drugs because of their potential to alter the responsivity of smooth muscles. Segments should be cut from the sital ileum, with care taken to avoid damaging the muscle layers with forceps. The location and length of these segments should be standardized in an experimental series; commonly, 1-cm segments are used, avoiding use of the most distal 10 cm of the ileum. The mesentery should be cut away from the gut segment, which then is placed in Tyrode's solution at 37°C. Once relaxed, the tissue should be gently flushed with solution at 37°C to remove the intraluminal contents. The segment will maintain its viability for several hours if held in Typrode's solution at just below 20°C. For studies in a single muscle type, the longitudinal muscle layer is readily dissected from the ileal segment [75]. When obtained from the guinea pig, the tissue retains the mesenteric plexus and is therefore useful for study of agents affecting neuronal conduction or neurotransmitter release, as well as smooth muscle contractility.

Threads are secured to the proximal andsital end of the segment without occluding the lumen. The tissue is mounted with proximal end up, in an organ bath containing Tyrode's solution, which is maintained at 37°C and gassed with 95% O_2/5% CO_2. The thread tied to the proximal end of the tissue is attached to a device for quantificating the contractile response of the tissue. Commonly, this apparatus consists of a Staham force displacement transducer connected to an amplifier and chard recorder permitting continuous monitoring of response under isometric conditions. Initially, tension, e.g., 1 g, should be applied to the tissue followed by an equilibration period, e.g., 30 minutes, before the addition of test substances to the bath. The tension applied to the muscle should be chosen to stretch the tissue to a length resulting in the greatest generated force on activation. This length, referred to as the optimum length, is determined by measuring the difference between passive and active tension for different lengths of the muscle with a maximally effective stimulus, such as high potassium-induced (110 mM) depolarization [56]. To reduce the contribution of spontaneous contractile activity and tone to "passive" tension, which may be confounding with intestinal muscle from some species, force-length relationships can be examined at a lower temperature (22°C), which minimizes this factor [76].

Modifications of this technique include electrical stimulation of the tissue and application of intraluminal pressure for eliciting peristaltic reflexes [77]. The parameters chosen for electrical stimulation of tissue determine the mechanism and characteristics of the contractile response. For electric field stimulation, muscle is hung between platinum electrodes connected to a stimulator. Stimuli of 1-Hz frequency for 1 millisecond at 100 V are likely to induce contractions mediated by neural stimulation. These contractions are sensitive to inhibition of neural conduction by sodium channel inactivators, e.g., tetrodotoxin (10^{-7} g/ml). Stimuli of higher frequency, e.g., 60 Hz, are likely to depolarize the muscle membrane directly, to be unaltered by tetrodotoxin, and to reflect the mechanical properties of the muscle itself.

Substances tested with this preparation should be added in a broad concentration range with washing of the tissue after each application. When a concentration of the test substance is found to cause contraction, additional concentrations should be tested to permit construction of a concentration-response curve that will facilitate analyses by use of such statistics as the EC_{50}, the affinity, and the intrinsic activity. An additional consideration that may require evaluation is the development of tachyphylaxis to the test substance to describe the concentration-response relationship adequately.

13.3.6   Neoplasia

The proliferation of mucosal cells is what is actually measured specific to the GI tract. *in vivo*, this proliferation can be measured by rates of incorporation of ^{3}H-thymidine. It can also now be measured using flow cytometry.

13.3.7   Symbiotic Population Alteration

The entire length of the GI tract is populated by a diversity of microorganisms. It is not uncommon to find 400 different species of bacteria in the feces of a single subject [43]. The species composition and quantity of the microflora differ markedly in the various regions of the GI tract. The stomach is relatively sparsely populated [$<10^3$ colony-forming units (CFU)/ml] because of the gastric juices destroying a majority of the organism, e.g., streptococci, staphylococci, and lactobacilli. The concentration of microorganisms in the small intestine increases from 10^3 to 10^4 CFU/ml in the duodenum from 10^6 to 10^7 CFU/ml in the distal ileum [43]. Along the small intestine, the concentration of gram-positive aerobes decreases as the concentration of gram-negative increases. The large intestine is densely inhabited by gram-negative anaerobes such that luminal contents may contain 10^{12} CFU/ml. Among the more prominent species are *Bacteroides*, *Bifidobacterium*, *Fusobacterium*, *Clostridium*, and *Eubacterium* [43]. Studies with animals delivered and reared under germ-free conditions have demonstrated clearly that the indigenous microflora influences the structure and function of the GI tract [22,36].

The gut microflora may mediate toxic effects on this organ by a variety of mechanisms. Pathogenic strains, such as *Shigella*, may elicit damage in part by their capacity to invade the mucosal epithelium and produce necrosis and hemorrhagic effects. Microorganisms are also capable of elaborating toxins causing diarrhea by perturbing mucosal cell electrolyte and water flux. The toxin of *Vibrio cholerae*, for example, has become an important tool for elucidation of biochemical mechanisms regulating cyclic adenosine monophosphate (AMP)-dependent chloride efflux. Overgrowth of the indigenous bacterial population in the small intestine also may produce malabsorption, most notably of fats and vitamin B_{12} [66].

Another mechanism of toxicity mediated by intestinal microorganisms is an indirect one. The bacterial population is capable of catalysis of numerous chemical reactions that can alter the biologic activity of dietary substances and xenobiotics, especially in more distal portions of the gut. Many examples exist of bacterial activation of chemicals to a form exerting toxicity in the gut. A classic case is that of cycasin, the BETA-glucosidase of methylzoxymethanol (MAM), which is contained in nuts of cycad plants [78]. The BETA-glucosidase activity of indigenous bacteria hydrolyzes cycasin and releases MAM, a mutagenic carcinogen causing tumor formation in the colon, liver, and kidney [68].

Methodologies exist for evaluating bacterial metabolism of gut flora *in vitro* (with gut contents having been collected and incubated) and *in vivo*, and the *in situ* bacterial flora of the intestine. The actual study of effects on the flora is best done *in vitro* using variations on traditional microbiology techniques, but it is difficult to model the complex interactions between these organisms and a host organism using an *in vitro* methodology.

A conventional technique for assessing the potential of a microorganism to perturb electrolyte and water flux in the gut is the use of the rabbit ileal segment. Toxins of bacteria, e.g., *V. cholerae*, *Clostridium difficile*, and *Escherichia coli*, which produce diarrhea in humans and experimental animals, cause fluid accumulation when injected into a closed segment of the small intestine. This technique provides a simple experimental approach for identifying the potential of a toxin for inducing intestinal antiabsorptive or secretory activity and estimating its potency [69,70]. Young (12–24-week-old) rabbits are fasted overnight. The ileum, which is more sensitive and gives more consistent results than does the jejunum, is tied into three or four closed segments each 3 in long. Segments are separated by 6 in. Cultures or supernatant factions are injected into the segments with a 22-gauge needle. The animal is allowed to recover from anesthesia and then reanesthetized to examine the ileal contents after 4–6 hours. The experimenter can distinguish toxicity induced by an exotoxin from that resulting from bacterial invasion of the mucosa.

Methods for bacterial metabolism studies. An experimental approach for assessing the role of intestinal bacteria in mediating GI or systemic toxicity is the use of germ-free animals. Rats, for example, that have been aseptically delivered by cesarean section and reared under germ-free conditions are available from animal suppliers. Procedures required for establishing and maintaining a germ-free animal colony are described by Foster [71]. To test whether bacteria are responsible for metabolic alteration of a compound, comparison can be made of its urinary and fecal elimination in germ-free and conventional animals of the same strain. A difference in the composition of the metabolites of the compound in the urine or the feces may suggest participation of bacterially mediated reactions in the biotransformation process. Absence of a particular metabolite of an administered compound in the urine or feces from the germ-free animal is presumptive evidence that the reaction is only catalyzed by bacterial and not mammalian enzymes.

Support for such a hypothesis can be obtained by examining the *in vitro* metabolic capability of defined bacterial cultures or of homogenates of the luminal gut contents. In experiments of this type, the culture conditions can markedly affect the experimental outcome. With studies of large intestinal contents or bacterial isolates form this source, it is especially important to carry out experiments using anaerobic conditions because most organisms from this site are obligate anaerobes highly sensitive to even brief exposure to aerobic conditions. Demonstration of a particular enzymatic activity in a bacterial strain *in vitro* provides additional evidence of a possible bacterial role. However, it has frequently been difficult to predict *in vivo* routes of intestinal metabolism from *in vitro* studies. One possible reason for such discrepancies is the synergistic interaction between an extract from *Bacteroides fragilis*, which populates the distal intestine, and microsomal enzyme activity of the colonic mucosa. Each component alone had minimal capacity to transform 2-aminoanthracene into a mutagen (with respect to a *Salmonella* tester strain). However, in combination, a substantial increase in mutagen formation occurred. The mechanism of this activating effect of a heat-liable component of the bacteria on the colonic microsomes is not known. This observation illustrates the complex interaction between bacterial, intestinal tissue, and xenobiotics, and the potential limitations of *in vitro* studies.

Table 13.2 presents an overview of eight different *in vitro* test systems available (and used for) evaluating GI toxicity. Each system focuses on one or two specific functional endpoints of toxicity, with either perfused/superfused tissues or organs or cultured cells as models. These endpoints are categorized, somewhat arbitrarily, as either screening or mechanistic systems, depending on the emphasis of their use to date.

13.4 SPECIFIC PROBLEMS/LIMITATIONS OF *IN VITRO* TEST SYSTEMS

Though components of many of the functions of the GI tract are expressions of the function of a specific cell or tissue type, the overall functions of interest are largely expressions of an integrated physiologic and biochemical function. The chief limitation of *in vitro* assessment of GI toxicity is that it is either not possible or only possible on a limited basis to model the organ-system–side nature of such functions as motility (mass transport) and absorption. Similarly, effects on the microbial populations/ecologies of the GI tract regions are complex. Aspects of these effects can be studied *in vitro*, as Table 13.2 establishes, but not the entire functionality.

Similarly, modeling the *in vivo* modes (and characteristics) of exposure to environmental agents in *in vitro* models is particularly difficult. In part, this difficulty is because so many sources of potential toxicants exist—ingestion (which can include inhaled, particularly water, contaminants, dietary components and contaminants, drugs, etc.), tract secretions (both plasma and biliary toxicants), microbial products, and the products of GI biotransformation of environmental agents.

The present-day state-of-the-art system allows maintenance of isolated perfused

TABLE 13.2. *In vitro* test systems for gastrointestinal toxicity

System	Endpoint	Evaluation	References
Isolated perfused intestines (M)	Functional: biochemical and metabolic	Correlation with *in vivo* finding for methylprednisolone	Mehendale [16] Hohenleitner and Senior [20]
Isolated perfused intestines (M)	Functional: biochemical and metabolic	Limited	Mehendale [16]
Isolated superfused ileum (S)	Functional: pharmacologic responses and biochemical	Correlation with *in vivo* findings for antioxidants and receptor-mediated agents	Gad et al. [13]
Stomach wall in a Ussing chamber (M)	Functional: gastric secretion (acid secretion), electrolyte flux	Correlation for pharmacological agents	Soll [9] Sachs and Berglingh [8]
Parietal cells from gastric gland (M, S)	Functional: hydrogen ion secretion measurement	Good correlation for ulcerogens	Soll [9] Sachs and Berglingh [8] Soll and Berglingh [12]
Inverted intestinal sac (M)	Functional: energy-mediated carrier	Well studied	Wilson and Wiseman [6]
Isolated rabbit jejunum (M)	Functional: motility measurement	Well studied and correlated	University of Edinburgh [14]
Cultured myocytes (M)	Functional motility	Limited	Bitar and Makhlouf [15]
Culture of GI mucosa (M)	Functional; metabolite and biochemical	Well studied	Howdle [21]

NOTE: Letters in parentheses indicate primary employment of system: (S), screening system; (M), mechanistic tool.

organ preparations with adequate physiologic and biochemical integrity for only short periods of time. Clinically, advances have allowed maintenance of the kidney for several days for later physical transplantation in patients. These procedures require subambient temperatures to preserve organ function. Such techniques are generally not useful in toxicologic studies, as maintenance of the organ at optimal functional levels is a prerequisite for most toxicologic studies. Hence, the principal limitation imposed by the isolated perfused organ preparations is the short duration of study. Critical and vital organ functions deteriorate in isolated perfused organs with time. For example, isolated perfused lung preparations can be maintained for a maximum of 4 hours [72]. Often, it is not possible to determine the effect of therapeutic agents on lung tissue in such a short period of time. Similarly, isolated perfused liver preparations cannot be maintained for longer periods without compromising liver function [73]. A practical consideration of interest in this connection may also be the level of expertise required for setting up perfusion experiments. Setting up and conducting successful perfusion experiments requires specially trained personnel in all aspects of the surgical procedures as well as the technical aspects of associated instrumentation.

Often, a principal argument in favor of isolated perfused organ studies is the maintenance of natural membrane barriers, the integrity of intact cells, and the complex and dynamic interrelations between individual and groups of cells. For certain studies, this argument may represent a principal limitation. The complexity of a whole organ deprives the toxicologist of access to individual reactions occurring within the organ. Compartments and permeability barriers may prevent substrates and test drugs from exerting effects known to manifest when the particular toxic agent is allowed direct access to the enzyme or organelle of interest. *in vitro* experiments with homogenate preparations and tissue slice preparations would be the obvious choice of technique when dissection of individual transport processes and biotransformation reactions is the principle objective. The size of an experiment and the time required to perform it may make organ perfusion far less efficient from the perspective of time and resources than it may make *In vitro* preparations demonstrating the same effects with less investment of time and resources. Another consideration is the availability of the experimental tissue of organs. Although access to valuable human tissues might be available, such access might be infrequent; and, in any event, the available tissue would be limited. Clearly, isolated cell techniques or other *in vitro* techniques have the advantages of maximizing the use of such invaluable experimental material when designing and carrying out such studies. Schimmel and Knobil [74] pointed out the greater efficiency of establishing an experimental fact with tissue slices than with isolated perfused organ. Finally, despite all of the refined techniques of maintaining the organ *in vitro* in as near a normal state as possible, the resulting preparation may differ in some highly significant manner from the organ *in vivo*, limiting the interpretation and application of results obtained in the organ perfusion system.

References

1. Walsh CT. *Mechanisms of Gastrointestinal Toxicology*. Reston, VA: Society of Toxicology, 1988.
2. Walsh T. Methods in gastrointestinal toxicology. In: Hayes AW, ed. *Principles and Methods of Toxicology*. New York: Raven. 1989:659–675.
3. Schiller CM. *Intestinal Toxicology*. New York: Raven, 1984.
4. Betton GR. The digestive system I: the gastrointestinal tract and exocrine pancreas. In: Turton J, Hooson J, eds. *Target Organ Pathology*. Bristol, PA: Taylor and Francis, 1998:29–61.
5. Rowland IR. *Role of the Gut Flora in Toxicity and Cancer*. New York: Academic Press, 1988.

6. Wilson TH, Wiseman G. The use of sacs of everted small intestine for the study of the transference of substances from the mucosal to the serosal surface. *J Physiol (Lond)* 1954;123:116–125.

7. Kimmich GA. Intestinal absorption of sugar. In: Johnson LR, ed. *Physiology of the Gastrointestinal Tract*, Vol. 2. New York: Raven, 1981:1,035–1,062.

8. Sachs G, Berglingh T. Physiology of the parietal cell. In: Johnson LR, ed. *Physiology of the Gastrointestinal Tract*. New York: Raven, 1981:570–574.

9. Soll AH. Secretagogue stimulation of ^{14}C-aminopyrine accumulation by isolated canine parietal cells. *Am J Physiol* 1980;238:G366–G375.

10. Szabo S, Reynolds ES, Lichtenberger LM, Haith LR, Dzau VJ. Pathologenesis of duodenal ulcer: gastric hyperacidity caused by propionitrile and cysteamine in rats. *Res Commun Chem Pathol Pharmacol* 1977;16:311–323.

11. Segal HL, Miller LL, Plumb EJ. Tubeless gastric analysis with an azure A ion-exchange compound. *Gastroenterology* 1955:28:402–408.

12. Soll AH, Berglingh T. Physiology of isolated gastric land and parietal cells: receptors and effectors regulating function. In: Johnson LR, ed. *Physiology of the Gastrointestinal Tract*. New York: Raven, 1987:883–909.

13. Gad SC, Leslie SW, Acosta D. Inhibiting actions of butylated hydroxytoluene (BHT) on isolated rat ileal, atrial and perfused heart preparations. *Toxicol Appl Pharmacol* 1979;48:45–52.

14. Department of Pharmacology, University of Edinburgh. *Pharmacological Experiments on Isolated Preparations*. Edinburgh: E & S Livingstone, 1970.

15. Bitar KN, Makhlouf GM. Receptors on smooth muscle cells: characterization by contraction and specific antagonists. *Am J Physiol* 1982;242(4):G400–G407.

16. Mehendale HM. Application of isolated organ techniques in toxicology. In: Hayes AW, ed. *Principles and Methods of Toxicology*. New York: Raven, 1989:699–740.

17. Crane RK, Wilson TH. *In vitro* method for the study of the rate of intestinal absorption of sugars. *J Appl Physiol* 1958;12:145–146.

18. Ussing HH, Zerahn K. Active transport of sodium as the source of electric current in the short-circuited isolated frog skin. *Acta Physiol Scand* 1951;23:110–127.

19. Crane RK, Mandelstam P. The active transport of sugars by various preparations of hamster intestine. *Biochim Biophys Acta* 1960;45:460–476.

20. Hohenleitner FJ, Senior JR. Metabolism of canine small intestine vascularly perfused *in vitro*. *J Appl Physiol* 1969;26:119–128.

21. Howdle PD. Organ culture of gastrointestinal mucosa. *Postgrad Med J* 1984;60:645–652.

22. Kotsonis FN, Mackey M, Hjelle JJ. *Nutritional Toxicology*. New York: Raven, 1994.

23. Mehendale HM. Aldrin epoxidase activity in the developing rabbit lung. *Pediatr Res* 1980;14:282–285.

24. Itakura N, Fisher AB, Thurman RG. Cytochrome P450-linked p-nitroanisole O-demethylation in the perfused lung. *J Appl Physiol* 1984;43:238–245.

25. Thurman RG, Marazzo DP, Jones LS, Kauffman FC. The continuous kinetic determina-

tion of p-nitroanisole O-demethylation in hemoglobin-free perfused rat liver. *J Pharmacol Exp Ther* 1977;201:498–506.

26. Israili ZH, Dayton PG. Enhancement of xenobiotic elimination: role of intestinal excretion. *Drug Metab Rev* 1984;15:1,123–1,159.

27. Pfeiffer CJ. Gastroenterologic response to environmental agents: absorption and interactins. In: Lee DHK, Falk HL, Murphy SD, Geiger SR, eds. *Handbook of Physiology, Section 9, Reactions to Environmental Agents.* Bethesda, MD: American Physiology Society, 1977:349–374.

28. Reinhardt MC. Macromolecular absorption of food antigens in health and disease. *Ann Allergy* 1984;53:597–601.

29. Acra SA, Ghishan FK. Methods of investigating intestinal support. *Jpn J Parenter Enterol Nutr* 1991;15:93S–98S.

30. Kroes R, Wester PW. Forestomach carcinogens: possible mechanisms of action. *Food Chem Toxicol* 1986;24:1,083–1,089.

31. Kimmich GA. Intestinal absorption of sugar. In: Johnson LR, ed. *Physiology of the Gastrointestinal Tract.* New York: Raven, 1981:1,035–1,062.

32. Thompson ABR, Dietschy JM. Intestinal lipid absorption: major extra cellular and intracellular events. In: Johnson LR, ed. *Physiology of the Gastrointestinal Tract.* New York: Raven, 1981:1,147–1,220.

33. Smithson KW, Millar DB, Jacobs LR, Gray GM. Intestinal diffusion barrier: unstirred water layer or membrane mucous coat. *Science* 1981;214:1,241–1,244.

34. Mayer RM, Treadwell CR, Gallo LL, Vahouny GV. Intestinal mucins and cholesterol uptake *in vitro. Biochem Biophys Acta* 1985;833:34–43.

35. Thompson ABR, O'Brien BD. Uptake of homologous series of saturated fatty acids into rabbit intestine using three in vitro techniques. *Dig Dis Sci* 1980;25:209–215.

36. Kellet GL, Barker ED. The effect of vandadate on glucose transport and metabolism in rat small intestine. *Biochim Biophys Acta* 1989;979:311–315.

37. Hutchison JDM, Undrill VJ, Porteous JW. The vascularly and luminally perfused small intestine *in vitro*: dissection technique and model system. *Lab Anim* 1997;25:168–183.

38. Levine RR, McNary WF, Kornguth PJ, LeBlanc R. Histological reevaluation of everted gut technique for studying intestinal absorption. *Eur J Pharmacol* 1970;9:211–219.

39. Aw TY, Bai C, Jones DP. Small intestinal enterocytes. In: Tyson CA, Frazier JM, eds. *Methods in Toxicology. In Vitro Biological Systems.* Boston: Academic Press, 1993:193–201.

40. Harrison DD, Webster HL. The preparation of isolated intestinal crypt cells. *Exp Cell Res* 1969;55:257–260.

41. Bevan C, Foulkes EC. Interaction of cadmium with brush border membrane vesicles from the rat small intestine. *Toxicology* 1989;54:297–309.

42. Hopfer U. Isolated membrane vesicles as tools for analysis of epithelial transport. *Am J Physiol* 1977;233:E445–E449.

43. Stevens BR, Kavnitz JD, Wright EM. Intestinal transport of amino acids and sugars: advances using membrane vesicles. *Annu Rev Physiol* 1984;46:417–433.

44. Shirazi-Beechey SP, Davies AG, Tebbutt K, et al. Preparation and properties of brush-border membrane vesicles from human small intestine. *Gastroenterology* 1990;98:676–685.

45. Douglas AP, Kerley R, Isselbacher KJ. Preparation and characterization of the lateral and basal plasma membranes of the rat intestinal epithelial cell. *Biochem J* 1972;128:1,329–1,338.

46. Bell GI, Kayano T, Buse JB, et al. Molecular biology of mammalian glucose transporters. *Diabetes Care* 1990;13:198–208.

47. Elsas LJ, Longo N. Glucose transporters. *Annu Rev Med* 1982;43:377–393.

48. Kayano T, Burant CF, Fukumoto H, et al. Human facilitative glucose transporters. Isolation, functional characterization, and gene localization of cDNAs encoding an isoform (GLUT5) expressed in small intestine, kidney, muscle, and adipose tissue and an unusual glucose transporter pseudogene-like sequence (GLUT6). *J Biol Chem* 1990;265:13,276–13,282.

49. Stevens BR, Fernandez A, Hirayama B, Wright EM, Kempner ES. Intestinal brush border membrane Na^+/glucose cotransporter functions in situ as a homotetramer. *Proc Natl Acad Sci U S A* 1990;87:1,456–1,460.

50. Dudeja PK, Wali RK, Harig JM, Brasitus TA. Characterization and modulation of rat small intestinal brush-border membrane transbilayer fluidity. *Am J Physiol* 1991;260:G586–G594.

51. Schultz SG, Curran PF, Chez RA, Fuisz RE. Alanine and sodium fluxes across mucosal border of rabbit ileum. *J Gen Physiol* 1967;50:1,241–1,260.

52. Rivière PJM, Sheldon RJ, Malarchik ME, Burks TF, Porreca F. Effects of bombesin on mucosal ion transport in the mouse isolated jejunum. *J Pharm Exp Ther* 1980;253:778–783.

53. Sheldon RJ, Malarchik ME, Fox DA, Burks TF, Porreca F. Pharmacological characterization of neural mechanisms regulating mucosal ion transport in mouse jejunum. *J Pharmacol Exp Ther* 1989;249:572–582.

54. Howdle PD. Organ culture of gastrointestinal mucosa. *Postgrad Med J* 1984;60:645–652.

55. Dharmsathaphorn K, McRoberts JA, Mandel KG, Tisdale LD, Masui H. A human colonic tumor cell line that maintains vectorial electrolyte transport. *Am J Physiol* 1984;246:G204–G208.

56. Hecht G, Koutsouris A, Pothoulakis C, LaMont JT, Madara JL. *Clostridium difficile* toxin B disrupts the barrier function of T84 monolayers. *Gastroenterology* 1992;102:416–423.

57. Lencer WI, Reinhart FD, Neutra MR. Interaction of cholera toxin with cloned human goblet cells in monolayer culture. *Am J Physiol* 1990;258:G96–G102.

58. Hiraishi H, Terano A, Ivey KJ. Gastric mucosal cell culture for toxicology study. In: Tyson CA, Frazier JM, eds. *Methods in Toxicology. In Vitro Biological Systems*. Boston: Academic Press, 1993:182–192.

59. Sanders MJ, Soll AH. Cell separation by elutriation: major and minor cell types from complex tissues. *Methods Enzymol* 1989;171:482–497.

60. Derblom H, Johansson H, Nylander G. A simple method of recording quantitatively certain gastrointestinal motility functions in the rat. *Acta Clin Scand* 1966;132:154–165.

61. Walsh CT, Ryden EB. The effect of chronic ingestion of lead on gastrointestinal transit in rats. *Toxicol Appl Pharmacol* 1984;75:485–495.

62. Miller MS, Galligan JJ, Burks TF. Accurate measurement of intestinal transit in the rat. *J Pharmacol Methods* 1981;6:211–217.

63. Staff of the Department of Pharmacology, University of Edinburgh. *Pharmacological Experiments on Isolated Preparations*. Edinburgh: E & S Livingstone, 1970.

64. Bitar KN, Makhlouf GM. Receptors on smooth muscle cells: characterization by concentration and specific antagonists. *Am J Physiol* 1982;242:G400–G407.

65. Makhlouf GM. Isolated smooth muscle cells of the gut. In: Johnson LR, ed. *Physiology of the Gastrointestinal Tract*. New York: Raven, 1987:555–569.

66. Mathias JR, Clench MH. Review: pathophysiology of diarrhea caused by bacterial overgrowth of the small intestine. *Am J Med Sci* 1985;289:243–248.

67. Murthy K, Grider J, Maklouf G. Receptor-coupled G proteins mediate contractions and Ca++ mobilization in isolated intestinal muscle cells. *J Pharmacol Exp Ther* 1992;260:90–97.

68. Laqueur GL, McDaniel EG, Matsumoto H. Tumor induction in germfree rats with methylazoxymethanol (MAM) and synthetic MAM acetate. *J Natl Cancer Inst* 1967;39:355–371.

69. Jenkin CR, Rowley D. Possible factors in the pathogenesis of cholera. *Br J Exp Pathol* 1959;40:474–481.

70. Triadafilopoulos G, Pothoulakis C, O'Brien MJ, LaMont JT. Differential effects of *Clostridium difficile* toxins A and B on rabbit ileum. *Gastroenterology* 1988;93:273–279.

71. Foster HL. Gnotobiology. In: Baker HJ, Lindsey JR, Weisbroth SH, eds. *The Laboratory Rat, Vol 2, Research Applications*. New York: Academic Press, 1980:43–57.

72. Niemeier RW, Bingham E. An isolated perfused lung preparation for metabolic studies. *Life Sci* 1972;11:807–820.

73. Kilbinger H, Krieglstein J. Applicability of the isolated perfused rat brain for studying central cholinergic mechanisms. *Naunyn Schmiedebergs Arch Pharmacol* 1974;285:407–411.

74. Schimmel RJ, Knobil E. Role of free fatty acid in stimulation of gluconeogenesis during fasting. *Am J Physiol* 1969;217:1,803–1,808.

75. Walsh CT, Harnett KM. Inhibitory effect of lead acetate on contractility of longitudinal smooth muscle from rat ileum. *Toxicol Appl Pharmacol* 1986;83:62–68.

76. Gordon AR, Siegman MJ. Mechanical properties of smooth muscle. I. Length-tension and force-velocity relations. *Am J Physiol* 1971;221:1,243–1,249.

77. Van Neuten JM, Geivers H, Fontaine J, Janssen PAJ. An improved method for studying peristalsis in the isolated guinea-pig ileum. *Arch Int Pharmacodyn* 1973;203:411–414.

78. Mickelsen O. Introductory remarks, symposium on cycads. *Fed Proc* 1972;31:1,465–1,546.

14

In Vitro
Immunotoxicology

Marc Pallardy,[1] Herve Lebrec,[1] Saadia Kerdine,[1]
Florence G. Burleson,[2] and Gary R. Burleson[2]

[1]Immunotoxicology group, INSERM U461, Faculté de Pharmacie
Paris XI, Chatenay-Malabry, France
[2]BRT-Burleson Research Technologies, Raleigh, North Carolina

Immunotoxicology can best be defined as the discipline concerned with the study of the adverse effects resulting from the interactions of xenobiotics, i.e., chemicals, drugs, and biologicals, with the immune system. These adverse effects may result as a consequence of (1) a direct or indirect action of the compound (or its metabolites) on the immune system; (2) an immunologically based host response to the compound or its metabolites, or to host antigens modified by the compound or its metabolites [6].

A testing battery has been developed and validated to evaluate the potential adverse effects of compounds on the immune system during development of drugs [24,25]. This methodology is currently used to characterize immune alterations occurring after *in vivo* chemical exposure in rodents. This approach, although important for the detection of chemical-induced immunotoxicity, provides little mechanistic information.

At present, minimal information is available regarding molecular events associated with chemical-induced immunotoxicity. Progress in understanding the actions of xenobiotics on cellular and molecular targets will permit predictions regarding structure-activity relationships and potential human health risks to the immune system. Development of useful, reproducible *in vitro* models should facilitate these studies.

14.1 GENERAL ORGANIZATION OF THE IMMUNE SYSTEM

Lymphoid organs are generally divided into two categories: The primary lymphoid organs are those in which the maturation of T and B lymphocytes into antigen-recognizing lymphocytes takes place. The secondary lymphoid organs are those organs in which antigen-driven proliferation and differentiation takes place. Progenitor cells from bone marrow migrate to the thymus, where they differentiate into T lymphocytes. T lymphocytes then leave the thymus to enter the peripheral blood circulation, through which they are transported to the secondary lymphoid organs. Only 5–10% of lymphocytes leave the thymus, whereas 90–95% of all thymocytes die in the thymus. B lymphocytes differentiate from hematopoietic stem cells in the fetal liver during embryonic life and in the bone marrow after birth. The major secondary lymphoid organs are the spleen, where 50% of spleen cells are B lymphocytes and 30–40% T lymphocytes, and the lymph nodes. In addition, the tonsils, the appendix, the clusters of lymphocytes distributed in the lining of the small intestine (Peyer's patches), as well as the lymphoid aggregates spread throughout mucosal tissue in the body serve as secondary lymphoid organs.

Two pathways of differentiation exist from the pluripotent stem cells: the myeloid progenitor giving rise to erythrocytes, thrombocytes, granulocytes, and monocytes/macrophages, and the lymphoid progenitor giving rise to T and B lymphocytes. The neutrophil granulocyte plays a role in the innate immunity in phagocyting microorganisms, whereas eosinophil and basophil granulocytes play a role in parasitic infections and allergic diseases. Monocytes/macrophages are unique cells capable of phagocytosis with bactericidal and tumoricidal activities. These cells are also efficient antigen-presenting cells (APCs). B lymphocytes are the cells concerned with the synthesis of antibody, and it is generally agreed that B cells are mature when IgM and IgD are coexpressed. The second pathway of differentiation of bone marrow stem cells differentiating into lymphocytes passes through the thymus. The result of thymic selection and maturation is the formation of a repertoire of antigen-recognizing T cells of different specificity: helper T cells, cytotoxic T cells, and suppressor T cells. Another cell type mainly involved in the killing of tumors cells or cells infected with viruses is the natural killer (NK) cell. NK cells are lymphoid cells found in the spleen, the lymph nodes, the bone marrow, and the peripheral blood of nonimmune animals and normal humans. These cells can lyse a variety of target cells lacking class I major histocompatibility molecules (MH°C I) without prior antigen recognition.

The immune cells are constantly recirculating through the body, entering lymph nodes by the afferent lymphatics and leaving the lymph nodes by efferent lymphatics converging in the thoracic duct emptying into the vena cava. The traffic of lymphocytes between lymphoid and nonlymphoid tissue ensures that after exposure to an antigen, the antigen and lymphocytes specific to that antigen are sequestrated in the lymphoid tissue, where lymphocytes undergo proliferation and differentiation.

14.2 TRIGGERING THE IMMUNE RESPONSE

The term immunity refers to all mechanisms used by the body as protection against environmental agents foreign to the body. These agents may be microorganisms, tumors, food, chemicals, or drugs. Such immunity may be innate (nonspecific of antigen) or acquired (specific of antigen).

14.2.1 Innate Immunity

Innate immunity includes physical barriers (skin, mucous membrane), chemical barriers (pH), proteins (enzymes, complement), and cells (phagocytic cells, NK cells). Phagocytic cells consist of polymorphonuclear leukocytes (PMNs, granulocytes), phagocytic monocytes, and macrophages. These cells produce peroxide and superoxide radicals toxic to many microorganisms and contain granules (lysosomes) filled with hydrolytic enzymes. PMNs represent the primary line of defense against infectious agents. In the event that PMNs either cannot contain or are destroyed by the infectious agent, macrophages are recruited to the site.

NK activity or "non major histocompatibility complex (MHC)-restricted cytotoxicity" is mediated by lymphocytes, distinct from T and B lymphocytes and sometimes called N lymphocytes, that do not recognize their target via the T-cell receptor (TCR) antigen or Ti/CD3 complex. NK cells have an antiviral activity and patients with deficient NK activity show recurrent viral infections [3].

14.2.2 Acquired Immunity

It is now well known that T cells recognize fragments of antigen (Ag), rather than native intact molecules, and that these fragments must be presented in the context of major histocompatibility complex gene products. T-dependent antigens must first be unfolded, degraded, and "processed" by APCs. Antigen fragments bind to MHC class II (MHC II) molecules and are transported to the plasma membrane of the APC. APCs include Langerhans cells (LCs), macrophages, monocytes, B cells, Kupffer cells, dendritic cells (DCs), and glial cells. The antigen–MHC complex, on the surface of the APC, interacts with the TCR for the antigen to form a trimolecular complex (Ag-MHC-TCR). In addition, a number of adhesion structures stabilize the trimolecular complex (CD2/LFA-3, LFA-1/ICAM-1, CD28/CD80, CTLA4/CD86). These adhesive interactions not only stabilize cell–cell interactions, but also serve to transduce regulatory signals to the T cell.

Occupation of the TCR by antigen and MHC, stabilized by interactions between adhesion molecules, is accompanied by transmission of an activation signal across the T-cell plasma membrane. The T cell must receive a second signal in the form of cytokines, such as interleukin-1 (IL-1) and -6 (IL-6), produced by the APC to complete the process of activation.

Stimulation of the TCR induces virgin or naive T cells to progress from the G0 to the G1 stage of the cell cycle and to express high-affinity IL-2 binding sites (IL-2R). Once T cells are stimulated to secrete IL-2, this cytokine binds to its receptor, thus inducing cell progression through the cell cycle from G1 to S, from G2 to M. Hence, antigen-activated T CD4$^+$ cells producing IL-2 can (1) promote their own clonal expansion, and (2) promote the expansion of other T cells activated by the same antigen, but they may not produce IL-2. As a consequence of T-cell activation, naive CD4$^+$ T-helper lymphocytes or T-helper precursors (THp) produce only IL-2 and proliferate in response to this cytokine. After antigen stimulation, these cells differentiate into TH0 cells producing IL-2, IL-4, interferon-gamma (γIFN), IL-5, IL-6, IL-10, and IL-13. Ultimately, TH1 and TH2 cells will develop from these TH0 lymphocytes. TH1 lymphocytes produce IL-2 and γIFN, whereas TH2 lymphocytes synthetize IL-4, IL-5, IL-10, and IL-13.

The next step of the immune response to an antigen is the expansion and differentiation of B cells into plasma cells and CD8$^+$ cells into cytotoxic T cells. These cells will serve as effector cells.

B cells can bind the same antigen as T cells. B cells, however, recognize native antigen through immunoglobulins present on their surface. Therefore, this process does not require processing of the antigen. Cytotoxic T-lymphocytes are T cells possessing a cytolytic activity directed against cells infected by a pathogen (virus, bacteria). An important difference between CD4$^+$ (helper T cells) and CD8$^+$ (cytotoxic T cells) is that CD4$^+$ cells are activated by signals provided both by an antigen–MHC II molecule complex and IL-1 provided by an APC, whereas activation of CD8$^+$ cells requires a signal provided by the antigen–MHC I molecule complex. All nucleated cells possess MHC I molecules.

Termination of the immune response leads to the generation of memory T cells and the destruction of the excess lymphocytes. These cells undergo apoptosis either by cytokine deprivation or by activation-induced cell death (AICD) using the Fas/Fas ligand pathway.

14.3   CHEMICAL-INDUCED IMMUNOTOXICITY

14.3.1   Immunosuppression

Immunosuppression induced by environmental chemicals or drugs carries the risk of increased susceptibility to exogenous infectious agents or reactivation of latent infection by microorganisms already present in the host. The experience of iatrogenic immunodeficiencies, for example, in transplanted patients, reveals that long-term immunosuppression is associated with a higher incidence of malignancies. Skin and lip lesions (39%), solid lymphomas (27%), and in situ carcinoma of the cervix (8%) were the most frequent cancers observed compared with the normal population and were induced by oncogenic viruses (lymphomas and Epstein–Barr virus, lip lesions and herpesvirus, skin lesions, and cytomegalovirus or herpesvirus) [30].

14.3.2 Hypersensitivity Reactions

Under some circumstances, immunity, rather than protection, produces damaging and sometimes fatal results. These immunoallergic reactions, known collectively as hypersensitivity, are mediated by specific mechanisms triggered by an immunogen. These phenomenona are genetically restricted, meaning that only a few individuals will be affected and are not dose dependent. Most xenobiotics have a molecular weight less than 1000 d and are not immunogenic. They become immunogenic if they are conjugated to high molecular weight carriers that are mainly proteins. In addition, xenobiotics need to be reactive enough to form a stable complex with proteins, which is usually not the case. So, it is believed that reactive metabolites are implicated in the formation of hapten–protein conjugates. This mechanism has been demonstrated for penicillin, in which antipenicilloyl, antipenaldate, and antipenicillenate antibodies are generally found in patients allergic to penicillin. Hypersensitivity reactions are characterized by a period of sensitization, meaning that clinical symptoms appear only in people sensitized to the antigen or hapten. Importantly, the appearance of allergic reactions after primary exposure, or subsequent reexposure, to an immunogen is dependent on T lymphocyte activation. This dependence applies not only to contact hypersensitivity (DTH), which is a cell-mediated immune response, but also to immediate hypersentivity reactions mediated by specific antibodies.

14.3.3 Autoimmunity

Under certain circumstances, xenobiotics can destroy the integrity of self-tolerance with the development of an immune response directed toward an autoantigen, thus leading to an autoimmune disease. These manifestations are genetically restricted and observed only in a small percentage of individuals, are developed in sensitized people, and are not dose dependent. The major difference between hypersensitivity and autoimmunity is that the former is the consequence of immune responses to exogenous determinants, such as xenobiotics, whereas the latter is dealing with a reaction directed to an autoantigen generally modified by the xenobiotic. Clinical manifestations linked to autoimmunity may be organ-specific (hepatitis, nephropathy, cytopenia) or systemic (lupus disease). Drug-induced autoimmune hepatitis is often associated with treatment by dihydralazine or tienilic acid [2,3]. It is assumed that a reactive metabolite is produced that reacts with the enzyme, cytochrome P-450 1A2 for dihydralazine and P-450 2C9 for tienilic acid, and modifies the structure resulting in an immunogenic autoantigen [15]. Diphenylhydantoin, hydralazine, procaïnamide, gold salts, or D-penicillamine can lead to the appearance of lupus-like syndrome with fever, joint pain, and damage to the central nervous system, kidneys, and heart as major clinical signs.

14.4   CULTURING IMMUNE CELLS

14.4.1   Origin of the Cells

Immune cells, except granulocytes, can be divided in two categories: APCs or accessory cells (monocytes, macrophages, LCs, DCs) and lymphoid cells (T and B lymphocytes, thymocytes, NK cells). For an immune response to develop *in vitro*, the two populations of cells are required, except in the case of NK activity. Immune cells can be obtained from different sources: primary lymphoid organs (bone marrow, thymus), secondary lymphoid organs (spleen, lymph nodes, Peyer patches, tonsils), and peripheral blood. In the adult, the bone marrow and the thymus are the main sites of generation of new lymphocytes, and these organs contain immature cells as well as mature components [21]. In this chapter, only tests using mature lymphoid cells will be described and secondary lymphoid organs and peripheral blood compartment are indicated as sources of immune cells.

In most species, the spleen is the largest aggregate of secondary lymphoid tissue in the body. The total lymphoid content reaches adult levels at 8–10 weeks of age in rats and mice. The spleen is composed of white pulp, rich in lymphoid cells, and red pulp, which contains many sinuses as well as large quantities of erythrocytes and macrophages, some lymphocytes, and a few other cells. In rodents, approximately 50% of spleen cells are B lymphocytes and 30–40% are T lymphocytes. After antigenic stimulation, the germinal centers contain a large number of B cells and plasma cells. These cells synthesize and release antibodies.

Lymph nodes are small ovoid structures found in various regions throughout the body. They are composed of a medulla with many sinuses and a cortex, which is surrounded by a capsule of connective tissue. The cortical region contains primary lymphocytic follicules that after antigenic stimulation form germinal centers containing dense populations of lymphocytes, mostly B cells, undergoing mitosis. The deep cortical area or paracortical region contains T cells and macrophages. The medullary area contains antibody-secreting plasma cells that have traveled from germinal centers into the cortex.

Peripheral blood is the most accessible source of lymphoid cells in humans, and it contains accessory cells (monocytes). Human blood contains 5–10×10^6/ml leukocytes and 1,000-fold more erythrocytes. About 30% of leukocytes are lymphocytes and 1–3% monocytes (referred to jointly as peripheral blood mononuclear cells, PBMC), whereas the rest are granulocytes.

14.4.2   Collection of Cells from Secondary Lymphoid Organs

Most lymphoid tissue can be dissociated with relatively little difficulty, because the cells are not tightly bound to each other by tight junctions or substantial amounts of connective tissue. The spleen is easily accessible by an incision with a scalpel on the left flank of the animal. The organ is then poured into a Petri dish containing culture media with fetal bovine serum (FBS) and cut into pieces. Simply crushing or teasing the spleen will release a large number of lymphocytes with good viability. The cell yield from a mouse is approximately $5-10 \times 10^7$ and from a rat approximately $2-4 \times 10^8$. Approximately 2×10^6 nucleated cells can be obtained per milligram fresh wet weight of spleen, with a viability of 80%. A reasonable number of cells can be obtained from the mesenteric, inguinal, and para-aortic lymph nodes, which are relatively accessible. The procedure to release cells from nodes is the same as the one used for spleen. The yield of lymph node cells from a young adult mouse may be $3-5 \times 10^7$ and in a rat, $1-2 \times 10^8$, with a viability of 80–85%. Lymph nodes may be visualized by injecting trace amounts of india ink subcutaneously (SC).

14.4.3 Collection of Cells from Peripheral Blood

Peripheral blood is obtained by cardiac puncture or from the retro-orbital sinus in mice and rats, and from veins in larger animals and humans. The best anticoagulant to use is EDTA. Do not use citrate or heparin because they may cause clumping of leukocytes. It is necessary to use a density-gradient procedure for rapid preparation of PBMC by centrifugation from whole blood in good yield (50%, 106 PBMC/ml of blood). These 1,000-fold purifications work because the giant polymers in the separation media (Ficoll for human cells and Lympholyte M for rodent cells) cause the formation of a high-density gradient. Erythrocytes in these gradients pellet to the bottom, along with granulocytes, which are presumably dense because of their granules, leaving PBMC at the interface between polymers and plasma. Among PBMC obtained through this procedure, T cells form most cells (70%).

14.4.4 Conditions of Culturing Immune Cells

For manipulations lasting a few hours, suspensions of peripheral lymphocytes tolerate room temperature well. Once the cells are isolated, they are cultured in tissue culture media at 37°C, 5% CO_2 in an incubator. The normal pH of the blood is 7.4. Handling media for cells may be buffered with phosphate, which provides a strong buffering without change of pH caused by exposure to air. For culture, the relatively high phosphate cannot be tolerated: Bicarbonate-buffered Hanks's bovine salt serum (BSS), Earle's BSS, minimal essential medium (MEM), or RPMI 1640 are used instead. We used RPMI 1640 with 25-mM HEPES and 2-g/l bicarbonate. These buffers lose CO_2 to the air gradually, and the culture are therefore incubated in an atmosphere of 5% CO_2 in air. It is essential to add protein in the culture media: 5 to 10% FBS in general when using human or rodent cells. In the case of human lymphocytes, it is possible to add autologous serum or a pool of AB serum (5 to 10%). Finally, when the cells are stimulated by an antigen or a mitogen, it is necessary to add 2⁻mercaptoethanol (2-ME) to the media (2×10^{-5} M, final concentration) to avoid cytotoxicity from free radicals.

14.4.5 Differences Between Cells from Human and Animal Origins

Results comparing effects of xenobiotics on animal and human models need to be interpretated with caution. Most studies conducted on animals use splenocytes, whereas peripheral blood lymphocytes are generally employed for human studies. However, lymphocyte populations in spleen and periphery are qualitatively and quantitatively different for humans and animals. In the Fisher 344 rat, the peripheral blood compartment contains 6.6% B cells in comparison to 63% in the spleen [26]. In addition, the percentage among T cells of CD4$^+$ and CD8$^+$ cells in spleen and peripheral blood is not the same (36% CD4$^+$ and 78% CD8$^+$ cells in the spleen, 83% CD4$^+$ and 22% CD8$^+$ in the blood). (Numbers found are greater than 100% because of double-positive cells.)

14.5 *IN VITRO* MODELS IN IMMUNOTOXICOLOGY: DESCRIPTION, ADVANTAGES, AND LIMITATIONS OF THE DIFFERENT *IN VITRO* MODELS CURRENTLY USED IN IMMUNOTOXICOLOGY

14.5.1 Dendritic Cells as a Tool to Evaluate the Sensitizing Potential of Hapten

The nature of hypersensitivity immune responses provoked by chemical allergens is essentially no different than that characterizing protective immunity and providing host resistance against infectious disease. Actually, *in vitro* models are composed of "primed" T lymphocytes, from animals having received hapten–protein conjugate, in the presence of accessory cells (macrophages, DCs). However, although "primed" T cells are not dependent on the quality of APCs required, the presence of DCs is a prerequisite for "naive" T lymphocytes to be activated *in vitro*. In addition, DCs found within the epidermis that are LCs play a key role in cutaneous immune responses. Substantial evidence exists that DCs play a key role in hapten-induced hypersensitivity reactions [18].

DCs are professional APCs found in all tissues. They can be classified into interstitial DCs of the heart, kidney, gut, and lung; LCs in the skin and mucous membrane; and interdigitating DCs (IDCs) in the thymic medulla [8]. Immature DCs are efficient in phagocytizing and in processing antigen, whereas mature DCs are less effective in processing antigen but present antigen efficiently to T lymphocytes [1]. Mature DCs are characterized by their dendritic morphology, expression of CD1α, high levels of MHC II molecules and other molecules, such as CD54 (ICAM-1), CD80 (B7-1), and CD86 (B7-2) [36]. Human DCs can be obtained either from peripheral blood using an enriched fraction of CD14$^+$ monocytes or from cord-blood–isolated CD34$^+$ cells [9,35]. Cord blood CD34$^+$ cells are isolated using immunomagnetic beads and separation columns and then cultured with GM-CSF and tumor necrosis factor (TNF-α) for 10 days. The addition of stem cell factor

(SCF) and Flt3 ligand enhances the proliferation of CD34+ cells [33]. Adherent human PBMCs are cultured with genetically modified-colony stimulating factor (GM-CSF) and IL-4 for 5 days and then cultured with GM-CSF and TNF-α for another 2–4 days to allow maturation of these DC-like cells [41]. These cells are then ready to be stimulated by the hapten.

Recently, it has been shown that epicutaneous application of sensitizing agents, such as 2,4 dinitrolfuorobenzene (DNFB), 2,4 dinitrochlorobenzene (DNCB), and nickel (Ni), but not irritants, reduced the total number of LCs in the epidermis [31].

IL-1B was produced as early as 4 hours after skin challenge by LCs, and IL-1B neutralization inhibited LC migration and prevented hypersensitivity induced by a chemical agent, such as trinitrochlorobenzene [17].

Assessing IL-1B production or IL-1B mRNA production using reverse transcriptase-polymerase chain reaction (RT-PCR) by DC after hapten stimulation could be a tool for predicting the sensitizing potential of xenobiotics. This approach is still at the level of fundamental research, and validation protocols have just started using well-known sensitizers.

14.5.2 Models to Assess Lymphoproliferation

Xenobiotics-induced immunosuppression can be the result of two types of mechanisms that are not mutually exclusive: (1) cytotoxicity with a diminution of the number of immunocompetent cells (see apoptosis section below) or (2) alteration of lymphocyte functions resulting in an inhibition of lymphocyte proliferation and differentiation.

Lymphoproliferation of T cells can be measured after activation with several reagents: lectins [concanavalin A (Con A), parahydroxyamphetamine (PHA), anti-CD3 antibodies, or allogeneic cells (mixed lymphocyte response, MLR). PHA and Con A bind to a number of carbohydrate residues on glycoproteines and induce a similar cascade of events as an antigen. Similar strong polyclonal proliferative responses of T cells can be induced with anti-TCR or anti-CD3 monoclonal antibody. Antibodies directed to the E subunit of the CD3 complex (145-2C11 in murine models, OKT3 in human models) are the most widely used protocols. T-cell responses to these artificial ligands, i.e., mitogens or monoclonal antibody, generally require the presence of accessory cells. It is believed that these cells have two main roles. First, through nonspecific binding, accessory cells can present the ligand to T cells in cross-linked form; notably, it has been well documented for an anti-CD3 monoclonal antibody (mAb), in which immobilization of these mAbs via the Fc receptor to accessory cells is critical for their mitogenic effects on T cells [37]. Secondly, accessory cells provide "second signals" through the interaction of surface molecules on T cells with ligands on accessory cells or cytokines, at least in the case of lectins. A degree of caution, however, must be exercised in the interpretation of studies solely relying on these reagents, especially monoclonal antibody. Binding of these reagents may not truly mimic the physiologic antigen-binding event with respect to epitope specificity, avidity, or valency. In addition, the extensive cross linking of all molecules reactive with a particular monoclonal antibody is not a situation likely to be mimicked by physiologic ligands of such cell surface molecules.

When lymphoid cells from genetically distinct animals of the same species are mixed together in tissue culture, they interact with each other and undergo blast transformation (MLR). Because both populations of lymphoid cells in the MLR are capable of recognition and subsequent response, a unidirectionnal MLR is preferred, whereby the stimulator cells are inactivated by mitomycin C or radiation treatment before addition to culture. It has been shown in numerous species that this property of stimulating an MLR is a function of disparity at the MHC, and alloreactive T cells recognize genetically disparate MHC molecules. Proliferating cells in this model are mainly CD4+ lymphocytes in response to class II determinants. A primary cell-mediated response to allogeneic MHC antigens can be achieved *in vitro*, whereas an *in vitro* response to "nominal" antigens requires *in vivo* priming. T cells reactive with an allogeneic MHC determinant may represent as many as 2% of the total cell population, whereas T cells reactive with an exogenous immunogen generally represent approximatly one in 10,000 of the same cell pool.

B cells can also be induced to proliferate after treatment with polyclonal B-cell activators, such as lipopolysaccharide (LPS), a major constituent of the outer leaflet of the outer membrane of gram-negative bacteria. LPS is directly mitogenic for B lymphocytes, independent of antigen specificity, and can activate a large fraction of B cells (as high as one in three). In addition, LPS triggers the production of cytokines from macrophages: IL-1, IL-6, IL-8, TNF-α, αIFN, and macrophage inflammatory proteins 1 and 2 (MIP-1 and MIP-2). However, results obtained with LPS should be interpreted with caution because this reactant activates B cells in a calcium-independent manner, which is not the case after antigenic stimulation through the B-cell antigen receptor.

Numerous xenobiotics can modulate T- or B-cell proliferation. These xenobiotics include environmental chemicals, such as 7,12-dimethylbenz(a)anthracene (DMBA) [10] or 2,3,7,8-tetrachlorodibenzo-p-dioxin (TCDD), and drugs, such as cyclosporine A or glucocorticoids (GC). Recently, we have studied the effects of several drugs on murine T and B lymphocyte proliferation using a panel of selected tests (Con A, PHA, anti-CD3, MLR, LPS) [29]. Two groups of drugs were used: immunosuppressive drugs (azathioprine, cyclosporin A, hydrocortisone, dexamethasone,) and nonimmunosuppressive drugs (furosemide, indomethacin, cimetidine). The results show that none of the nonimmunosuppressive drugs were able to modify B- or T-cell proliferation, confirming the specificity of these tests. Cyclosporin A significantly inhibited T lymphocyte proliferation with no effect on LPS-induced B-cell proliferation, as previously described [14]. Inhibition of T-cell proliferation was not dependent on the activation signal used in contrast to GC. This difference may be because GC inhibit T lymphoproliferation by a mechanism involving both inhibition of cytokine production and apoptosis, and this could be modulated depending on the stimulus employed (mitogen, allogeneic cells).

14.5.3 Cytokine Measurement to Assess T Helper or T Helper 2 Lymphocyte Differentiation

A key characteristic of T lymphocytes resides in their cytokine secretion profile. Thus, differentiated helper T lymphocytes can be characterized by their pattern of cytokine secretion and be divided into T helper (Th1) and T helper 2 (Th2) subpopulations. Th1 cells mainly produce IL-2 and γIFN, whereas Th2 lymphocytes produce IL-4, IL-5, IL-10, and IL-13 [26,28]. Similarly, CD8+ T cells can be differentiated *in vitro* into T cytotoxic 1 (Tc1) or T cytotoxic 2 (Tc2) lymphocytes [16]. The involvement of Th1- or Th2-like cytokines in contact sensitization has mainly been described in the mouse, with reference compounds, such as dinitrochlorobenzene (TNCB), dinitrofluorobenzene, or oxazolone [34]. All of these data clearly demonstrate that strong contact sensitizers activate γIFN, producing Tc1 effector lymphocytes, and that γIFN plays a central role in the inflammatory cutaneous reaction [40]. In contrast, trimellitic anhydride (TMA) and diisocyanates, which are known respiratory sensitizers, trigger a Th2-like response after cutaneous treatment in the mouse. This Th2-like response is characterized by production of IL-4 and IL-10 [11-13]. The assessment of cytokine production can be performed at the protein level, and cytokines can be measured in cell culture supernatants using commercial enzyme-linked immunosorbent assays (ELISAs). *in vitro* bioassays for cytokines [20] and interferons [7] are well established. Flow cytometric analysis of intracellular staining with fluorescent conjugated monoclonal antibodies can also be used, *in vitro* or *ex vivo*. However, measuring cytokine production at the protein level, by ELISA, for example, is not always possible. Indeed, measurement of a detectable *ex vivo* IL-4 production in the mouse after topical exposure to TMA is only possible after an additional nonspecific stimulation of sensitized lymphocytes by lectins, such as Con A [12].

Cytokine mRNAs are detectable by RT-PCR followed by liquid hybridization to labeled internal probes [4,32]. The RT-PCR is an *in vitro* synthesis of complementary DNA (cDNA) followed by amplification of specific cDNA sequences by simultaneous primer extension of complementary strands of DNA. An *in vitro* human skin model [SKIN² (Advanced Tissue Sciences, La Jolla, CA)], a three-dimensional model consisting of multilayered dermal fibroblasts and well-differentiated epidermal keratinocyte layers, including a stratum corneum, was studied. This human skin model produced constitutive levels of cytokine gene expression mRNA for IL-1α, IL-1B, IL-6, IL-8, IL-10, TNF-α, GM-CSF, TGFB1, and IL-12 p35 (but not IL-12 p40) [4]. The dermal component of this model, consisting of multilayered human dermal fibroblasts, constitutively produced mRNA for IL-1α, IL-1B, IL-6, IL-8, GM-CSF, TGFB1, and IL-12 p35, but not IL-10, TNF-α, or IL-12 p40 [4]. The main problems faced with RT-PCR have to deal with contamination and nonspecific amplification or poor PCR efficiency. Thus, cytokine mRNA measurement seems to be an appropriate endpoint for evaluation of sensitization to drugs, chemicals, protein allergens, or sensitizer-protein conjugates.

14.5.4 Evaluation of Apoptosis Induced by Xenobiotics

Programmed cell death, or apoptosis, is a highly regulated process by which an organism eliminates unwanted cells without eliciting an inflammatory response. However, xenobiotics are also able to trigger unwanted apoptosis or to alter the regulation of programmed cell death. Cytologic characteristics of apoptosis are generally different from those seen in acute pathologic cell death resulting from cell injury. The morphologic characteristics of apoptosis are unique, including cell shrinkage, membrane blebbing, chromatin condensation, DNA fragmentation, dissolution of the nuclear lamina, nuclear fragmentation, and emergence of apoptotic bodies. It is now established that apoptosis plays a critical role in both development and homeostasis of the immune system: thymic selection, cytotoxicity, deletion of autoreactive cells, and regulation of the size of the lymphoid compartment.

Assessment of apoptosis relies on the morphologic and biochemical modifications of the dying cells. As a rule, and because an apoptotic cell rarely displays all of the characteristic apoptotic features, several criteria should be monitored in parallel: analysis of cell morphology by microscopy, identification of DNA fragmentation, measurement of the subG1 peak using cell-cycle analysis, and detection of plasma membrane changes using fluorescent annexin V.

Changes in cell, and particularly in nuclear, morphology, which led to the recognition of apoptosis as a particular mode of cell death, remains a gold standard for identification. Fluorescence microscopy performed without cell fixation, thus allowing differential uptake of DNA binding dyes, is a method of choice that can be used for cells in suspension or adherent cells. Hoechst 33342, a vital dye, gives the clearest images, but when ultraviolet (UV) excitation is not available, it does not contrast well. Acridine orange can be used instead of Hoechst stain, and it requires a filter combination similar to that used for fluorescein (excitation at 488 nm). In addition to intercalating into DNA, acridine orange also diffuses slightly into the cytoplasm, allowing for a better visualization of apoptotic bodies formed by cytoplasmic budding and fragmentation and containing small amounts of condensed chromatin. Ethidium bromide or propidium iodide can also be used at the same time to visualize primary or secondary necrosis.

Digestion of DNA into oligonucleosomal fragments that can be resolved as a typical ladder pattern by agarose gel electrophoresis has long been considered as the best and most convincing evidence of apoptosis, but this is less the case now, because it has been clearly shown that not all cells are able to degrade their DNA in this manner. However, because most lymphoid and myeloid apoptotic cells undergo oligonucleosomal fragmentation, which was first identified in thymocytes, agarose gel electrophoresis can be useful in immunotoxicology studies.

Apoptotic cells analyzed with classic methods for cell-cycle determination show a reduced DNA stainability by DNA-intercalating dyes, such as propidium iodide, lower than that of G0/G1 cells (sub-G1 peak). This phenomenon is generally assumed to be from the generation of the small molecular weight DNA fragments that, after ethanol permeabilization of the plasma membrane, can diffuse through this altered membrane and lower total DNA content. It has also been suggested that this effect may be explained by a reduced accessibility of the dye to condensed chromatin. Whatever the exact mechanism, analysis of this sub-G1 peak by flow cytometry has the advantages of being simple and quantitative because the percentage of cells with reduced stainability in a given population can be measured accurately.

A consequence of chromatin fragmentation is the exposure of 3'-hydroxyl ends of DNA, which can be detected by repetitive enzymatic addition of modified nucleotides, such as fluorescein-labeled deoxyuridine triphosphate (dUTP). Thus, DNA breaks can be revealed *in situ* in individual cells. Two different enzymes can catalyse this reaction: terminal deoxynucleotidyl transferase (TdT) and DNA polymerase. When the former is used, the technique is called TdT-mediated dUTP nick end-labeling (TUNEL), whereas *in situ* nick translation (INST) or *in situ* end-labeling (ISEL) is used to refer to assays using DNA-polymerase. TUNEL has remained as the most widely used term, whatever the nature of the enzyme involved. TdT is generally perceived as more sensitive because it can catalyze template-independent nucleotide addition (oligonucleosomal fragments are assumed to be blunt ended) as well as recognize both 3'-recessed and 5'-recessed fragments (DNA-polymerase only recognizes 3'-recessed fragments). Aside from the nature of the enzyme, commercial kits offer a variety of staining methods: direct labeling (FITC-dUTP, BODIPY-dUTP, CY2-dUTP) or indirect (digoxigenin-conjugated dUTP, biotin-conjugated dUTP) followed by secondary detection systems based on fluorescein, peroxidase, alkaline phosphatase. The TUNEL method is, however, not routinely used for single cells given its high cost and the availability of numerous alternative, less-expensive techniques. The main application of TUNEL is for immunocytochemistry with tissue sections.

Preservation of plasma membrane integrity is one of the major characteristics of apoptosis. Uptake of propidium iodide appears in the late stages of apoptosis triggered *in vitro* and denotes the secondary necrosis of dying cells, which are not engulfed as they would be *in vivo*. However, loss of phospholipid asymmetry, with exposure of phosphatidylserine residues normally maintained exclusively in the inner leaflet is an early and widespread phenomenon. Annexin V, in the presence of Ca^{++}, preferentially binds to negatively charged phosphatidylserine, and this property is used to quantify apoptosis by flow cytometry.

14.5.5 Models to Assess Lymphocyte Cytotoxicity (Innate or Acquired)

Lymphocyte cytolytic activity can be either innate (NK cell activity) or acquired [cytoxic lymphocyte (CTL) activity] and involves a complex series of events: (1) rapid adhesion of a CTL or NK cell to its target (conjugate formation), (2) delivery of the "lethal hit" by the killer cell resulting in prelytic fragmentation of the target's DNA, and (3) target cell dissolution (lysis), detachment, and recycling of the killer cell to start another lytic interaction.

The generation of CTL represents an important acquired effector mechanism for resistance to viral infections and surveillance against neoplastically transformed cells. Induction of CTL by differentiation of precursor CTL cells into fully cytotoxic cells can be accomplished, *in vitro*, in a one-way mixed lymphocyte tumor cell interaction. Cytolytic activity of the CTL is assessed in a 4-hour ^{51}Cr-release assay after a 5-day *in vitro* sensitization to allogeneic-inactivated tumor cells expressing class I molecules. Both CD4+ and CD8+ cells need to be present at the initiation of the assay. P815 mastocytoma and WFUG1 lymphoma tumor cell lines are currently used for the generation of murine and rat CTL, respectively. The Jurkat leukemic T-cell line can be used for generation of human CTL *in vitro*.

CTL activity has proven to be a valuable tool in immunotoxicity testing for assessing activation, proliferation, and differentiation phases of the immune response. For example, it has been reported that CTL activity is suppressed by *in vitro* exposure to DMBA both in murine and human models (P. Saas, personnal communication) and to cyclosporin A and dexamethasone but not azathioprine [23]. Because of the complexity of this cytotoxic mechanism, one xenobiotic can act on more than one step of CTL activity. Cyclosporin A inhibits IL-2 and γIFN production by CD4^{+} cells, thus acting on the differentiation step, but it also inhibits target cell-induced granule-containing perforin or granzyme exocytosis [38].

Spontaneous natural cytotoxicity is an important cytolytic effector mechanism in resistance to tumors and viral diseases. A compilation of xenobiotics affecting NK activity has been published [5]. NK tumoricidal activity is assessed in a 4-hour microcytotoxicity assay by culturing splenocytes or peripheral blood lymphocytes with a ^{51}Cr-labeled cell line sensitive to NK lysis. Tumor cells commonly used are K562, a cell line derived from a patient with chronic myelogenous leukemia in blast crisis, for assessment of human NK cell activity and YAC, a murine lymphoma, for assessment of murine and rat NK cell activity. The NK assay should be performed using at least four effector-to-target ratios, to accurately assess the dose-response relationships between target lysis and a number of effector cells present [39]. Limited studies exist on *in vitro* evaluation of NK activity after exposure to xenobiotics. DMBA inhibits rat, murine, and human NK activity with comparable 50% inhibitory concentrations (IC$_{50}$s) (P. Saas, personnal communication). Human, murine, and rat NK cell activities are inhibited after *in vitro* exposure to cyclosporin A with IC$_{50}$s of 2.6 x 10^{-5} M, 1 x 10^{-7} M, and 1.7 x 10^{-6} M, respectively [23]. Azathioprine has no effect on human and murine NK activity, and dexamethasone poorly inhibits human NK while leaving murine NK activity unaffected [23].

14.5.6 *In Vitro* Antibody Production

The antibody plaque-forming cell (PFC) response to sheep red blood cells (SRBCs) was first described in 1963 by Jerne and Nordin [22]. The PFC to SRBC assay along with cell surface marker analysis resulted in the highest association with immunotoxic potential when 51 different chemicals were studied using the National Toxicology Program (NTP) panel [24]. The PFC response in immunotoxicology has been described in elegant detail [19]. Four basic approaches are available for use of the PFC in immunotoxicologic studies: (1) One approach is *in vivo–in vivo*, (2) another approach is *in vivo–in vitro*, (3) a third approach is *in vitro–in vitro*, and (4) a fourth approach is *in vitro–in vitro* coupled with a metabolic activation system [19]. Three major components exist to the IgM PFC response to SRBC: the B cell for production of antibody, the macrophage because of its essential role in antigen processing, and the T lymphocyte because SRBCs are T-dependent antigens. The importance of the PFC for detecting immunotoxicity is a reflection of this dependence on three cell types, and suppression of any one cell type will cause an alteration in the PFC response.

14.6 RELEVANCE OF *IN VITRO* MODELS

The relevance of *in vitro* models in immunotoxicology is an important aspect even if, at this time, these models are more "comprehensive" than they are "predictive." Three criteria of *in vitro* models need to be examined:

1. Establish that the cultured cells retain the functional activity(ies), whose impairment *in vivo* would lead to adverse effects

2. Establish a correlation between the measured *in vitro* function and the known *in vivo* effects

3. Assess the usefulness of the chosen *in vitro* test as a predictor for known immunotoxic xenobiotic

The first criterion applies primarily to xenobiotic-induced modulation of immune functions, i.e., immunosuppression. All tests proposed for *in vitro* immunotoxicity assessment use either splenocytes or peripheral blood lymphocytes. These cell populations contain all effectors of innate or acquired immune responses, and it is clearly established that functional impairment of these cell populations can lead to an increase in tumors and infections.

The second criterion is clarified by the results obtained in transplant patients, where long-term immunosuppression is associated with impairment of lymphocyte functions (MLR, PHA, CTL, NK), whereas currently measurable immunologic parameters (lymphocyte subsets, serum immunoglobulin levels) are not affected [32]. However, a lot of work still needs to be done regarding metabolism, species differences (mouse versus rat), comparison of results obtained with human and animal cells, and clinical relevance of the changes observed *in vitro*.

The third criterion is important because the initiation of a specific immune response is dependent on a set of events, i.e., activation, proliferation, and differentiation. It is likely that molecules will act solely on one event or one mechanism.

For example, azathioprine, although inhibiting proliferation of murine, rat, and human T lymphocytes, *in vitro*, after activation by different stimuli, does not affect NK or CTL assays in the same conditions. In other words, for predictive purposes, several tests, evaluating the different functions of the immune system, are needed. Continued progress in the evaluation of *in vitro* models will allow these tests to be more and more useful in predicting the immunotoxicity of xenobiotics.

References

1. Banchereau J, Steinman RM. Dendritic cells and the control of immunity. *Nature* 1998;392:245–247.

2. Beaune PH, Dansette PM, Mansuy D, Kiffel L, Finck M, Amar C, Leroux JP, Homberg JC. Human anti-endoplasmic reticulum autoantibodies appearing in a drug-induced hepatitis are directed against a human liver cytochrome P-450 that hydroxylates the drug. *Proc Natl Acad Sci U S A* 1987;84:551.

3. Bourdi M, Larrey D, Nataf J, Bernuau J, Pessayre D, Iwasaki M, Guenguerich FP, Beaune P. Anti-liver endoplasmic reticulum autoantibodies are directed against human cytochrome P-450 IA2. A specific marker of dihydralazine-induced hepatitis. *J Clin Invest* 1990;85:1,967–1,970.

4. Burleson FG, Limardi LC, Sikorski EE, Rheins LA, Donnelly TA, Gerberick GF. Cytokine mRNA expression in an *in vitro* human skin model, SKIN². *Toxicology In Vitro* 1996;10:513.

5. Burleson GR. Alteration of cellular interactions in the immune system: Natural killer activity and N lymphocytes. In: Milman HA, Elmore E, eds. *Biochemical Mechanisms and Regulation of Intercellular Communication.* Princeton, NJ: Princeton Scientific, 1987:51–96.

6. Burleson GR, Dean JH. Immunotoxicology: past, present and future, In: Burleson GR, Dean JH, Munson AE, eds. *Methods in Immunotoxicology*, Vol. 1. New York: Wiley-Liss, 1995:3–10.

7. Burleson GR, Burleson FG. Bioassay of interferons. In: Burleson GR, Dean JH, Munson AE, eds. *Methods in Immunotoxicology*, Vol. 1. New York: Wiley-Liss, 1995:345–356.

8. Caux C, Liu YJ, Banchereau J. Recent advances in the study of dendritic cells and follicular dendritic cells. *Immunol Today* 1995;16:2.

9. Caux C, Vanbervliet B, Massacrier C, Dezutter-Dambuyant C, de saint-Vis B, Jacquet C, Yoneda K, Imamura S, Schmitt D, Banchereau J. CD34⁺ hematopoietic progenitors from human cord blood differentiate along two independent dendritic cell pathways in response to GM-CSF+TNF alpha. *J Exp Med* 1996;184:695.

10. Dean JH, Ward EC, Murray MJ, Lauer LD, House RV, Stillman WS, Hamilton TA, Adams DO. Immunosuppression following 7,12-dimethylbenz(a)anthracene exposure in B6C3F1 mice. II-Altered cell-mediated immunity and tumoral resistance. *Int J Immunopharmac* 1986;8:189.

11. Dearman RJ, Kimber I. Differential stimulation of immune function by respiratory and contact chemical allergens. *Immunology* 1991;72:563.

12. Dearman RJ, Basketter DA, Kimber I. Characterization of chemical allergens as a function of divergent cytokine secretion profiles induced in mice. *Toxicol Appl Pharmacol* 1996;138:308.

13. Dearman RJ, Mitchell JA, Basketter DA, Kimber I. Differential ability of occupational chemical contact and respiratory allergens to cause immediate and delayed dermal hypersensitivity reactions in mice. *Int Arch Allergy Immunol* 1992;97:315.

14. Di Padova FE. Pharmacology of cyclosporin (Sandimmum). Pharmacological effects on immune function: *in vitro* studies. *Pharmacol Rev* 1989;41:373.

15. Druet P, Pelletier L. Chemical-induced autoimmunity. *Eur Cytokine Net* 1991;2(suppl 3):43.

16. Dutton RW. The regulation of the development of CD8 effector T cells. *J Immunol* 1996;157:4,287.

17. Enk AH, Angeloni VL, Udey MC, Katz SI. An essential role for Langerhans cell-derived IL-1 beta in the initiation of primary immune responses in skin. *J Immunol* 1993;150:3,698.

18. Grabbe S, Schwarz T. Immunoregulatory mechanisms involved in elicitation of allergic contact hypersensitivity. *Immunol Today* 1998;19:37.

19. Holsapple MP. The plaque-forming cell (PFC) response in immunotoxicology: An approach to monitoring the primary effector function of B lymphocytes, In: Burleson GR, Dean JH, Munson AE, eds. *Methods in Immunotoxicology*, Vol. 1. New York: Wiley-Liss, 1995:71–108.

20. House RV. Cytokine bioassays for assessment of immunomodulation: Applications, procedures, and practical considerations. In: Burleson GR, Dean JH, Munson AE, eds. *Methods in Immunotoxicology*, Vol. 1. New York: Wiley-Liss, 1995:251–276.

21. Hunt SV. Preparation of lymphocytes and accessory cells. In : Klaus GGB, ed. *Lymphocytes, A Practical Approach*. Washington, DC: IRL Press, 1987:ch. 1.

22. Jerne NK, Nordin AA. Plaque formation in agar by single antibody producing cells. *Science* 1963;140:405.

23. Lebrec H, Roger R, Blot C, Burleson GR, Bohuon C, Pallardy M. Immunotoxicological investigation using pharmaceutical drugs. *In vitro* evaluation of immune effects using rodent or human immune cells. *Toxicology* 1995;96:147.

24. Luster MI, Munson AE, Thomas PT, Holsapple MP, Fenters JD, White KL, Jr, Lauer LD, Germolec DR, Rosenthal GJ, Dean JH. Developpment of a testing battery to assess chemical-induced immunotoxicity: National Toxicology Program's guidelines for immunotoxicity evaluation in mice. *Fundam Appl Toxicol* 1988;10:2.

25. Luster MI, Portier C, Pait DG, White KL, Gennings C, Munson AE, Rosenthal G. Risk assessment in immunotoxicology I. Sensitivity and predictability of immune tests. *Fundam Appl Toxicol* 1992;18:200.

26. Mosmann TR, Cherwinski H, Bond MW, Giedlin MA, Coffman RL. Two types of murine helper T cell clone. I. Definition according to profiles of lymphokine activities and secreted proteins. *J Immunol* 1986;136:2,348.

27. Munson AE, Wei Cao, Kimber I, White KL, Jr. Practical aspects of immunotoxicology. *Eur Cytokine Net* 1991;2:21.

28. O'Garra A. Cytokines induce the development of functionally heterogeneous T helper cell subsets. *Immunity* 1998;8:275.

29. Pallardy M, Lebrec H, Blot C, Burleson G, Bohuon C. *In vitro* evaluation of drug-induced toxic effects on the immune system as assessed by proliferative assays and cytokine production. *Eur Cytokine Net* 1991;2:201.

30. Penn I. Tumors in the immunocompromised patients. *Annu Rev Med* 1998;39:63.

31. Rambukkana A, Pistoor FH, Bos JD, Kapsenberg ML, Das PK. Effects of contact allergens on human Langerhans cells in skin organ culture: migration, modulation of cell surface molecules, and early expression of interleukin-1 beta protein. *Lab Invest* 1996;74:422.

32. Revillard JP. Iatrogenic immunodeficiencies. *Curr Opin Immunol* 1990;2:445.

33. Rosenzwajg M, Camus S, Guigon M, Gluckman JC. The influence of interleukin (IL)-4, IL-13, and Flt3 ligand on human dendritic cell differentiation from cord blood CD34+ progenitor cells. *Exp Hematol* 1998;26:63.

34. Ryan CA, Dearman RJ, Kimber I, Gerberick F. Inducible interleukin 4 (IL-4) production and mRNA expression following exposure of mice to chemical allergens. *Toxicol Lett* 1998;94:1.

35. Sallusto F, Lanzavecchia A. Efficient presentation of soluble antigen by cultured human dendritic cells is maintained by granulocyte/macrophage colony-stimulating factor plus interleukin 4 and downregulated by tumor necrosis factor a. *J Exp Med* 1994;179:1,109.

36. Steinman RM, Pack M, Inaba K. Dendritic cells in the T-cell areas of lymphoid organs. *Immunol Rev* 1997;156:25.

37. Tax WJM, Willems HW, Reekers PPM, Capel PJA, Koene RAP. Polymorphism in mitogenic effects of IgG1 monoclonal antibodies against T3 antigen on human T cells. *Nature* 1983;307:445.

38. Trenn G, Taffs S, Hohman R, Kincaid R, Shevach EM, Sitkovsky M. Biochemical characterization of the inhibitory effect of CsA on cytolytic T lymphocyte effector functions. *J Immunol* 1989;142:3,796.

39. Whiteside TL, Bryant J, Day R, Herberman RB. Natural killer cytotoxicity in the diagnosis of immune dysfunction: criteria for a reproducible assay. *J Clin Lab Anal* 1990;4:102.

40. Xu H, Dilulio NA, Fairchild RL. T cell populations primed by hapten sensitization in contact sensitivity are distinguished by polarized patterns of cytokine production: interferon gamma-producing (Tc1) effector CD8+ T cells and interleukin (IL)-4/IL-10-producing (Th2) negative regulatory CD4+ T cells. *J Exp Med* 1996;183:1,001.

41. Zhou LJ, Tedder TF. CD14+ blood monocytes can differentiate into functionally mature CD83+ dendritic cells. *Proc Natl Acad Sci U S A* 1996;93:2,588.

15

Strategy and Tactics for Employment

Shayne Cox Gad

Gad Consulting Services, Raleigh, North Carolina

The test methods designed and used to evaluate the potential of man-made materials to cause harm to the people making, transporting, using, or otherwise coming into contact with them hold a unique and ambivalent place in our society. On the one hand, our society not only is critically dependent on technologic advances to improve or maintain standards of living, but it is also intolerant of risks (real or potential) to life and health—for example, diseases like acquired immunodeficiency syndrome (AIDS) and multiple sclerosis and illness or disability caused by household products, pesticides, or waste products. At the same time, the traditional tests (with both their misuse and misunderstanding of their use) have served as the rallying point for those individuals concerned about the humane and proper use of animals. This campaign has caused the field of testing for the potential to cause irritation or damage to the eyes to become both the most active area for the development of alternatives and innovations and the most sensitive area of animal testing and use in research. But all testing using animals has come under question.

In recent years, tremendous progress has been made in our understanding of biology down to the molecular level. This progress has translated into many modifications and improvements in *in vitro* testing procedures that now give us tests that (1) are more reliable, reproducible, and predictive of potential hazards in humans; (2) use fewer animals; and (3) are considerably more humane than were earlier test forms. A number of *in vitro* test systems have been proposed, developed, and validated, and through the Interagency Committee for Alternative Models (ICAM) process, they have been accepted by regulatory agencies and incorporated into use.

It is the intent of this volume to make more scientists aware of the full range of new available techniques. But more importantly, it is hoped that the whole process involved in testing can be modified so that only what needs to be done will be and that such tests will answer the desired questions in a manner maximizing efficiency, effectiveness, and animal welfare.

The entire product safety assessment process, in the broadest sense, is a multistage process in which none of the individual steps is overwhelmingly complex, but the integration of the whole process involves fitting together a large complex pattern of pieces. This volume as a whole seeks to address the questions of the use of *in vitro* test methods. How the data generated by the various test systems and models described elsewhere in this volume could be integrated into programs for government and corporate enterprise to provide for a safe product life cycle is the subject of this chapter. As will be seen, this objective calls for a significant conceptual modification of the approach to the general product safety assessment problem, and it will be addressed by starting with the current general case and progressing to the plans and the means for changing the process in an interactive fashion. Along the way, limitations of current models and approaches and places where testing and research data could be made more practically useful will be pointed out.The integration of *in vitro* methodologies into the product safety assessment process has become essential, particularly with an understanding of mechanisms becoming increasingly important in both product design and evaluating the relevance of findings.

The entire safety assessment process, which supports new product research and development, is a multistage process in which none of the individual steps is overwhelmingly complex, but for which the integration of the whole process involves fitting together a large complex pattern of pieces. In this chapter, an approach is proposed in which integration of *in vitro* test systems calls for a modification of the approach to the general product safety assessment problem. This modification can be addressed by starting with the current general case and progressing to a means for changing the process in an iterative fashion. The integration of *in vitro* methodologies, especially into the pharmaceutical safety assessment process, has become essential, particularly with an understanding of mechanisms becoming increasingly important in both product design and evaluating the relevance of findings.

The most important part of any product safety evaluation program is, in fact, the initial overall process of defining and developing an adequate data package on the potential hazards associated with the product life cycle (the manufacture, sale, use, and disposal of a product and associated process materials) [2]. To do this, one must ask a series of questions in an interactive process, with many of the questions designed to identify or modify their successors. The first question is, what information is needed?

Determining what information is needed calls for understanding the way in which the chemical is to be made and used, as well as understanding the potential health and safety risks associated with exposure of humans who will be either using the drug or associated with the processes involved in making it, on the basis of a hazard and toxicity profile. Once such a profile is established (as illustrated in Fig. 15.1), the available lit-

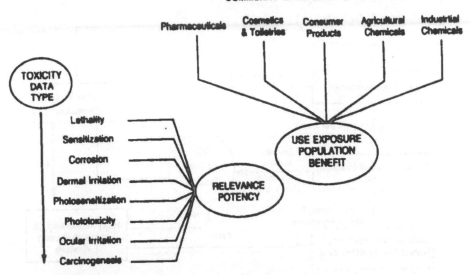

FIG. 15.1. Multidimensional matrix for hazard assessment. The hazards associated with a new product are a multidimensional problem depending on (1) the product's intended use, its innate toxicity, and its physiochemical properties and (2) the potential human and environmental exposure. This matrix illustrates the key questions involved in developing the final hazard assessment profile.

erature is searched to determine what is already known. Much of the necessary information for support of safety claims in registration of a new drug is mandated through regulation. This is not the case at all, however, for those safety studies done (1) to select candidate products for development, or (2) to design pivotal safety studies to support registration, or (3) to pursue mechanistic questions.

Taking into consideration this literature information and the previously defined exposure profile, a tier approach (Fig. 15.2) has traditionally been used to generate a list of tests or studies to be performed.[1] What goes into a tier system is determined by (1) regulatory requirements imposed by government agencies, (2) the philosophy of the parent organization, (3) economics, and (4) available technology. How such tests are actually performed is determined on one of two bases. The first basis (and most common) is the menu approach: selecting a series of standard design tests as "modules" of data. The second is an interactive/iterative approach, in which strategies are developed and studies are designed based both on needs and on what has been learned to date about the product.

[1] The special case of pharmaceutical and pesticide products also exists, in wich mandated minmum test batteries.

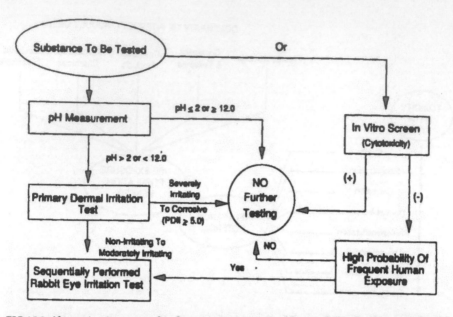

FIG 15.2. Alternative tier approaches for eye irritation testing. The usual plan for characterizing the toxicity of a compound or product is to develop information in a tier approach. More information is required (a higher tier level is attained) as the volume or production and the potential for exposure increase. A common scheme is shown.

15.1 DEFINING TEST OBJECTIVES

The initial and most important aspect of a product safety evaluation program is the series of steps leading to an actual statement of the problem or of the objectives of any testing and research program. This definition of objectives is essential and, as proposed here, consists of five steps: (1) defining product or material use, (2) estimating or quantitating exposure potential, (3) identifying potential hazards, (4) gathering baseline data, and (5) designing and defining the actual research program to answer outstanding questions. Later, we will look at the specific application of this approach to dermal and ocular toxicity cases, in which the concept of communities of interest (that is, an understanding of the needs of the organization producing the material and the users) will become essential. This concept will be discussed in detail later.

15.1.1 Objectives Behind Data Generation and Utilization

To understand how product safety and toxicity data are stored, and how the data generation process might be changed to better meet the product safety assessment needs of society, it is essential to understand that different regulatory organizations have different answers to these questions. The ultimate solution is in the form of a multidimensional matrix, with the three major dimensions of the matrix being (1) toxicity data type (lethality, sensitization, corrosion, irritation, photosensitization, phototoxity, etc.), (2) exposure characteristics (extent, population size, population characteristics, etc.), and (3) stage in the research and development process we are dealing with.

A careful, zero-based consideration of what the optimum product safety assessment strategy for a particular problem should be is what is needed. A framework for such a strategy integrating both *in vitro* and *in vivo* tests is as shown in Fig. 15.1. Before formulating such a strategy and deciding what mix of tests should be used, it is first necessary to decide criteria for what would constitute an ideal (or acceptable) test system.

Any useful test system must be sufficiently sensitive that the incidence of false negatives is low. Clearly, a high incidence of false negatives is intolerable. In such a situation, a large number of dangerous chemical agents would be carried through extensive additional testing only to find that they possess undesirable toxicologic properties after the expenditures of significant time and money. On the other hand, an overly sensitive test system will give rise to a high incidence of false positives, which could have the deleterious consequence of rejecting potentially valuable drugs. The "ideal" test will fall somewhere between these two extremes and, thus, provide adequate protection without unnecessarily stifling development.

The "ideal" test should have an endpoint measurement providing data such that dose-response relationships can be obtained. Furthermore, any criterion of effect must be sufficiently accurate in the sense that it can be used to reliably resolve the relative toxicity of two compounds producing distinct (in terms of hazard to humans), yet similar, responses. In general, it may not be sufficient to classify compounds into generic toxicity categories, such as "intermediate" toxicity, because a candidate chemical falling into a given category, yet borderline to the next, more severe toxicity category, should be treated with more concern than a second candidate falling at the less-toxic extreme of the same category. Therefore, it is useful for a test system to be able to rank compounds with potentially similar uses accurately within any general toxicity category.

The endpoint measurement of the "ideal" test system must be objective, which is important so that a given compound will give similar results when tested using the standard test protocol in different laboratories. If it is not possible to obtain reproducible results in a given laboratory over time or between various laboratories, the historic database against which new compounds are evaluated will be time/laboratory dependent. Along these lines, it is important for the test protocol to incorporate internal standards to serve as quality controls. Thus, test data could be represented using a reference scale based on the test system response to the internal controls. Such normalization, if properly documented, could reduce interest variability.

The test results for any given compound should be reproducible both intrinsically (within the same laboratory over time) and extrinsically (between laboratories). If this condition is not satisfied, significant limitations will be placed on the application of the test system because it would potentially produce conflicting results. From a regulatory point of view, this possibility would be highly undesirable (and perhaps indefensible).

Alternatives to current *in vivo* test systems basically should be designed to evaluate the subject toxic response in a manner as closely predictive of that in humans as possible while also reducing animal use and avoiding inhumane treatments.

From a practical point of view, several additional features of the "ideal" test should be satisfied. The test should be rapid so that the turnaround time for a given compound is reasonable. Obviously, the speed of the test and the ability to conduct tests on several chemicals simultaneously will determine the overall productivity. The test should be inexpensive so that it is economically competitive with current testing practices. And finally, the technology should be easily transferred from one laboratory to another without excessive capital investment for test implementation. It should be kept in mind that although some of these practical considerations may appear to present formidable limitations for a given test system at the present time, the possibility of future developments in testing technology could overcome these obstacles.

This brief discussion of the characteristics of the "ideal" test system provides a general framework for evaluation of alternative test systems in general. No test system is likely to be "ideal."

Therefore, it will be necessary to weigh the strengths and weaknesses of each proposed test system to reach a conclusion on how "good" is a particular test. The next section will present the basis for the specific test evaluation.

In both theory and practice, both *in vivo* and *in vitro* tests have potential advantages. Tables 15.1 and 15.2 summarize these advantages.

15.2   DESIGNING THE RESEARCH PROGRAM

The next step, given that no data are found from any literature sources (and that it has been determined that data are needed), is to perform appropriate predictive tests. The bulk of this volume addresses the specifics of performing such tests using *in vitro* models. Before considering how to design and conduct a testing program, we must first consider how the practice of safety assessment came to its current state in the employment of such tests.

To understand how product safety and toxicity data are used, and how the data generation process might be changed to better meet the product safety assessment needs of society, it is essential to understand that different commercial and regulatory organizations have different answers to these questions. The ultimate answer is a multidimensional matrix, with the three major dimensions of the matrix being (1) toxicity data type (lethality, sensitization, corrosion, irritation, photosensitization, phototoxicity, etc.), (2) exposure characteristics (extent, population size, population characteristics, etc.), and (3) type of commercial organization (or organizations) regulated (which we will call the "community of interest"). This matrix was shown in Fig. 15.1.

TABLE 15.1. Rationale for using *in vivo* test systems

1. Provides evaluation of actions/effects on intact animal and organ/tissue interactions.

2. Either pure chemical entities or complete formulated products (complex mixtures) can be evaluated.

3. Either concentration or diluted products can be tested.

4. Yields data on the recovery and healing processes.

5. Required statutory tests for agencies, such as the Food and Drug Administration (FDA) (for "pivotal" safety studies) and the European Economic Community (EEC).

6. Quantitative and qualitative tests with scoring system, generally capable of ranking materials as to relative hazards.

7. Amenable to modifications to meet the requirements of special situations (such as multiple dosing or exposure schedules).

8. Extensive available database and cross-reference capability for evaluation of relevance to human situation.

9. The ease of performance and relative low capital costs in many cases.

10. Tests are generally both conservative and broad in scope, providing for maximum protection by erring on the side of overprotection of hazard to humans.

11. Tests can be either single endpoint (such as lethality, pyrogenicity, etc.) or shotgun (also called multiple endpoint, including such tests systems as a 13-week oral toxicity study).

TABLE 15.2. Rationale for seeking *in vitro* alternatives for toxicity tests

1. Avoid complications (and potential confounding or masking findings) or animal and tissue/organ *in vivo* evaluation.

2. *In vivo* systems may only assess short-term site of application or immediate structural alterations produced by agents. Note, however, that tests may only be intended to evaluate acute local effects.

3. Technician training and monitoring are critical (particularly if the evaluation called for it to be subjective in nature).

4. If our objective is either the total exclusion of a particular type of agent or the identification of truly severe-acting agents on an absolute basis (that is, without false positives or false negatives), *in vivo* tests in animals do not perfectly predict results in humans.

5. Clearly, structural and biochemical differences exist between test animals and humans making extrapolation from one to the other difficult.

6. Lack of standardization of *in vivo* systems.

7. Variable correlation with human results.

8. Large biologic variability between more complex experimental units (i.e., individual animals).

9. Large, diverse, and fragmented databases that are not readily comparable.

Communities of interest are defined by how the products are to be used, who regulates their use, and what benefits are expected for the consumer. A number of ways of classifying such communities exist, but for our purposes, we will divide and define them as follows.

Pharmaceuticals. Materials of concern are agents intended as therapeutics (or medical devices), in which the production worker or health-care provider (doctor, nurse, or pharmacist) may have a significant chance of exposure, but the major concern is for those patients receiving or using the drug or device. The FDA is the primary U.S. regulator.

Cosmetics and toiletries. The materials are cosmetics, fragrances, shampoos, hand and body soaps, hair dyes, and other materials intended to improve appearance and personal presentation. These materials are intended to be applied to the skin (or other body surface) or to be applied or used in a manner making dermal or ocular exposure (at least) unavoidable. The major U.S. regulators are the FDA and the Consumer Product Safety Commission (CPSC).

Consumer products. Products intended to be used by the average person in and around their home can be divided into those having a high potential for exposure (dish and laundry detergents, for example), those having a low potential for such exposure (drain cleaners, oven cleaners, etc.), or those somewhere in a wide range in between (such as window and carpet cleaners). The primary regulators are the CPSC and the Department of Transportation (DOT), but the Environmental Protection Agency (EPA) also is important in terms of new chemical entities and disposal and waste management.

Agricultural products. These products are pesticides, herbicides, fertilizers, and other international food additives (such as preservatives, sterilants, etc.). The extent of dermal, inhalation, oral, and other routes of exposure will vary widely in use. Note that these products could be subdivided into those agents used in the field (that is, actually used in agriculture), those used for home or inside use, and those used in the storage and processing of foods. Regulatory oversight is vested primarily in the EPA [under the Federal Insecticide, Fungicide, and Rodenticide Act (FIFRA)], with secondary considerations by the DOT and the FDA. The third group (those used on foods) has the FDA as the primary driving force, with EPA and DOT concerns secondary.

Industrial chemicals. These chemicals are materials to which the major exposure is to workers involved in the manufacture and transportation of products. In a sense, all materials (e.g., the above categories) fall into this group at some time, plus a number of other chemicals never appearing (as such) in those categories (such as hydrofluoric acid and plasticizers). These materials are handled by a smaller population relative to most other products. Direct contact is never intended; in fact, active measures are taken to prevent it. The use of safety assessment data in these cases is to fulfill labeling requirements for shipping (DOT) and protecting workers [Occupational Safety and Health Act (OSHA)] and to provide hazard assessment information for accidental exposure and its treatment (DOT and OSHA). The results of such tests do not directly affect the economic future of a material.

Each of these communities has different needs and uses for each of the kinds of data produced, and these must be considered independently.

A careful consideration of what the optimum product safety assessment strategy should be is what is needed. A framework for such a strategy is as shown in Fig. 15.3. The components constititing each of the data generation toolboxes shown (screens, confirmatory tests, higher tier tests, and mechanistic evaluations) are common to all safety assessment programs in some form. But what is actually used for each of these tasks is not common to all of these programs, nor is how the judgment ovals (here labeled acceptance criterion and risk/benefit judgment) operate. The selection of these details is what constitutes the actual formation of a strategy. Before formulating such a strategy and deciding what mix of tests should be used, it is first necessary to decide criteria for what would constitute an ideal (or acceptable) test program.

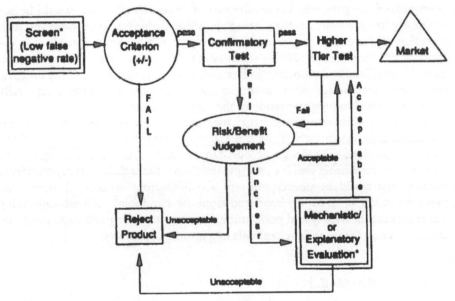

*Points For Initial Employment of In Vitro Tests

FIG 15.3. Iterative approach to a tiered product safety assessment. Within a tiered product safety assessment approach, it is possible to ask each question to a different degree of detail. If an early evaluation provides sufficient information, testing may be complete with relatively low expense. Conversely, it may also be determined that a more detailed (or specific) assay must be performed. Shown here is such an interactive approach identifying points where *in vitro* systems could readily be used in place of *in vivo* systems.

15.2.1 Considerations in Adopting New Test Systems

Conducted toxicologic investigations in two or more species of laboratory animals is generally accepted as being a prudent and responsible practice in developing a new chemical entity, especially one expected to receive widespread use and have exposure potential over human lifetimes. Adding a second or third species to the testing regimen offers an extra measure of confidence to the toxicologist and the other professionals responsible for evaluating the associated risks, benefits, and exposure limitations or protective measures. Although it undoubtedly broadens and deepens a compound's profile of toxicity, the practice of enlarging on the number of test species is, as has been demonstrated in multiple points in the literature (ref. 5, for example), an indiscriminate scientific generalization. Moreover, such a tactic is certain to generate the problem of species-specific toxicoses; that is, a toxic response or an inordinately low biologic threshold for toxicity is evident in one species or strain, whereas all other species examined are either unresponsive or strikingly less sensitive. The investigator confronting such findings must be prepared to address the all-important question, "Are humans likely to react positively or negatively to the test agent under similar circumstances?"

Assuming that numeric odds prevail and that humans automatically fit into the predominant category, whether on the side of being safe or at risk, would be scientifically irresponsible. Far from being an irreconcilable nuisance, however, such a confounded situation can be an opportunity to advance more quickly into the heart of the search for predictive information. Species-specific toxicosis can frequently contribute toward better understanding of the general case if the underlying biologic mechanism either causing or enhancing toxicity is defined, especially if it is discovered to uniquely reside in the sensitive species.

A mention of species-specific toxicoses usually implies that either different metabolic pathways for converting and excreting xenobiotics or anatomical differences are involved. The design of our current safety evaluation tests appear to serve society reasonably well (i.e., significantly more times than not) in identifying hazards that would be unacceptable in a confirmatory manner. However, the process can just as clearly be improved from the standpoints of both improving our protection of society and performing necessary screening and exploratory research in a manner using fewer animals in a more humane way.

15.2.2 *In Vitro* Models

In vitro models, at least as screening tests, have been with us in toxicology for some 20 years now. The last 5–10 years have brought a great upsurge in interest in such models. This increased interest is from economic and animal welfare pressures and technologic improvements.

It should be noted that, in addition to potential advantage, *in vitro* systems *per se* also have a number of limitations contributing to them not being acceptable models. Some of these reasons are detailed in Table 15.3.

TABLE 15.3. Possible interpretations when *in vitro* data do not predict results of *in vivo* studies

1. Compound is not absorbed at all or is poorly absorbed in *in vivo* studies.

2. Chemical is well absorbed, but it is subject to first-pass effect in liver.

3. Compound is distributed so that less (or more) reaches the receptors than would be predicted on the basis of its absorption.

4. Chemical is rapidly metabolized to an active or inactive metabolite that has a different profile of activity or different duration of action than does the parent drug.

5. Compound is rapidly eliminated (e.g., through secretory mechanisms).

6. Species of the two test systems used are different.

7. Experimental conditions of the *in vitro* and *in vivo* experiments differed and may have led to different effects than were expected. These conditions include factors such as temperature, age, sex, and strain of animal.

8. Effects elicited *in vitro* and *in vivo* by the particular case in question differ in their characteristics.

9. Tests used to measure responses will probably differ greatly for *in vitro* and *in vivo* studies, and the types of data obtained may not be comparable.

10. The *in vitro* study did not use adequate controls (e.g., pH, vehicle used, volume of test agent given, samples taken from sham-operated animals).

11. *In vitro* data cannot predict the volume of distribution in central or in peripheral compartments.

12. *In vitro* data cannot predict the rate constants for movement of drug agent between compartments.

13. *In vitro* data cannot predict the rate constants of chemical elimination.

14. *In vitro* data cannot predict whether linear or nonlinear kinetics will occur with specific dose of a drug *in vitro*.

15. Pharmacokinetic parameters (e.g., bioavailability, peak plasma concentration, half-life) cannot be predicted based solely on *in vitro* studies.

16. *In vivo* effects of chemical are caused by an alteration in the higher order of an intact animal system, which cannot be reflected in a less-complex system.

At the same time, as demonstrated throughout this volume, substantial potential advantages exist in using *in vitro* system. These advantages of using cell or tissue culture in toxicologic testing are (1) isolation of test cells or organ fragments from homeostatic and hormonal control, (2) accurate dosing, and (3) quantitation or results. It is important to devise a suitable model system related to the mode of toxicity of the compound. Tissue and cell culture had the immediate potential to be used in two different ways by industry. First, it has been used to examine a particular aspect of the toxicity of a compound in relation to its toxicity *in vivo* (i.e., mechanistic or explanatory studies). Second, it has been used as a form of rapid screening to compare the toxicity of a group of compounds for a particular form of response. Indeed, the pharmaceutical industry has used *in vitro* test systems in these two ways for years in the search for new potential drug entities.

The author has already addressed the theory and use of screens in toxicology [3]. Mechanistic and explanatory studies are generally called for when a traditional test system gives a result that is unclear or whose relevance to the real-life human exposure is doubted. *In vitro* systems are particularly attractive for such cases because they can focus on defined aspects of a problem or pathogenic response, free from the confounding influence of the multiple responses of an intact higher level organism. Note, however, that first one must know the nature (indeed, the existence) of the questions to be addressed.

15.2.3 Short-Term Advances: A Mixed Battery

1. Product safety assessment should not continue to be performed as it traditionally has been.
2. No generally accepted *in vitro* test systems are immediately available to completely replace all (or, indeed, any) of the *in vivo* testing requirements.
3. Some steps can be taken to move development and acceptance of *in vitro* systems along.
4. Some modifications to current *in vivo* testing methods can and should be adopted.

Before developing these points, however, one must consider the needs of the communities of interest responsible for testing and understand the basic concept of a screen (as opposed to a definitive test).

15.2.4 Concept of Screens

Screens are simple tests trying to answer questions with great sensitivity (but not necessarily marked specificity). Many of the currently proposed *in vitro* systems show immediate promise as screens. As such, they would be employed to rapidly and efficiently identify those materials that were clearly strong irritants or corrosives—for example, those that would, therefore, not need further evaluation.

Until *in vitro* tests are further developed and accepted, the appropriate strategy (or mix or tests) for industrial safety assessment should be mixed tier approach (such as Fig. 15.3) using *in vitro* tests as screens. The tier approach presented in Table 15.4 is one form that can be (and is being) used currently (for ocular irritation), but this can be improved. For the short term, a better system could be used employing a mixed series of screening steps. Such a system would have the following stages (using the particular case of evaluation of potential eye irritants as an example):

I. *In Vitro* Screen

a. Extremely active compounds in any assay (such as one of the cytotoxicity test systems) should be considered strong irritants (or worse) and classified/handled as such. Note that this process should also serve to identify (among others) the same compounds as the current pH screen, and therefore, that step is not required.

b. Less active or inactive compounds would pass on, unless the testing need was only to identify I(a)-type compounds (in which case, testing is complete).

II. Primary Dermal Irritation

a. Severely irritating-to-corrosive compounds should be treated as I(a).

b. Mild-to-nonirritant compounds should be treated as I(b).

III. Staggered Eye Irritation Test

Using a low-volume–type test, materials should be evaluated using an animal. If no irritation was found at 24 hours, a second (and 24 hours later, a third) rabbit would be added to the test. Clear positive findings would stop the test.

15.2.5 Far Horizons: How to Get Them

Clearly, great progress has been made both in practices regarding the conduct of safety assessment tests in intact animals and in developing an array of promising *in vitro* candidates for replacement of the *in vivo* test.

TABLE 15.4. Tier testing[a,b]

Tier testing	Mammalian toxicology	Genetic toxicology	Remarks
0	Literature review	Literature review	After initial identification of a problem, a database of existing information and of particulars of use of materials is established
1	Cytotoxicity screens Dermal sensitization Acuted systemic toxicity Lethality screens	Ames tests *In vitro* SCE *In vitro* cytogenetics	R&D materials and low-volume chemicals with severe limited exposure
2	Subacute studies Metabolism Primary dermal irritation Eye irritation	Forward mutation/CHO *In vivo* SCE *In vivo* cytogenetics	Medium-volume materials or those with a significant chance of human exposure
3	Subchronic studies Reproduction Teratology Chronic studies Mechanic studies		Any material with a high volume or a potential for widespread repeated human exposure or one giving indications of specific long-term effects

[a]Shown here is a now widely adopted example of an iterative/tier approach to toxicity testing using a range of test systems (physicochemical, *in vitro*, and *in vivo*) to minimize both cost and animal usage and distress.

[b]Abbreviations: CHO, Chinese hamster ovary; R&D, research and development; SCE, Sister chromatid exchange.

We would like to have in place (that is, accepted and used by industry and reg- ulation agencies) one or a battery of *in vitro* systems that would reduce the need for intact animal testing to necessary cases. And we would also like to have dupli- cate or unnecessary testing of materials reduced to a minimum. These goals are dictated as much by economic reasons and the need to do better science as they are by ethical and humane concerns. The efficient and effective safety assess- ment/toxicology laboratory of the very near future will have as its "front door" an *in vitro* screening shop that will "draw" validated specific target organ screens from a library, as needed, to perform the initial go/no-go evaluations on new com- pounds (or at least provide guidance as to where further evaluation is required). This same shop would also provide (again, from its established collection) *in vitro* system models to elucidate mechanistic questions later in the assessment process. Some people would say that this process is the current state-of-the-art model. Clearly, much of the necessary library could be assembled [4].

Though substantial progress has been made in improving the design and con- duct of *in vivo* tests, it does not appear that we are currently closer to achieving the one optimum case than we were in 1980 (though we clearly understand it better), and minimal progress toward the specific objectives of improved strategy is imme- diately apparent. How does the science and practice of toxicology go about getting to this point?

Two critical steps must be taken for the fulfillment of these objectives. The first of these steps is the need for acceptance of a scientific approach to the problem of safety assessment. The second step is to develop an operative validation and ac- ceptance process for new test procedures.

A scientific approach to safety assessment, such as the one presented in this chapter, does have proponents and adherents. Such an approach required those individuals involved in the management and conduct of the safety assessment process to continually question (and test) the efficiency and the validity of their evaluation systems and processes. More to the point, it requires recognition of the fact that "we have always done it this way" is not a reason for continuing to do so. This approach asks first what is the objective behind the testing and, then, how well is our testing meeting this objective. Such questions are exemplified by efforts such as those questioning if the information given by using six rabbits in an eye ir- ritation test is more predictive than that given by using two or three [1,7].

The second necessary step that is currently absent is that a collaborative process is needed, involving industrial, academic, and regulatory agencies for the valida- tion and "acceptance" of new test systems. The general model of peer recognition leading to acceptance by the scientific community [6] is not working in this case, as should have been expected from a situation in which politics, social policy, and litigation have as much influence as science.

15.3 CONCLUSION

The first principle in hazard assessment is to have your real data as near as possible to the real-life situation you are concerned about. That is, the nearer the model to humans, the better the quality of the prediction of any potential hazards.

The second principle should now also be clear: To be able to translate toxicity to hazard, and to be able to manage such hazards, it is essential to know how the agent is to be used and the marketplace it is to be a part of. It is hoped that this chapter has made these relationships clear.

Finally, alternatives of both *in vitro* and *in vivo* types are in the process of development for almost all of the different endpoints of concern in safety assessment. Many of these alternatives have great promise, and they could be used as screens for many of the uses presented here or as mechanistic tools. But complete replacement is clearly not near at hand, particularly for the more complicated endpoints. How these alternatives, then, can (and should) be integrated into strategies for product safety assessment is the key scientific and managerial challenge for the next decade, for not only do strong reasons exist, which make it adverse to continue where we are, but also the potentials for tremendous competitive advantage exist to those individuals successfully managing to integrate *in vitro* tools as both efficient screens and effective means of isolating and understanding the mechanistic underpinning for toxic and pathogenic processes. At the same time, each practicing toxicologist should feel both a moral and ethical compulsion to reduce the number of animals used in research and testing to the fullest extent possible and to ensure that those animals used are maintained and used in as humane a manner as possible.

References

1. DeSousa DJ, Rouse AA, Smolon WJ. Statistical consequences of reducing the number of rabbits utilized in eye irritation testing: data on 67 petrochemicals. *Toxicol Appl Pharmacol* 1984;76:234–242.

2. Gad SC. *Product Safety Evaluation Handbook*, 2nd ed. New York: Marcel Dekker, 1999.

3. Gad SC. An approach to the design and analysis of screening data in toxicology. *J Am Coll Toxicol* 1988;7(2):127–138.

4. Gad SC. A tier testing strategy incorporating *in vitro* testing methods for pharmaceutical safety assessment. *Hum Innov Alter Anim Exp* 1989;1:75–79.

5. Gad SC, Chengelis CP. *Acute Toxicology*, 2nd ed. San Diego, CA: Academic Press, 1997.

6. Greim H, Andrae U, Forster U, Schwarz L. Application, limitation and research requirements of *in vitro* test systems in toxicology. *Arch Toxicol* 1986;9(suppl):225–236.

7. Talsma DM, Leach CL, Hatoum NS, Gibbons RD, Roger JC, Garvin PJ. Reducing the number of rabbits in the Draize eye irritancy test: a statistical analysis of 155 studies conducted over 6 years. *Fundam Appl Toxicol* 1988;10:146–153.

16

Scientific and Regulatory Considerations in the Development of *In Vitro* Techniques in Toxicology

Patricia D. Williams

SRA Life Sciences, Falls Church, Virginia

The application of *in vitro* techniques in toxicology has received much attention over the past 15 years. The impetus for this growing interest has come from many sources. The *in vitro* "movement" has been driven by scientific, economic, and societal demands. Scientifically, the state of technologic developments in biochemistry and pharmacology (e.g., cellular biology, receptor pharmacology) has had an influence on the model systems available. Economic and societal pressures within academic and industrial institutions have also contributed to the interest in reducing animal use. The rising costs of drug development within industry and increasingly lean sources of funding for academic research have had an impact on limiting the use of animals from an economic perspective. At the same time, public opinion and involvement in animal welfare and use issues have provided additional impact and impetus to the practicing toxicologist/pharmacologist to consider alternatives to animal use.

Although considerable activity in *in vitro* toxicology has occurred and is still occurring, significant hurdles remain in the evolution of alternative procedures. The most significant issues are of a scientific and regulatory nature. Although scientific and regulatory areas are often considered exclusive, the present discussion will address the need for scientific and regulatory acceptance of *in vitro* alternatives to proceed in parallel. Indeed, the separation of scientific and regulatory issues poses a major obstacle to the continued advancement and integration of *in vitro* techniques in toxicologic assessment.

16.1 SHARED GOALS: PREDICTABILITY AND RESPONSIBILITY

We will first discuss the scientific and regulatory issues relative to the proximate and ultimate goals of the toxicologist, both within industry and within the various regulatory agencies. Scientifically, challenges remain in defining what specific *in vitro* systems can do in terms of modeling events occurring *in vivo*. Thus, in the short term, the goal of fully characterizing the *in vitro* models by their morphologic, biochemical, and functional capacity is paramount. Ultimately as well, the development of systems that can more accurately predict *in vivo* events is a key scientific goal and challenge. From a regulatory perspective, the issue of what will be acceptable in terms of alternative procedures remains unresolved. However, the regulatory goal of providing data predictive of *in vivo* toxic liabilities is similar to the ultimate scientific challenge and objective of *in vitro* test development. Thus, the goals of the scientist working in the alternative areas are congruent with the objectives of the regulatory agencies—that is, to develop model systems accurately predicting events occurring *in vivo*. The congruence between scientific and regulatory issues in alternative test development has also been noted in a publication from the Division of Toxicology of the Food and Drug Administration (FDA), in which Dr. Sidney Green stresses the need for test systems correlating with *in vivo* endpoints and objectives [3].

16.2 ISSUES: WHEN AND HOW TO USE *IN VITRO* SYSTEMS

Assuming, as discussed above, that the scientific and regulatory needs to develop accurate and relevant *in vitro* models for safety assessment are congruent, what measure, if any, can be taken at present to use existing techniques?

16.2.1 The Drug Discovery Process

Toxicology is an important participant in the early stages of drug discovery and selection. *In vitro* systems designed to monitor adverse or toxic properties are used alone and in concert with *in vivo* testing to evaluate and select against specific target organ effects and liabilities. These test systems are often similar to those used in examining pharmacologic properties, employing subcellular and cellular models. The use of *in vitro* systems as prescreens for toxicity assessments is an application that is generally well accepted by the scientific and regulatory communities [4]. At these early stages of chemical synthesis and evaluation, *in vitro* toxicologic models can also assist in drug design via structure-activity relationship studies. Secondly, the identification of structures potentially improving the toxicity profile of existing drugs (e.g., toxicity inhibitors) is facilitated by the use of *in vitro* systems. In this fashion, toxicology can play an intimate and key role in the discovery process. A broader recognition and acceptance of *in vitro* toxicologic techniques at the early stages of compound discovery and selection is certainly evolving within the drug industry. The need to make early assessments of toxicologic properties before commitment of resources to develop new drugs provides continuing impetus to apply *in vitro* toxicologic techniques in parallel with discovery efforts (which also frequently employ *in vitro* pharmacologic techniques).

At the preliminary stages of compound selection, *in vitro* systems can be and are used to make such decisions on the fate of a compound. Typically, compounds selected via *in vitro* screening procedures are subsequently tested in definitive *in vivo* testing before project commitment, or within the safety assessment process (Fig. 16.1). Thus, the evidence or decision of product safety before the clinical introduction of new drug entities would not be made on the basis of an *in vitro* test alone. An exception would be in the area of genotoxicity, in which positive *in vitro* results might deter the development of a new product.

16.2.2 The Safety Assessment Process

During the conduct of preclinical safety assessment studies, *in vitro* systems are most commonly used to assist in resolving issues that develop. In this capacity, *in vitro* systems offer unique opportunities in exploring mechanisms of toxicity and in performing species comparisons. Such problem solving can also occur at the clinical phase of drug development, in which issues develop in humans. To a limited extent, *in vitro* testing also contributes directly to the product safety profile at the preclinical state of development. The use of genotoxicity assays is an example of this application. An *in vitro* test that has received approval for direct use in safety assessment is Corrositex, a biochemical procedure predicting chemical corrosivity [6]. This test can be used to replace the rabbit test for corrosivity and has been accepted by regulatory bodies, including the Department of Transportation (DOT) [6] and the Environmental Protection Agency (EPA) [7]. In the future, it is possible that *in vitro* systems will play an even greater role in the preclinical safety assessment process; however, the extent will depend on the level of scientific advancement and validation of *in vitro* techniques in toxicology.

16.2.3 Selection of Test Procedures

The most reliable tests are those for which a large database of experimental and mechanistic data exists on a variety of drugs/chemicals. The isolated hepatocyte, for example, has emerged as a powerful tool for evaluating drug-induced hepatotoxicity and for establishing metabolic profiles, and it has also emerged as an ad-

SCIENTIFIC AND REGULATORY CONSIDERATIONS

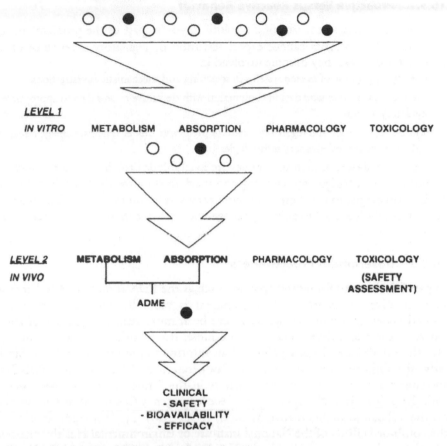

FIG. 16.1. Preclinical evaluation of new chemical entities.

junct to other *in vitro* systems (e.g., metabolic supplementation). In addition, numerous *in vitro*/biochemical assays have multiple uses in pharmacology and toxicology. Many of these test systems have not been as extensively applied to toxicology as they could be. Examples of such systems would include (1) central nervous system/cardiovascular receptor assays in isolated membranes, cells, and tissues; (2 enzyme assays (e.g., cholinesterase); and (3) simple *in vitro* assays measuring specific processes (e.g., histamine release in mast cells). It must also be noted that reliability and acceptance of test will be determined to some degree by the specific application. For example, a vast difference exists in the acceptability of tests for screening purposes compared with tests designed to potentially replace an *in vivo* procedure.

16.3 ISSUES: WHERE DO WE GO FROM HERE?

16.3.1 Proactive Versus Reactive Behavior

The saying "if you're not part of the solution, you're part of the problem" never rang truer. Toxicologists cannot expect scientific or regulatory acceptance of *in vitro* systems unless they become involved in:

The appropriate use of *in vitro* systems in screening and mechanistic investigations

The submission of *in vitro* data in conjunction with definitive *in vivo* data to appropriate regulatory agencies

Addressing scientific and regulatory needs for predictive model systems, namely, research and development of advanced technologies

As discussed above, numerous avenues are available for the immediate reduction in animal use in toxicology screening and mechanistic investigations. But it is up to the toxicologist to use these tools more extensively and creatively. Toxicologists cannot simply "default" to government agencies to make these decisions for them.

16.3.2 Government Proactiveness

In parallel, the need for federal agencies, such as the FDA and the EPA to become more involved in *in vitro* test development is painfully evident. Outside the United States, governmental agencies have been more visibly supportive of alternative research and development. For example, the Health Protection Branch of Health and Welfare Canada published an information report in 1991 [2]. Similarly, the Organisation for Economic Cooperation and Development (OECD) sponsored a review [1]. Recently, progress in the United States has been made with the publication of a report of an *ad hoc* Interagency Coordinating Committee on the Validation of Alternative Methods (ICCVAM) [5]. This report represents the combined efforts of the National Institute of Environmental Health Sciences, the National Institutes of Health, the U.S. Public Health Service, and the Department of Health and Human Services. Thus, governmental agencies can contribute directly to the continuing dialogue and discussion of alternative procedures.

16.3.3 Scientific/Regulatory Collaboration

Perhaps most critical to the advancement of alternative procedures is the establishment of mechanisms whereby active communication, debate, and guidance can occur between industrial and governmental agencies. Furthermore, the evolution of true collaborative efforts between scientists and regulators, in designing and supporting testing strategies, in the present and future, offers the potential to accelerate the integration of alternatives into the safety assessment process.

An example of the type of collaborative effort that could be initiated might involve conducting retrospective *in vitro* analyses of key compounds identified in historic databases at FDA, EPA, or industry. Such analyses would provide information on the predictive value of proposed test systems on compounds, for which a great deal of data exist on target organ effects *in vivo*.

16.4   VALIDATION

Validation, it is said, has different meanings to different people. One aspect, however, is clear: Validation by any definition is meaningless without sound science. An alternative that is simple, reproducible, and transferable means *nothing* unless it is relevant to the physiologic or biochemical endpoint one is seeking to predict or model. Thus, a key factor in the validation of *in vitro* techniques in toxicology involves the degree of correlation between events occurring *in vitro* and those the toxicologist evaluates in the intact animals. This correlation determines (1) the ultimate scientific value of the techniques and (2) the level of confidence associated with a particular test in terms of its predictability from a safety perspective. The scientific criteria determining the degree of correlation or level of confidence in a given test are predictability, identity, mechanisms of injury, and compensatory factors (Fig. 16.2). Unfortunately, test reproducibility, simplicity, and transferability are frequently viewed as the critical ingredients to test validation, at the expense of any mechanistic or scientific validity. It can be argued that such components, though important in test standardization and acceptance, have little bearing on the true scientific validation and rigor of new test procedures.

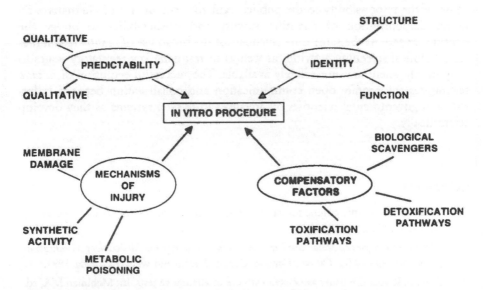

FIG. 16.2. Scientific and regulatory criteria for establishing validity of *in vitro* test procedures.

The degree to which an *in vitro* test fulfills the scientific and regulatory criteria outlined in Fig. 16.2 will determine its level of utility in the drug discovery and safety assessment process. The more empirical the test procedure, the greater the risk that such a test will fail to accurately predict *in vivo* events. Thus, the usefulness of certain tests may be limited to early selection and screening of new chemical entities before *in vivo* evaluation. More mechanistically sound and scientifically valid models can be used as adjuncts to resolve *in vivo* issues developing preclinically or clinically.

16.5 SUMMARY

Considerable interest and activity has evolved over the past decade in the development and application of *in vitro* techniques for risk assessment. At the same time, issues of a scientific and regulatory nature persist and must be addressed in parallel for the continued evolution and acceptance of *in vitro* techniques. From a scientific perspective, challenges remain in the development and application of model systems accurately reflecting events occurring *in vivo*. Although some of the scientific challenges involve advancing the existing technologies to meet the needs of mechanistically relevant systems, the potential to apply long-established biochemical and pharmacologic techniques has not been fully exploited. From a regulatory perspective, confidence that *in vitro* tests can either reduce or replace existing *in vivo* procedures presents both scientific and political challenges. Because of the responsibility to the public, regulatory acceptance of alternatives requires demonstration of scientific validity and predictability to bridge the confidence gap. At the same time, avenues for the broad use of *in vitro* techniques for preclinical screening activities as well as in resolving key regulatory issues in drug development are immediately available. The successful evolution of *in vitro* techniques will require open communication and collaboration between industrial and governmental scientists on advances in *in vitro* systems as they develop scientifically.

References

1. Frazier JM. Scientific criteria for validation of *in vitro* toxicity tests. *OECD Environ Monogr* 36, 1990.

2. Gilman JPW. *Report on Status and Trends in In Vitro Toxicology and Methodology Modifications for Reducing Animal Use.* Ottawa, Ontario, Canada:Health and Welfare Canada, 1991.

3. Green S. Regulatory issues associated with use of alternative tests. In: Mehlman MA, ed. *Advances in Modern Environmental Toxicology, Vol. X. Safety Evaluation: Toxicology, Methods,*

Concepts and Risk Assessment. Princeton, NJ: Princeton Scientific, 1987:107–116.

4. Green S. Animal alternatives in toxicology. *J Am Coll Toxicol* 1988:7(4):459–462.
5. ICVAM. *Validation and Regulatory Acceptance of Toxicological Test Methods: A Report of the Ad Hoc Interagency Coordinating Committee on the Validation of Alternative Methods.* Research Triangle Park, NC: NIEHS, March 1997. NIH Publication 97-3981.
6. DOT Publication DOT-E 10904. April 28, 1993. Renewal March 22, 1995.
7. EPA. Dermal corrosion, method 1120. *Fed Register* 1995;60(142).

17

Safety Issues in the Use of Human Tissue by *In Vitro* Metabolism and Toxicology Laboratories

Carol E. Green,[1] Janis Teichman,[1] and
Katherine L. Allen[2]

[1]SRI International, Menlo Park, California
[2]IDEC Pharmaceuticals, La Jolla, California

The use of human tissue in toxicology and metabolism studies has increased dramatically in recent years [1,2]. Just a few years ago, these studies were conducted primarily in university laboratories and research institutions. However, now human tissue preparations, particularly liver microsomes, hepatocytes, and slices, are used routinely in many pharmaceutical and chemical company research facilities and contract research organizations. Medical personnel, who have been trained to properly protect themselves, are more aware of the dangers of viral infections and other human pathogens than are most toxicology researchers. With every use of human tissue, researchers incur risk of infection and illness. To minimize these risks, every research laboratory using human tissue should establish strict policies and standard procedures. The Occupational Safety and Health Administration (OSHA) has promulgated guidelines for procedures protecting against exposure to hepatitis B virus (HBV) and human immunodeficiency virus (HIV) [3]. Other published documents provide reference for safe practices with biohazardous materials and information on the regulatory requirements [4–6]. Additionally, publications from other researchers provide practical advice that should be reviewed before introducing human tissues into a laboratory [7,8].

All human tissue should be treated as potentially infectious, because it is impossible to identify all of the hazardous agents associated with each specimen. This process is called "universal precaution," and all human blood, blood products, tissues, primary cell, and tissue cultures should be treated as if they were contaminated with HBV and HIV, at a minimum. Herpes B, SV-40, and ebola viruses have been known to be a problem in rhesus kidney cell cultures. Some of the viruses that can be associated with human cells and tissue are described below. Additionally, products derived from human tissue, such as human serum and other bodily fluids, are also a potential risk to researchers.

This chapter is designed to promote precautions that should be taken when using human tissue. It is not intended to replace standard procedures appropriate for each laboratory, which must be established based on the type of research being performed. Many of the suggestions are common sense. Those suggestions appearing overzealous actually are designed to be convenient, and to ensure worker safety without jeopardizing the laboratory efficiency.

Perhaps the most significant risk in the use of human tissue is the researcher's lack of perception of the danger involved in this work. Most accidents occur during routine procedures when an individual's concentration may stray. The laboratory and the procedure must be organized in such a way that the rules of worker safety are convenient. Every supervisor and technician working in laboratories performing experiments with human tissue must be constantly aware of the importance of proper attitude in the prevention of exposure accidents

For more than 15 years, SRI International has used human liver in many types of *in vitro* studies, including hepatotoxicity evaluations, metabolic activation for genotoxicity studies, and comparative metabolism of xenobiotics in numerous species. During this time, we have established a workable routine ensuring technician safety when processing human tissue.

17.1 SEROLOGY

The three common sources of human tissue, biopsy specimens, cadaver donor, and organ donor, should have extensive serology tests performed on them before they are used in research. Tests routinely performed on organ donor tissue include those for HIV, hepatitis B and C (HCV), syphilis, and cytomegalovirus (CMV). Many human tissue specimens will test positive for CMV but are still acceptable for use. Any tissue testing positive for HIV, HBV, HCV, or syphilis, however, should not be used unless specifically required by the project.

Many laboratories now use products provided by commercial supplies, such as human liver microsomes. Because these preparations can be characterized in advance of distribution to researchers, they should be stringently tested to assure safety. Nevertheless, these materials should still be considered a potential hazard.

In our laboratory, without exception, all personnel working with human tissue must be vaccinated against HBV before they are allowed to work in the laboratory. This policy is excellent because viral hepatitis is far more infectious than is HIV.

Regardless of the serology results, all human tissue should be treated as if it were infectious, which means following universal precautions. Serology tests are limited in scope and do not provide information on other infectious and rare diseases, such as kuru, Creutzfeldt–Jakob disease, scrapie, transmissible mink encephalopathy, measles, and papovavirus. Additionally false-negative test results are possible.

17.2 LABORATORY PROCEDURES AND GENERAL PRECAUTIONS

The next two sections briefly describe some recommended laboratory practices and policies. Most of these practices and policies can be readily implemented, with modifications in personal habits and laboratory procedures.

The door(s) to the laboratory designated for human tissue studies should always be properly labeled with the universal biohazard sign and kept closed while experiments are in progress. Incubators used to culture human cells, as well as refrigerators and freezers used to store samples from human cultures, should also be labeled with the universal biohazard sign.

Human tissues should be handled whenever possible in a biologic safety cabinet (class II) with a vertical laminar air flow hood (horizontal flow hoods provide absolutely no protection). In this way, spills and aerosols can be localized to an area that can be decontaminated easily and other personnel working in the laboratory area are not at risk. Also, all specimens of human tissue, including cell cultures and tissue homogenates, should be labeled prominently with a special warning alerting other laboratory personnel to potentially infectious material.

When transferring human tissue from one location to another, a secondary container should be used to prevent spills. For example, cell cultures that must be transferred between the incubator and the safety flow hood should be placed on a plastic tray with a lip.

Most viruses can remain infective in blood, body fluids, and tissues even after they have dried. For this reason, the laboratory area must be cleaned twice daily, e.g., before lunch and before leaving at the end of the day. Laboratory records should always be kept only in clean areas, away from human tissues. Papers contaminated with blood, media, or serum can transfer the infection to an unsuspecting technician who touches them [9].

When working with frozen human tissue, it is safest to wait until it has thoroughly thawed before cutting it. Alternatively, a piece can be broken from a frozen specimen wrapped in a plastic bag to contain fragments, so that tissue pieces are not scattered. Frozen tissue specimens should be transferred to a biologic safety cabinet, technicians should use barrier protection, and they should warn other people in the laboratory. Barrier protection includes disposable caps and hoods, face shields, water-impermeable laboratory coats, double gloves, and shoe covers.

17.3   LABORATORY PROCEDURES MINIMIZING RISK OF INFECTION

Numerous ways exist in which the researcher is exposed to infectious agents associated with human tissue and cells, including inhalation, ingestion, and needle sticks.

17.3.1   Inhalation

Aerosol. Although needle sticks and ingestion are the most direct routes of infection in clinical settings, inhalation of aerosols is probably more significant to researchers processing the human tissue into primary cell cultures, tissue homogenates, and microsomal fractions. Any procedure causing a break in a film of fluid generates aerosols and scatters the tiny droplets. The tiniest of these droplets dry out almost immediately, and the organisms they contain become airborne on the air currents in the laboratory. Many procedures, such as sonication and homogenization, will create droplets and aerosols that can be easily inhaled, not only by the technician performing the manipulation, but also by others in the laboratory. Extra care must be taken when performing procedures with human tissue that are known to produce aerosols. Most importantly, these procedures must be conducted in a biologic safety cabinet.

Pipetting. Ejection of fluids from pipettes or syringes must be performed in a controlled flow to produce a gentle stream. Forceful pressure causes aerosols. All pipetting of human tissue specimen fluids should be done in a biologic safety cabinet, because some aerosol formation always occurs. The discharge from the pipette should be released as close as possible to the plate, or the contents should be allowed to run down the wall of the tube or bottle, not dropped from a height above it (causing splashing or viral transfer to air components). Never use a syringe or needle as a substitute for a pipette when measuring or diluting biohazardous fluids.

Centifuging. Because centrifuging creates aerosols and droplets, samples should be contained in either capped centrifuge tubes or safety cups. The tubes should never be filled to the point that liquid is in contact with the lip of the tube, because the high G forces will drive the liquid past the cap seal and over the outside of the tube. Although many clinical laboratories maintain centrifuges in biologic safety cabinets, this precaution is not practical in some research laboratories, and so researchers must be careful to locate the centrifuge away from active areas of the laboratory. The lid of the centrifuge should always be closed when the unit is in motion. Serofuges, the type used in blood banks, should be oriented so that the air exhausting from the vent located at the base of the centrifuge is directed away from the operator. The centrifuge should be properly labeled with a universal biohazard sign and should be cleaned thoroughly after every use, as described below.

Homogenization. This process is most likely to create the largest volume of aerosols. When homogenizing with either a mortar-and-pestle–type apparatus or a Polytron homogenizer, this procedure must always be conducted in a biologic safety cabinet. Blenders also should be used only in a hood.

Cell sorters. Droplets are generated by fluorescent-activated cell sorters and represent potentially infectious material. Because these instruments are too large to be located in a hood, plastic shielding should be placed between the droplet-collecting area and the technician to reduce contamination.

17.4 SAFETY ISSUES

17.4.1 Other Routes of Aerosol Formation

Vacuum tubes used to collect blood specimens frequently retain vacuum, making it difficult to remove the rubber stopper. Aerosols are formed by "popping" the cork, so the preferred method of opening the tube is by first covering it with absorbent paper or cloth and then twisting the cork gently. Stirring and shaking also produce aerosols, particularly if the media contains material producing bubbles or foam, such as bovine serum albumin or fetal bovine serum. Sonication and lyophilization produce aerosols and should always be performed in a hood.

17.4.2 Ingestion

Ingestion is a route of exposure that is easy to avoid. Standard laboratory safety procedures, as well as common sense, should be followed to prevent ingestion of infectious agents. Never pipet by mouth. Do not eat, drink, smoke, or apply make up in the laboratory. Do not manipulate contact lenses. Many research facilities have limited office space, and frequently, desks are located in the laboratory. Although it is tempting to have a cup of coffee or eat a snack while doing calculations at the desk, these activities must not be performed in the laboratory. It is also imperative that no food be kept in the laboratory refrigerator, even if it is used only for storage of culture medium. Always remove gloves before leaving the laboratory. After removing gloves, technicians should thoroughly wash their hands and forearms.

17.4.3 Needle Sticks

Syringes. The easiest way to avoid needle sticks is to use needles and syringes only when absolutely necessary and then to use only disposable syringe–needle units. Fill the syringe carefully to minimize air bubbles and frothing of the inoculum, and expel excess air vertically into a cotton pledget moistened with disinfectant. Disposable syringe–needle units should be placed in a sharps (puncture-resistant) container after use. Needles should not be clipped, recapped, purposely bent, or broken before disposal or revised. Never force a needle into a sharps container. Fill containers for sharps only two-thirds full before disposal.

Several new products are also on the market, including blunt syringes for measuring and retractable needle syringes. These products reduce the likelihood of needle stick injuries.

Sharp items. Scalpel blades used to cut, slice, and apportion the human tissue are more commonly used by researchers than are needles. They must be handled with extraordinary care to prevent skin puncture and be disposed of in sharps container. Broken or chipped glassware should be discarded immediately in puncture-resistant containers (one for contaminated waste disposal and another for routine waste disposal). This source of exposure can be avoided by using only plasticware, rather than glass, when working with human tissue.

17.4.4 Contact with Mucous Membranes/Skin

Hand washing. Hands and wrists should be washed frequently and always immediately after removing gloves when working with potentially hazardous materials. Tests have shown that it is not unusual for microbial or chemical contamination to be present despite the use of gloves, because of unnoticed small holes, abrasions, tears, or entry at the wrist. Wearing two pairs of gloves reduces this risk. It is especially important to wash your hands and wrists before eating, smoking, drinking, or using the bathroom after working with human tissue. Hand washing should be done with gentle rubbing, not vigorous scrubbing.

Gloves. The importance of wearing gloves during all procedures requiring exposure to human tissue cannot be overemphasized. The selection of a glove appropriate for the work that is to be performed is crucial. Gloves made of polyethylene or polyvinyl chloride are ineffective barriers to virus particles [10]. Latex is a far superior barrier to virus permeation, but it does not provide a good chemical barrier. Gloves should be worn at all times while handling human tissue or equipment or materials that have come into contact with human tissue. The most difficult thing about wearing gloves is to resist touching any unprotected area of the body. It is important not to touch anything else when wearing gloves, such as the telephone or doorknobs (use a paper towel). All gloves should be disposed of properly immediately after use (into plastic biohazard waste) to avoid contaminating other surfaces in the laboratory. After removing gloves, wash your hands.

Eye/skin protection.   Personnel with open wounds should not work with human tissue until the wound has healed. A bandage is not a sufficient barrier against viruses!

Laboratory coats should always be worn in the laboratory and should be appropriately stored or discarded before leaving the laboratory. Disposable laboratory coats or coveralls are indispensable when working with human tissue because the contaminated garment can be discarded along with contaminated waste. It is also a good idea to wear disposable shoe covers in case blood or media are spilled on the floor. A common error is to leave the wrists uncovered, and this can be avoided by overlapping the laboratory coat sleeves with gloves.

17.5   PERSONAL HYGIENE

The importance of keeping hands away from mouth, nose, eyes, face, and hair has already been stressed. This habit must be developed. In some circumstances, a beard may be undesirable because it retains particulate contamination more persistently than does clean-shaven skin. Also, if the research requires a respirator, a clean-shaven face is essential for a proper fit. Those individuals with long hair should wear a head covering that can easily be decontaminated to protect the hair from fluid splashes and reduce facial contamination caused by adjusting the hair.

17.6   LABORATORY CLEANUP

Every laboratory bench top and the surface of every apparatus, chair, and floor area should be wiped after use with 0.525% sodium hypochlorite (10% household bleach). The outside of every sample tube and every container should also be wiped clean. All reusable surgical instruments and glassware should be soaked and rinsed thoroughly in 10% bleach, or some other disinfectant detergent for at least 10 minutes before washing and sterilization. Reusable pipettes should be laid flat in a container of the disinfectant (rather than being dropped vertically, which produces aerosols as the fluid rapidly rises to the pipette lumen). Every doorknob, switch, and surface that has been touched with a contaminated glove should also be wiped clean; this cleanup can be most conveniently done by maintaining a squirt bottle of 10% bleach (properly labeled with the concentration and date) at convenient locations in the laboratory. A routine, daily cleaning schedule, in addition to the cleanup immediately after an experiment, is also advised.

Note: Bleach must be used within 6 months after opening, because hypochlorite breaks down and chlorine escapes as a gas, causing the solution to slowly lose its potency after the container has been opened. Date and initial each bottle at the time of opening. Also, bleach reacts with proteins in general, so if an area is contaminated by blood, full-strength bleach must be used to decontaminate the area. Other disinfectant agents are also available. However, because of the toxic potential of these products, use of quaternary ammonium compounds, iodoform, or phenolic solutions is regulated by the Environmental Protection Agency (EPA) and requires specific record keeping and training.

17.7 WASTE DISPOSAL

Waste disposal is a critical issue, and special care should be taken by all researchers to dispose of human tissue correctly so that neither employees nor residents of the community are endangered by infectious human tissue. Two containers should be prepared: one for needles and sharp objects (this container must be puncture proof) and one for tissue, blood, gloves, garments, plastics, etc. Liquids, such as culture medium, may be decontaminated with 10% bleach for 10 minutes and then disposed in the sanitary sewer.

Sharp waste includes hypodermic needles, scalpel blades, needles with attached tubing, broken glass, Pasteur pipettes, blood vials, and anything else that could puncture the standard biohazard bag.

The waste containers should be kept separate from other waste in the laboratory and should be clearly labeled as biohazardous (use red bags labeled "Biohazard" with the universal biohazard emblem, which can be purchased from most large laboratory product suppliers). This area should be clearly marked with signs that can be read 25 ft away. The red biohazard bags must be kept in rigid outer containers for storage, handling, or transport and should be moved only after they have been tied. The rigid containers must be leak resistant. The containers should be properly washed and disinfected each time they are emptied unless they are disposable.

All human tissue waste is regulated under the Medical Waste Management Act. Consult your employer's Medical Waste Management plan for details relevant to the particular institution.

References

1. Rodrigues AD. Use of *in vitro* human metabolism studies in drug development. *Biochem Pharmacol* 48:2,147–2,156.

2. Obach RS, Baxter JG, Liston TE, Silber BM, Jones BC, MacIntyre F, Rance DJ, Wastall P. The prediction of human pharmacokinetic parameters from preclinical and *in vitro* metabolism data. *J Pharmacol Exp Ther* 1997;283:46–58.

3. Occupational Safety and Health Administration. *Enforcement Procedures for Occupational Exposure to Hepatitis B Virus and Human Immunodeficiency Virus. Compliance Assistance Guideline*. February 27, 1990. OSHA Instruction CPL 2-2.44B.

4. *Biosafety in Microbiological and Biomedial Laboratories*, 3rd ed. U.S. DHHS, PHS, CDC, May 1993.

5. Hensohn P, Jacobs R, Concoby B, eds. *Biosafety Reference Manual*, 2nd ed. Philadelphia, PA: ATHA Publications, 1995.

6. Nationl Academy of Sciences. *Biosafety in the Laboratory. Prudent Practices for the Handling and Disposal of Infectious Materials*. New York: National Academy Press, 1989.

7. Grizzle WE, Polt SS. Guidelines to avoid personnel contamination by infective agents in research laboratories that use human tissues. *J Tissue Cult Methods* 1988;11:191–200.

8. National Defense Research Interchange. *Guidelines for Handling Human Tissues and Body Fluids Used in Research*. Washington, DC: NCI Cooperative Human Tissue Network, 1987.

9. Pattison CP, Boyer DM, Maynard JE, Kelly PC. Epidemic hepatitis in a clinical laboratory: possible association with computer card handling. *JAMA* 1974;230:854–857.

10. Klein RC, Party E, Gershey EL. Virus penetration of examination gloves. *BioTechniques* 1990;9:196–199.

INDEX

T - #0079 - 101024 - C0 - 234/156/23 [25] - CB - 9781560327691 - Gloss Lamination